CHEMOMETRICS IN PRACTICAL APPLICATIONS

Edited by **Kurt Varmuza**

Chemometrics in Practical Applications
Edited by Kurt Varmuza

Published by InTech
Janeza Trdine 9, 51000 Rijeka, Croatia

Publishing Process Manager Maja Jukic
Technical Editor Teodora Smiljanic
Cover Designer InTech Design Team

Additional hard copies can be obtained from orders@intechopen.com

Chemometrics in Practical Applications, Edited by Kurt Varmuza
p. cm.
ISBN 978-953-51-0438-4

For EU product safety concerns:
IN TECH d.o.o., Prolaz Marije Krucifikse Kozulić 3, 51000 Rijeka, Croatia,
info@intechopen.com or visit our website at intechopen.com.

Meet the editor

Kurt Varmuza studied chemistry at the Vienna University of Technology in Austria. As one of the pioneers in chemometrics, his research activities in chemoinformatics and chemometrics involve development and applications of methods for spectra-structure relationships (MS and IR), structure-property relationships (QSPR), and classification of materials in archaeometry and cosmo chemistry. Since 1992 he has been a professor at the Vienna University of Technology, Austria.
In the book "Chemometrics in practical applications", various practical applications of chemometric methods in chemistry, biochemistry and chemical technology are presented, and selected chemometric methods are described in tutorial style. The book contains 14 independent chapters and is devoted to filling the gap between textbooks on multivariate data analysis and research journals on chemometrics and chemoinformatics.

Contents

Preface

Chemometrics has been defined as "a chemical discipline that uses statistical and mathematical methods to design or select optimum procedures and experiments, and to provide maximum chemical information by analyzing chemical data". Chemometrics can be considered as a part of the wider field chemoinformatics, and has close relationships to bioinformatics.

The start of chemometrics dates back to the 1960s, when multivariate data analysis methods - like for instance the "learning machine" - have been tried for solving rather complicated problems in chemistry, such as the automatic interpretation of molecular spectra. The name chemometrics was first used by Svante Wold in 1972 (in Swedish, kemometria) and it was established in 1974 by Bruce Kowalski. The first years of chemometrics were characterized by rather uncritical use of machine learning methods for complex - often too complex - tasks in chemistry and consequently sometimes accompanied by ignorance and refusal of many chemists. However, in this time also falls the presentation of the PLS regression method by chemometricians, which is now the most used method for evaluation of multivariate data, not only in chemistry. During the next decades chemometricians learned to use multivariate data analysis in a proper and safe way for problems with a realistic chance for success, and also found back to the underlying statistical concepts. Chemometrics contributed with valuable method developments and provided many stimulants in the area. Furthermore, commercial software became available and nowadays several basic chemometric methods, like principal component analysis, multivariate classification, and multiple regression (by PLS and other approaches) are routinely used in chemical research and industry. Admittedly, sometimes without the necessary elementary knowledge about the used methods.

Despite the broad definition of chemometrics, the most important part of it is still the application of multivariate data analysis to chemistry-relevant data. Chemical-physical systems of practical interest are often complicated and relationships between available (measurement) data and desired data (properties, origin) cannot be described by theory. Therefore, a typical chemometric approach is not based on "first principles" but is "data driven" and has the goal to create empirical models. A thorough evaluation of the performance of such models is essential for new cases. Multivariate statistical data analysis has been proven as a powerful tool for analyzing and structuring such data sets from chemistry and biochemistry.

This book is a collection of 14 chapters, divided into three sections. Assignment of the chapters to these sections only indicates the main contents of a chapter because most are interdisciplinary and contains theoretical as well as practical aspects.

In section "Methods" the topics comprise statistical model comparison, treatment of compositional data, methods for the estimation of kinetic parameters, and a new approach for exploratory data analysis. A comprehensive chapter presents an overview of experimental optimization.

Section "Biochemistry" deals with biomarker identification, kinetics of enzyme reactions, selection of substances with anticancer properties, and the use of an electronic nose for the identification of foodborne pathogens.

Section "Technology" focuses on chemometric methods used in food technology, for water quality estimation, for the characterization of nanocomposite materials by NMR spectra, and in chromatographic separation processes.

The topics of this book cover a wide range of highly relevant problems in chemistry and chemical/biological technology. The presented solutions may be of interest to the reader even if not working exactly in the fields described in the chapters. The book is intended for chemists, chemical engineers, and biotechnologists working in research, production or education. Students in these areas will find a source with highly diverse and successful applications of chemometric methods. In this sense, the major goal of this "mosaic of contributions" - presented in a book - is to promote new and adequate use of multivariate data analysis methods in chemistry and related fields.

March 2012

Kurt Varmuza
Vienna University of Technology,
Vienna,
Austria

Part 1

Methods

1

Model Population Analysis for Statistical Model Comparison

Hong-Dong Li[1], Yi-Zeng Liang[1] and Qing-Song Xu[2]
[1]*College of Chemistry and Chemical Engineering, Central South University, Changsha,*
[2]*School of Mathematic Sciences, Central South University, Changsha,*
P. R. China

1. Introduction

Model comparison plays a central role in statistical learning and chemometrics. Performances of models need to be assessed using a given criterion based on which models can be compared. To our knowledge, there exist a variety of criteria that can be applied for model assessment, such as Akaike's information criterion (AIC) [1], Bayesian information criterion (BIC) [2], deviance information criterion (DIC),Mallow's Cp statistic, cross validation [3-6] and so on. There is a large body of literature that is devoted to these criteria. With the aid of a chosen criterion, different models can be compared. For example, a model with a smaller AIC or BIC is preferred if AIC or BIC are chosen for model assessment.

In chemometrics, model comparison is usually conducted by validating different models on an independent test set or by using cross validation [4, 5, 7], resulting in a single value, *i.e.* root mean squared error of prediction (RMSEP) or root mean squared error of cross validation (RMSECV). This single metrics is heavily dependent on the selection of the independent test set (RMSEP) or the partition of the training data (RMSECV). Therefore, we have reasons to say that this kind of comparison is lack of statistical assessment and also at the risk of drawing wrong conclusions. We recently proposed model population analysis (MPA) as a general framework for designing chemometrics/bioinformatics methods [8]. MPA has been shown to be promising in outlier detection and variable selection. Here we hypothesize that reliably statistical model comparison could be achieved via the use of model population analysis.

2. Model population analysis

2.1 The framework of model population analysis

Model population analysis has been recently proposed for developing chemometrics methods in our previous work [8]. As is shown in **Figure 1**, MPA works in three steps which are summarized as (1) randomly generating N sub-datasets using Monte Carlo sampling (2) building one sub-model on each sub-dataset and (3) statistically analyzing some interesting output of all the N sub-models.

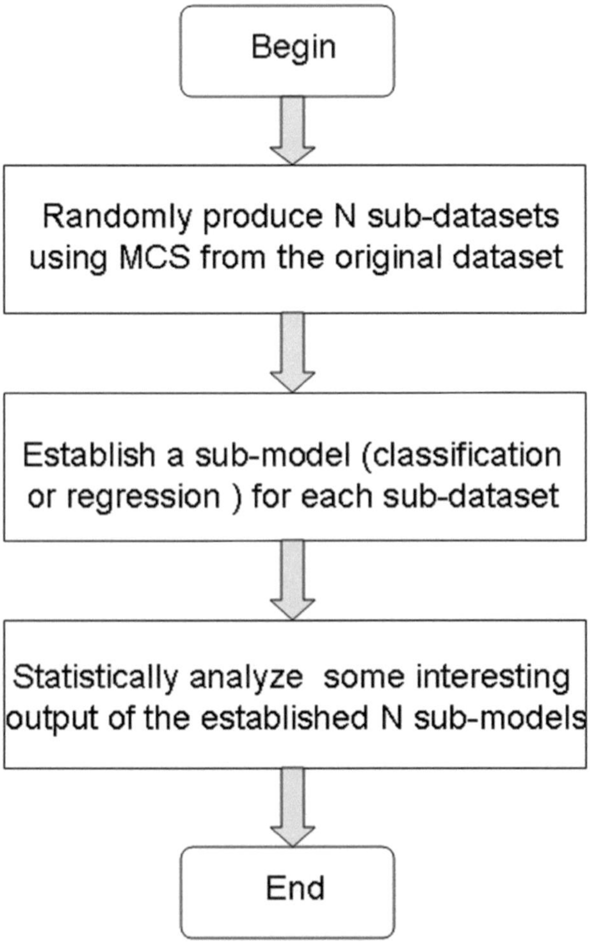

Fig. 1. The schematic of MPA. MCS is the abbreviation of Monte Carlo Sampling.

2.1.1 Monte Carlo sampling for generating a sub-dataset

Sampling plays a key role in statistics which allows us to generate replicate sub-datasets from which an interested unknown parameter could be estimated. For a given dataset (**X**, **y**), it is assumed that the design matrix **X** contains *m* samples in rows and *p* variables in columns, the response vector **y** is of size $m\times 1$. The number of Monte Carlo samplings is set to N. In this setting, N sub-datasets can be drawn from N Monte Carlo samplings with or without replacement [9, 10], which are denoted as $(\mathbf{X}_{sub}, \mathbf{y}_{sub})_i$, i = 1, 2, 3, …N.

2.1.2 Establishing a sub-model using each sub-dataset

For each sub-dataset $(\mathbf{X}_{sub}, \mathbf{y}_{sub})_i$, a sub-model can be constructed using a selected method, *e.g.* partial least squares (PLS) [11] or support vector machines (SVM) [12]. Denote the sub-model established as $f_i(\mathbf{X})$. Then, all these sub-models can be put into a collection:

$$C = (f_1(X), f_2(X), f_3(X), \ldots, f_N(X)) \quad (1)$$

All these N sub-models are mutually different but have the same goal that is to predict the response value **y**.

2.1.3 Statistically analyzing an interesting output of all the sub-models

The core of model population analysis is statistical analysis of an interesting output, *e.g.* prediction errors or regression coefficients, of all these sub-models. Indeed, it is difficult to give a clear answer on what output should be analyzed and how the analysis should be done. Different designs for the analysis will lead to different algorithms. As proof-of-principle, it was shown in our previous work that the analysis of the distribution of prediction errors is effective in outlier detection [13].

2.2 Insights provided by model population analysis

As described above, Monte Carlo sampling serves as the basics of model population analysis that help generate distributions of interesting parameters one would like to analyze. Looking on the surface, it seems to be very natural and easy to generate distributions using Monte Carlo sampling. However, here we show by examples that the distribution provided by model population analysis can indeed provide very useful information that gives insights into the data under investigation.

2.2.1 Are there any outliers?

Two datasets are fist simulated. The first contains only normal samples, whereas there are 3 outliers in the second dataset, which are shown in Plot A and B of **Figure 2**, respectively. For each dataset, a percentage (70%) of samples are randomly selected to build a linear regression model of which the slope and intercept is recorded. Repeating this procedure 1000 times, we obtain 1000 values for both the slope and intercept. For both datasets, the intercept is plotted against the slope as displayed in Plot C and D, respectively. It can be observed that the joint distribution of the intercept and slope for the normal dataset appears to be multivariate normally distributed. In contrast, this distribution for the dataset with outliers looks quite different, far from a normal distribution. Specifically, the distributions of slopes for both datasets are shown in Plot E and F. These results show that the existence of outliers can greatly influence a regression model, which is reflected by the odd distributions of both slopes and intercepts. In return, a distribution of a model parameter that is far from a normal one would, most likely, indicate some abnormality in the data.

2.2.2 Are there any interfering variables?

In this study, we first simulate a design matrix **X** of size 50 × 10, the response variable **Y** is simulated by multiplying X with a 10-dimensional regression vector. Gussian noises with standard deviation equal to 1 are then added to **Y**. That is to say, all the variables in **X** are "true variables" that collectively predict **Y**. This dataset (**X**, **Y**) is denoted SIMUTRUE. Then another design matrix **F** is simulated of size 50 × 10. Denote the combination of **X** and **F** as **Z**=[**X F**]. This dataset (**Z**, **Y**) is called SIMUINTF, which contains variables that are not predictive of **Y**. For both datasets, we randomly choose 70% samples to first build a regression model which is then used to make predictions on the remaining 30% samples, resulting in a RMSEP value. Repeating this procedure 1000 times, we, for both datasets,

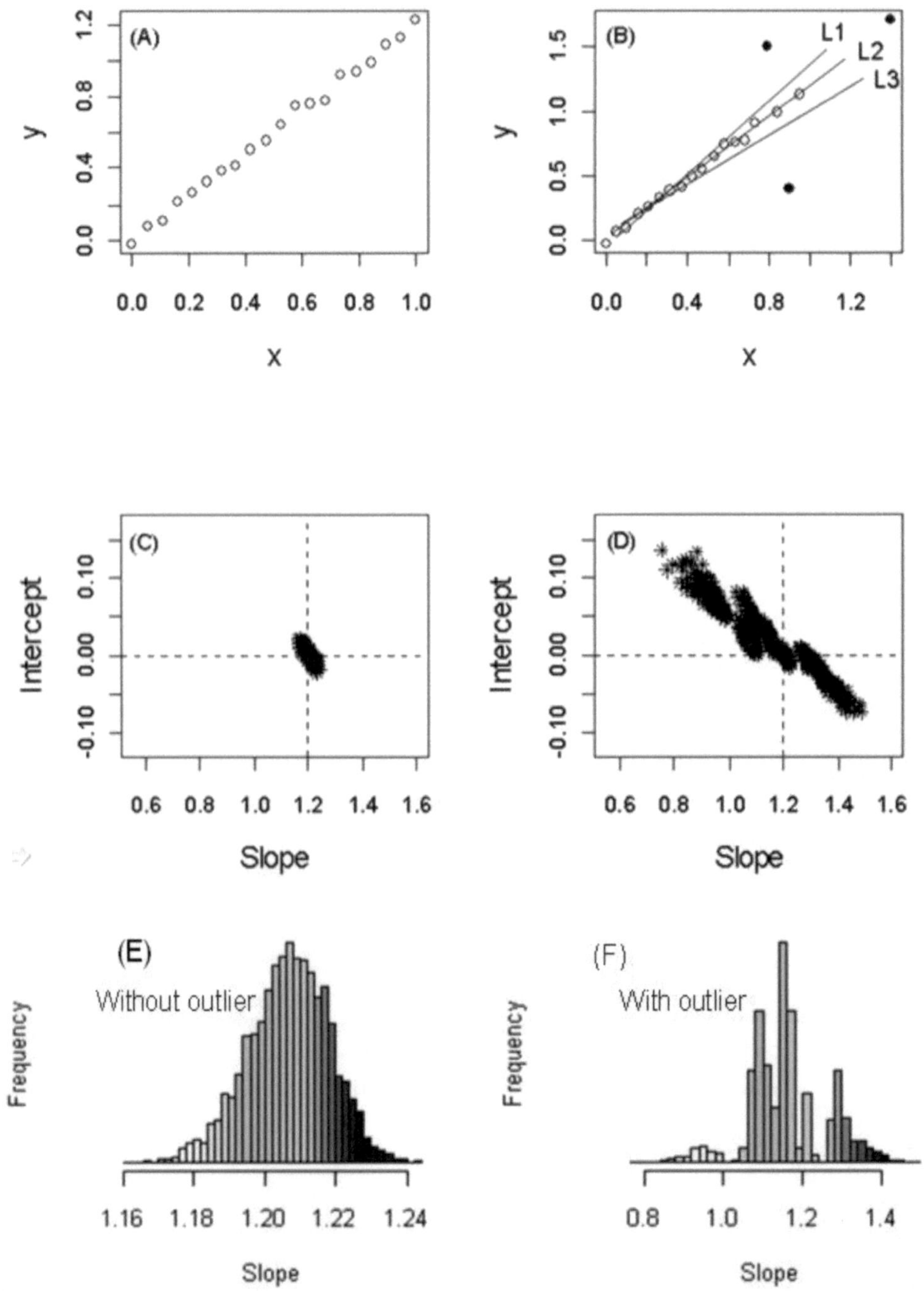

Fig. 2. A simulation study illustrating the use of model population analysis to detect whether a dataset contains outliers. Plot A and Plot B shows the data simulated without and with outliers, respectively. 1000 linear regression models computed using 1000 sub-datasets randomly selected and the slope and intercept are presented in Plot C and D. Specifically, the distribution of slope for these two simulated datasets are displayed in Plot E and Plot F.

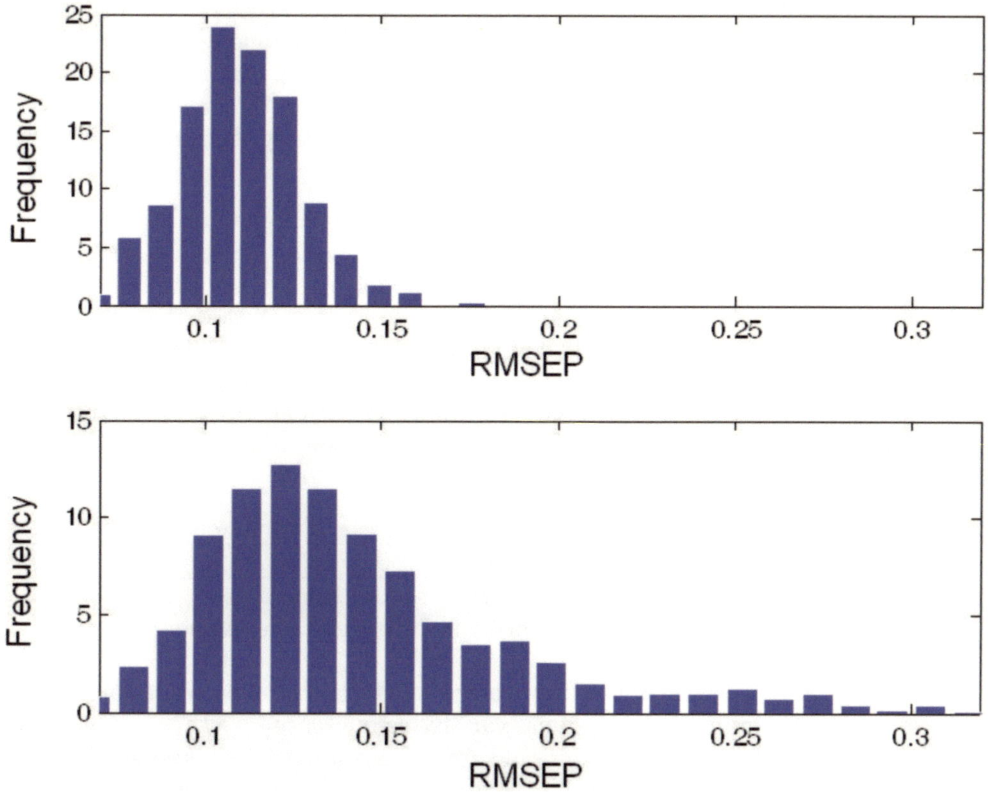

Fig. 3. The distribution of RMSEPs using the variable set that contains only "true variables" (upper panel) and the variable set that includes not only "true variables" but also "interfering variables" (lower panel).

obtain 1000 RMSEP values, of which the distributions are given in **Figure 3**. Clearly, the distribution of RMSEP of the SIMUINTF is right shifted, indicating the existence of variables that are not predictive of Y can degrade the performance of a regression model. We call this kind of variables "interfering variables". Can you tell whether a dataset contains interfering variables for a real world dataset? Curious readers may ask a question like this. Indeed, we can. We can do replicate experiments to estimate the experimental error that could serve as a reference by which it is possible to judge whether interfering variables exist. For example, if a model containing a large number of variables (with true variables included) shows a large prediction error compared to the experimental error, we may predict that interfering variables exist. In this situation, variable selection is encouraged and can greatly improve the performance of a model. Actually, when interfering variables exist, variable selection is a must. Other methods that use latent variables like PCR or PLS cannot work well because latent variables have contributions coming from interfering variables.

2.3 Applications of model population analysis

Using the idea of model population analysis, we have developed algorithms that address the fundamental issues in chemical modeling: outlier detection and variable selection. For outlier detection, we developed the MC method [13]. For variable selection, we developed subwindow permutation analysis (SPA) [14], noise-incorporated subwindow permutation analysis (NISPA) [15] and margin influence analysis (MIA) [16]. Here, we first give a brief description of these algorithms, aiming at providing examples that could help interested readers to understand how to design an algorithm by borrowing the framework of model population analysis.

As can be seen from **Figure 1**, These MPA-based methods share the first two steps that are (1) generating N sub-datasets and (2) building N sub-models. The third step "statistical analysis of an interesting output of all these N sub-models" is the core of model population analysis that underlines different methods. The key points of these methods as well as another method Monte Carlo uninformative variable elimination (MCUVE) that also implements the idea of MPA are summarized in Table 1. In a word, the distribution from model population analysis contains abundant information that provides insight into the data analyzed and by making full use of these information, effective algorithms can be developed for solving a given problem.

Methods*	What to statistically analyze
MC method	Distribution of prediction errors of each sample
SPA	Distribution of prediction errors before and after each variable is permuted
NISPA	Distribution of prediction errors before and after each variable is permuted with one noise variable as reference
MIA	Distribution of margins of support vector machines sub-models
MCUVE	Distribution of regression coefficients of PLS regression sub-models

*: The MC method, SPA, NISPA, MIA and MCUVE are described in references [13], [14], [15] [16] and [27].

Table 1. Key points of MPA-based methods.

2.4 Model population analysis and bayesian analysis

There exist similarities as well as differences between model population analysis and Bayesian analysis. One important similarity is that both methods consider the parameter of interest not as a single number but a distribution. In model population analysis, we generate distributions by causing variations in samples and/or variables using Monte Carlo sampling [17]. In contrast, in Bayesian analysis the parameter to infer is first assumed to be from a prior distribution and then observed data are used to update this prior distribution to the posterior distribution from which parameter inference can be conducted and predictions can be made [18-20]. The output of Bayesian analysis is a posterior distribution of some interesting parameter. This posterior distribution provides a natural link between Bayesian analysis and model population analysis. Taking Bayesian linear regression (BLR) [20] as an example, the output can be a large number of regression coefficient vectors that are sampled from its posterior distribution. These regression coefficient vectors actually represent a population of sub-models that can be used directly for model population analysis. Our future work will be constructing useful algorithms by borrowing merits of both Bayesian analysis and model population analysis.

2.5 Model population analysis and ensemble learning

Ensemble learning methods, such as bagging[21], boosting [22] and random forests [23], have emerged as very promising strategies for building a predictive model and these methods have found applications in a wide variety of fields. Recently, a new ensemble technique, called feature-subspace aggregating (Feating) [24], was proposed that was shown to have nice performances. The key point of these ensemble methods is aggregating a large number of models built using sub-datasets randomly generated using for example bootstrapping. Then ensemble models make predictions by doing a majority voting for classification or averaging for regression. In our opinion, the basic idea of ensemble learning methods is the same as that in model population analysis. In this sense, ensemble learning methods can also be formulated into the framework of model population analysis.

3. Model population analysis for statistical model comparison

Based on model population analysis, here we propose to perform model comparison by deriving an empirical distribution of the difference of RMSEP or RMSECV between two models (variable sets), followed by testing the null hypothesis that the difference of RMSEP or RMSECV between two models is zero. Without loss of generality, we describe the proposed method by taking the distribution of difference of RMSEP as an example. We assume that the data **X** consists of *m* samples in row and *p* variables in column and the target value **Y** is an m-dimensional column vector. Two variable sets, say V_1 and V_2, selected from the *p* variables, then can be compared using the MPA-based method described below.

First, a percentage, say 80%, from the *m* samples with variables in V_1 and V_2 is randomly selected to build two regression models using a preselected modeling method such as PLS [11] or support vector machines (SVMs) [12], respectively. Then an RMSEP value can be computed for each model by using the remaining 20% samples as the test set. Denote the two RMSEP values as $RMSEP_1$ and $RMSEP_2$, of which the difference can be calculated as

$$D = RMSEP_1 - RMSEP_2 \quad (2)$$

By repeating this procedure N, say 1000, times, N D values are obtained and collected into a vector **D**. Now, the model comparison can be formulated into a hypothesis test problem as:

Null hypothesis: the mean of **D** is zero.

Alternative hypothesis: the mean of **D** is not zero.

By employing a statistical test method, *e.g.* *t*-test or Mann-Whitney U test [25], a *P* value can be computed for strictly assessing whether the mean of **D** is significantly different from zero ($P<0.05$) or not ($P>0.05$). If $P<0.05$, the sign of the mean of **D** is then used to compare which model (variable set) is of better predictive performance. If $P>0.05$, we say two models have the same predictive ability.

4. Results and discussions

4.1 Comparison of predictive performances of variables subsets

The corn NIR data measured on *mp5* instrument is used to illustrate the use of the proposed method (http://software.eigenvector.com/Data/index.html). This data contain NIR spectra measured at 700 wavelengths on 80 corn samples. The original NIR spectra are shown in **Figure 4**. The chemical property modeled here is the content of protein. As was demonstrated in a large body of literature [26-30], variable selection can improve the predictive performance of a model. Here we would like to investigate whether the gain in predictive accuracy using variable subsets identified by variable selection methods is significant.

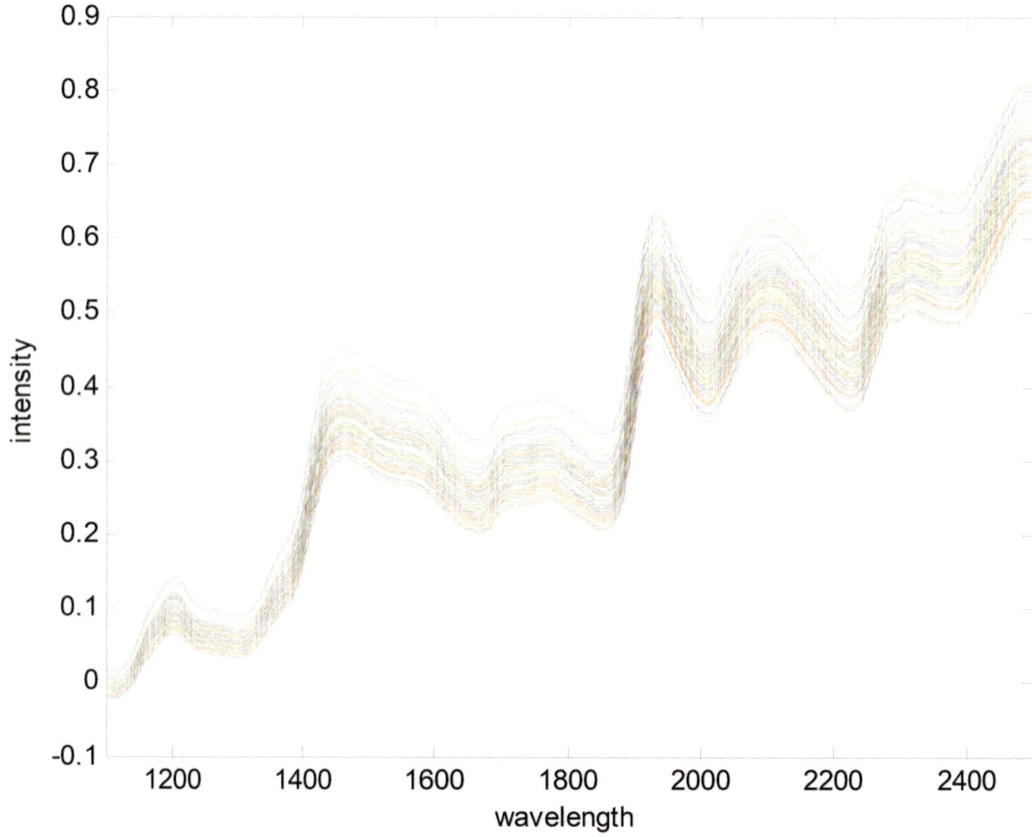

Fig. 4. Original near infrared spectra of corn on the mp5 instrument.

Uninformative variable elimination (UVE) is a widely used method for variable selection in chemometrics [26]. Its extended version, Monte Carlo UVE (MCUVE), was recently proposed [27, 31]. Mimicking the principle of "survival of the fittest" in Darwin's evolution theory, we developed a variable selection method in our previous work, called competitive adaptive reweighted sampling (CARS) [8, 28, 32, 33], which was shown to have the potential to identify an optimal subset of variables that show high predictive performances. The source codes of CARS are freely available at [34, 35].

In this study, MCUVE and CARS is chosen to first identify two variable sets, named V_1 and V_2, respectively. The set of the original 700 variables are denoted as V_0. Before data analysis, each wavelength of the original NIR spectra is standardized to have zero mean and unit variance. Regarding the pretreatment of spectral data, using original spectra, mean-centered

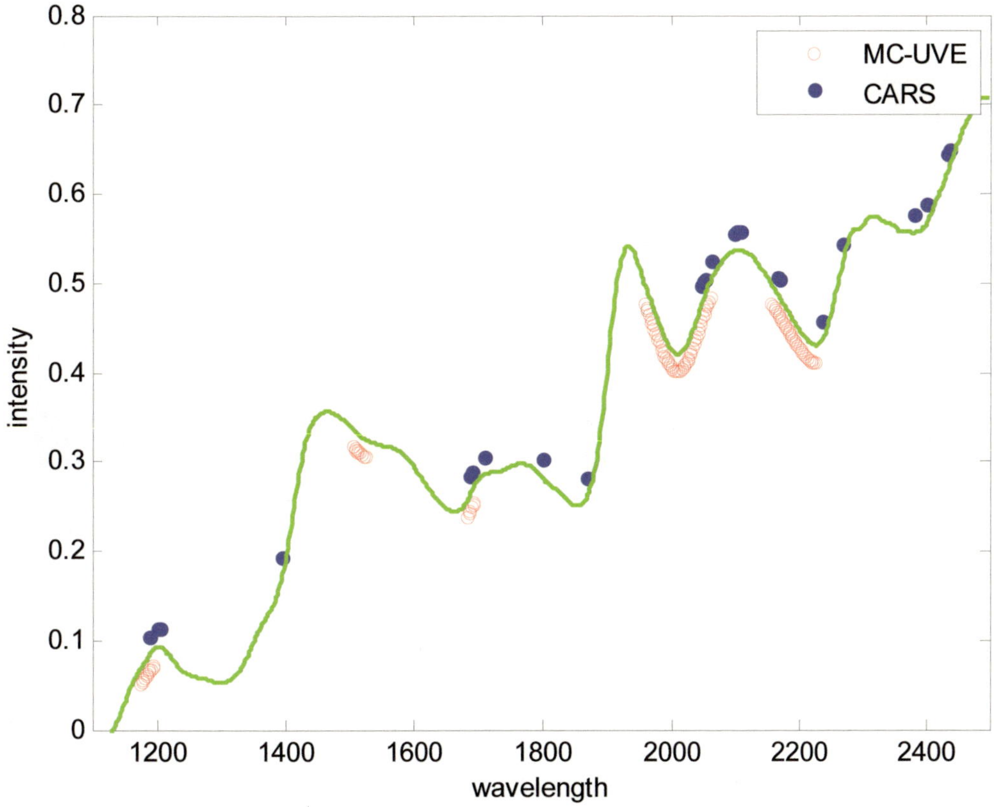

Fig. 5. Comparison of selected wavelengths using MC-UVE (red circles) and CARS (blue dots). The green line denotes the mean of the 80 corn NIR spectra.

spectra or standardized spectra indeed would lead to different results. But the difference is usually not big according to our experience. The reason why we choose standardization is to remove the influence of each wavelength's variance on PLS modeling because the decomposition of spectrum data X using PLS depends on the magnitude of covariance between wavelengths and the target variable Y. The number of PLS components are optimized using 5-fold cross validation. For MCUVE, the number of Monte Carlo simulations is set to 1000 and at each simulation 80% samples are selected randomly to build a calibration model. We use the reliability index (RI) to rank each wavelength and the number of wavelengths (with a maximum 200 wavelengths allowed) is identified using 5-fold cross validation. Using MCUVE, 115 wavelengths in 5 bands (1176-1196nm, 1508-1528nm, 1686-1696nm, 1960-2062nm and 2158-2226nm) are finally selected and shown in **Figure 5** as red circles. For CARS, the number of iterations is set to 50. Using CARS, altogether 28 variables (1188, 1202, 1204, 1396, 1690, 1692, 1710, 1800, 1870, 2048, 2050, 2052, 2064, 2098, 2102, 2104, 2106, 2108, 2166, 2168, 2238, 2270, 2382, 2402, 2434, 2436, 2468 and 2472 nm) are singled out and these variables are also shown in **Figure 5** as blue dots. Intuitively, MCUVE selects 5 wavelength bands while the variables selected by CARS are more diverse and scattered at different regions. In addition, the Pearson correlations variables selected by both methods are shown in **Figure 6**.

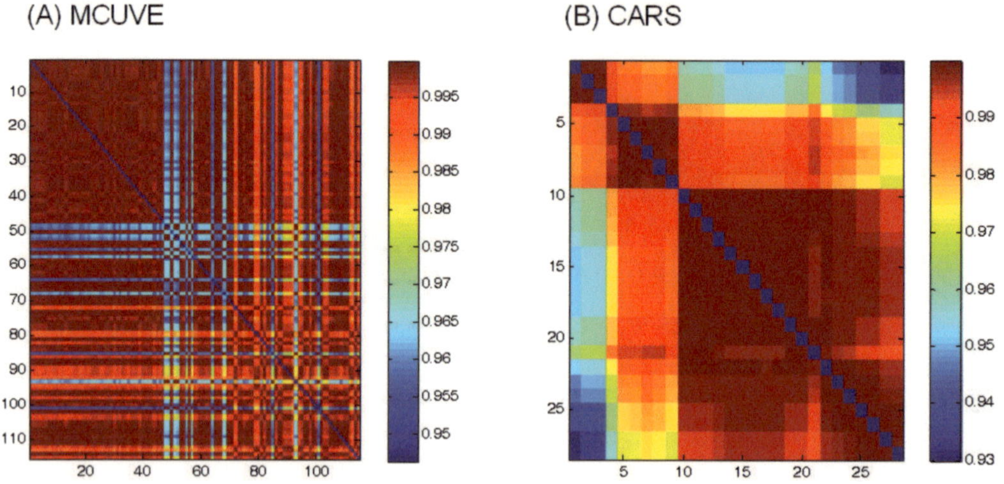

Fig. 6. The Pearson pair-wise correlations of variables selected using MCUVE (115 variables, left) and CARS (28 variables, right).

We choose PLS for building regression models. For the MPA-based method for model comparison, the number of Monte Carlo simulations is set to 1000 and at each simulation 60% samples are randomly selected as training samples and the rest 40% work as test samples. The number of PLS components is chosen based on 5-fold cross validation. In this setting, we first calculated 1000 values of $RMSEP_0$, $RMSEP_1$ and $RMSEP_2$ using V_0, V_1 and V_2, respectively. The distributions of $RMSEP_0$, $RMSEP_1$ and $RMSEP_2$ are shown in **Figure 7**. The mean and standard deviations of these three distributions are 0.169±0.025 (full spectra), 0.147±0.018 (MCUVE) and 0.108±0.015 (CARS). On the whole, both variable selection methods improve the predictive performance in terms of lower prediction errors and smaller standard deviations. Looking closely, the model selected by CARS has smaller standard deviation than that of MCUVE. The reason may be that CARS selected individual wavelengths and these wavelengths display lower correlations (see **Figure 6**) than those wavelength bands selected by MCUVE. The lower correlation results in better model stability which is reflected by smaller standard deviations of prediction errors. Therefore from the perspective of prediction ability, we recommend to adopt methods that select individual wavelengths rather than continuous wavelength bands.

Firstly, we compare the performance of the model selected by MCUVE to the full spectral model. The distribution of D values (MCUVE - Full spectra) is shown in Plot A of **Figure 8**. The mean of D is -0.023 and is shown to be not zero ($P < 0.000001$) using a two-side t test, indicating that MCUVE significantly improves the predictive performance. Of particular note, it can be observed that a percentage (83.1%) of D values are negative and the remaining (16.9%) is positive, which indicates model comparison based on a single split of the data into a training set and a corresponding test set may have the potential risk of drawing a wrong conclusion. In this case, the probability of saying that MCUVE does not improve predictive performances is about 0.169. However, this problem can be solved by the proposed MPA-based method because the model performance is tested on a large number of sub-datasets, rendering the current method potentially useful for reliably

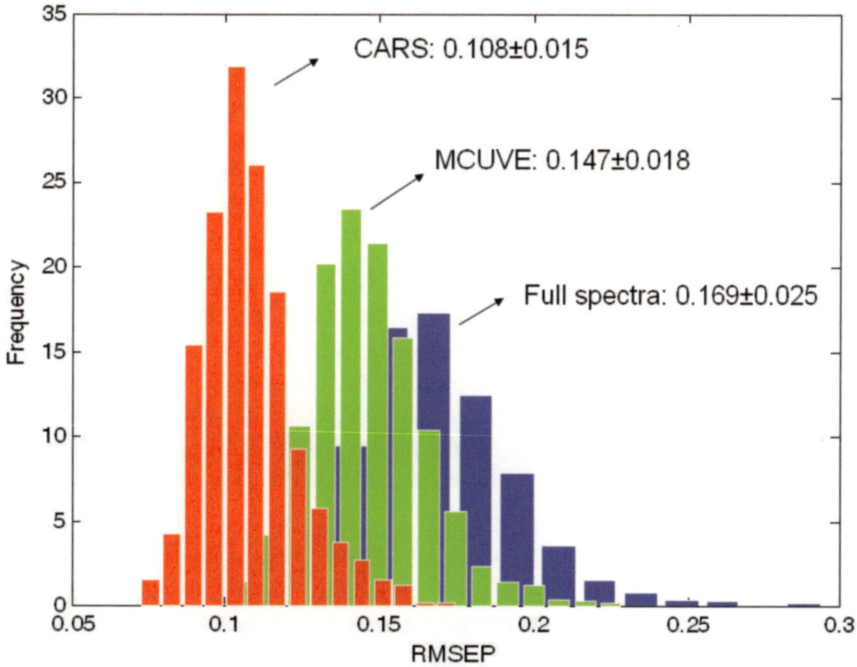

Fig. 7. Distributions of root mean squared errors of prediction (RMSEP) from 1000 test sets (32 samples) randomly selected from the 80 corn samples using full spectra and variables selected by MCUVE and CARS, respectively.

statistical model comparison. With our method, we have evidence showing that the improvement resulting from MCUVE is significant.

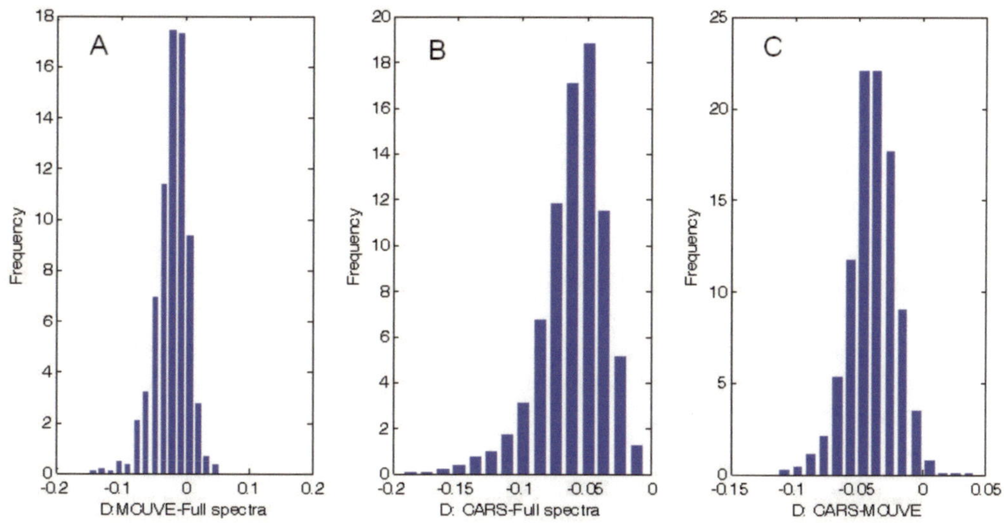

Fig. 8. The distributions of D values. The P values of t test for these three distributions are 8.36×10^{-120}, 0 and 0, respectively.

Further, the performance of the model selected by CARS is compared to the full spectral model. The distribution of D values (CARS – Full spectrum) is shown in Plot B of **Figure 8**. The mean of D is -0.061 which is much smaller that that from MCUVE (-0.023). Using a two-side t test, this mean is shown to be significantly different from zero (P = 0), indicating that the improvement over the full spectral model is significant. Interestingly, it is found that all the D values are negative, which implies the model selected by CARS is highly predictive and there is little evidence to recommend the use of a full spectral model, at least for this dataset.

Finally, we compare the models selected by MCUVE and CARS, respectively. The distribution of D values (CARS – MCUVE) is shown in Plot C of **Figure 8**. The mean of D values is -0.039. Using a two-side t test, this mean is shown to be significantly different from zero (P = 0), indicating that the improvement of CARS over MCUVE is significant. We find that 98.9% of D values are negative and only 1.1% are positive, which suggests that there is a small probability to draw a wrong conclusion that MCUVE performs better than CARS. However, with the help of MPA, this risky conclusion can be avoided, indeed.

Summing up, we have conducted statistical comparison of the full spectral model and the models selected by MCUVE and CARS based on the distribution of D values calculated using RMSEP. Our results show that model comparison based on a single split of the data into a training set and a corresponding test set may result in a wrong conclusion and the proposed MPA approach can avoid drawing such a wrong conclusion thus providing a solution to this problem.

4.2 Comparison of PCR, PLS and an ECR model

In chemometrics, PCR and PLS seem to be the most widely used method for building a calibration model. Recently, we developed a method, called elastic component regression (ECR), which utilizes a tuning parameter $\alpha \in [0,1]$ to supervise the decomposition of X-matrix [36], which falls into the category of continuum regression [37-40]. It is demonstrated theoretically that the elastic component resulting from ECR coincides with principal components of PCA when $\alpha = 0$ and also coincides with PLS components when $\alpha = 1$. In this context, PCR and PLS occupy the two ends of ECR and $\alpha \in (0,1)$ will lead to an infinite number of transitional models which collectively uncover the model path from PCR to PLS. The source codes implementing ECR in MATLAB are freely available at [41]. In this section, we would like to compare the predictive performance of PCR, PLS and an ECR model with $\alpha = 0.5$.

We still use the corn protein data described in Section 4.1. Here we do not consider all the variables but only the 28 wavelengths selected by CARS. For the proposed method, the number of Monte Carlo simulations is set to 1000. At each simulation 60% samples selected randomly are used as training samples and the remaining serve as test samples. The number of latent variables (LVs) for PCR, PLS and ECR (α = 0.5) is chosen using 5-fold cross validation.

Figure 9 shows the three distributions of RMSEP computed using PCR, PLS and ECR (α = 0.5). The mean and standard deviations of these distributions are 0.1069±0.0140, 0.1028±0.0111 and 0.0764±0.0108, respectively. Obviously, PLS achieves the lowest prediction errors as well as the smallest standard deviations. In contrast, PCR performs the worst. As a transitional model that is between PCR and PLS, ECR with α = 0.5 achieves the medium level performance.

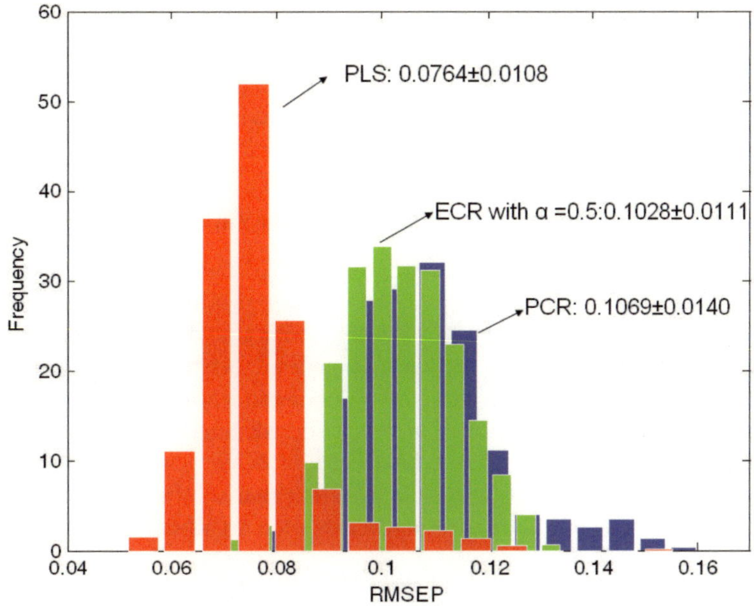

Fig. 9. The distributions of RMSEP from PCR, PLS and an ECR model with α =0.5

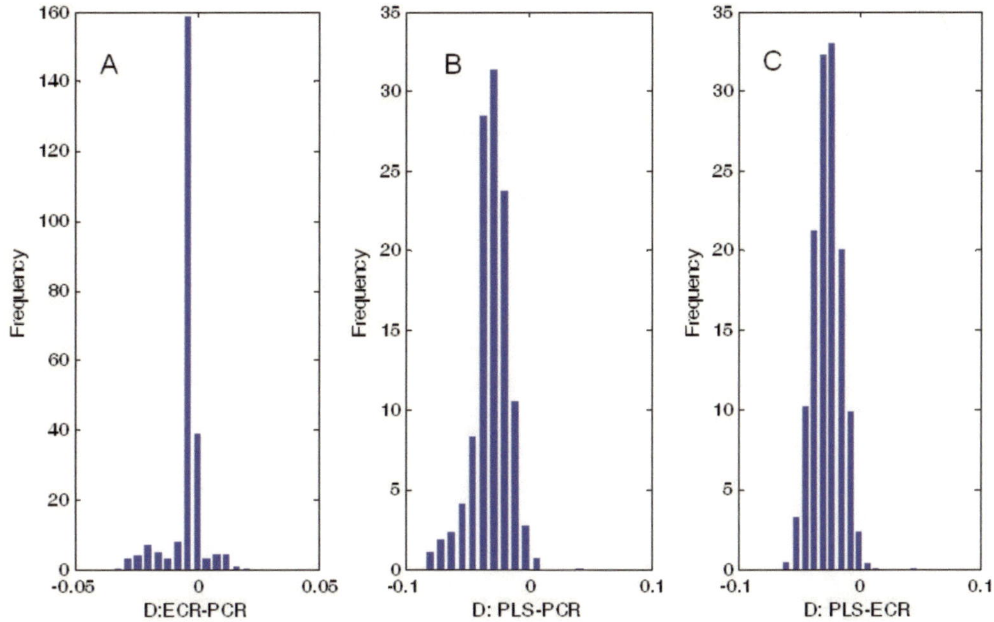

Fig. 10. The distributions of D values. The P values of t test for these three distributions are 0, 0 and 0, respectively.

The distributions of D values are displayed in **Figure 10**. The means of these three distributions are -0.0041 (Plot A), -0.0305 (Plot B) and -0.0264 (Plot C), respectively. Using a two-side t test, it is shown that all these three distributions of D values have a mean value that is significant not zero with P values equal to 0 , 0 and 0 for Plot A, Plot B and Plot C. To conclude, this section provides illustrative examples for the comparison of different modeling methods. Our example demonstrates that PLS (an ECR model associated with α = 1) performs better than PCR (an ECR model associated with α = 0) and a specific transitional ECR model associated with α = 0.5 has the moderate performance.

4.3 Comparison of PLS-LDA models before and after variable selection

Partial least squares-linear discriminant analysis (PLS-LDA) is frequently used in chemometrics and metabolomics/metabonomics for building predictive classification models and/or biomarker discovery [32, 42-45]. With the development of modern high-throughput analytical instruments, the data generated often contains a large number of variables (wavelengths, m/z ratios etc). Most of these variables are not relevant to the problem under investigation. Moreover, a model constructed using this kind of data that contain irrelevant variables would not be likely to have good predictive performance. Variable selection provides a solution to this problem that can help select a small number of informative variables that could be more predictive than an all-variable model.

In the present work, two methods are chosen to conduct variable selection. The first is t-test, which is a simple univariate method that determines whether two samples from normal distributions could have the same mean when standard deviations are unknown but assumed to be equal. The second is subwindow permutation analysis (SPA) which was a model population analysis-based approach proposed in our previous work [14]. The main characteristic of SPA is that it can output a conditional P value by implicitly taking into account synergistic effects among multiple variables. With this conditional P value, important variables or conditionally important variables can be identified. The source codes in Matlab and R are freely available at [46].We apply these two methods on a type 2 diabetes mellitus dataset that contains 90 samples (45 healthy and 45 cases) each of which is characterized by 21 metabolites measured using a GC/MS instrument. Details of this dataset can be found in reference [32].

Using t-test, 13 out of the 21 variables are identified to be significant ($P < 0.01$). For SPA, we use the same setting as described in our previous work [14]. Three variables are selected with the aid of SPA. Let V_0, V_1 and V_2 denote the sets containing all the 21 variables, the 13 variables selected by t-test and the 3 variables selected by SPA, respectively. To run the proposed method, we set the number of Monte Carlo simulations to 1000. At each simulation 70% samples are randomly selected to build a PLS-LDA model with the number of latent variables optimized by 10-fold cross validation. The remaining 30% samples working as test sets on which the misclassification error is computed.

Figure 11 shows the distributions of misclassification errors computed using these three variable sets. The mean and standard deviations of these distributions are 0.065±0.048 (all variables), 0.042±0.037 (t-test) and 0.034±0.034 (SPA), respectively. It can be found that the models using selected variables have lower prediction errors as well as higher stability in terms of smaller standard deviations, indicating that variable selection can improve the

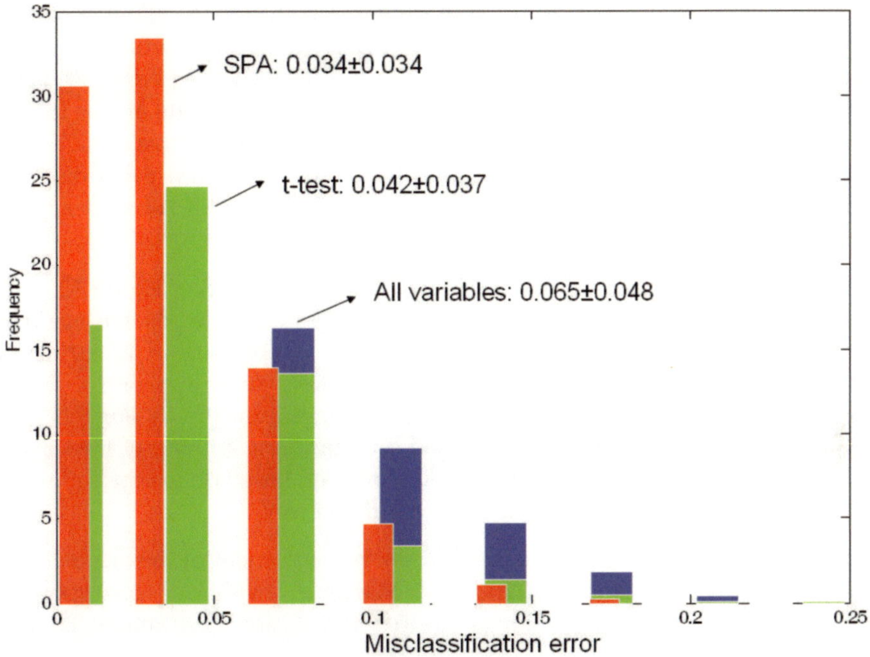

Fig. 11. The distributions of misclassification error on 1000 test sets using all variables and variables selected by t test and SPA, respectively.

performance of a classification model. The reason why SPA performs better than t-test is that synergistic effects among multiple variables are implicitly taken into account in SPA while t-test only considers univariate associations.

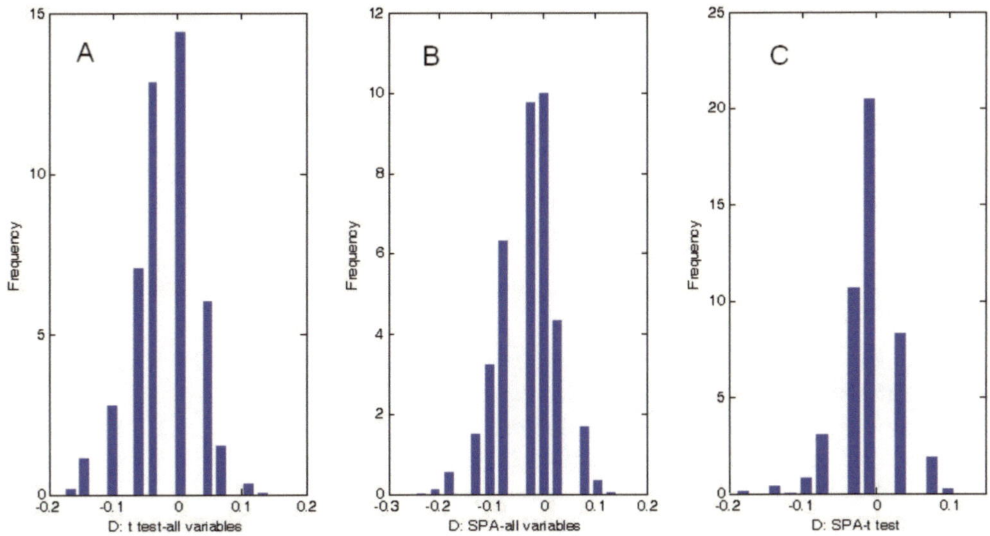

Fig. 12. The distributions of D values. The P values of t test for these three distributions are 1.66×10^{-46}, 1.02×10^{-57} and 1.27×10^{-8}.

We conducted pair-wise comparison of performances of the three variable sets described above. The distribution of D values (t-test – all variables) is shown in Plot A of **Figure 12**. The mean of D is -0.023 and is demonstrated to be significantly not zero (P=0) using a two-side t test, suggesting the improvement of variable selection. In spite of this improvement, we should also notice that a percentage (17.3%) of D values is positive, which again imply that model comparison based on a single split of the data into a training set and a corresponding test set is risky. However, with the aid of this MPA-based approach, it is likely to reliably compare different models in a statistical manner.

The distribution of D values (SPA – all variables) is shown in Plot B of **Figure 12**. The mean of D is -0.031 and is shown to be not zero (P = 0). Also, 171 D values are positive, again indicating the necessity of the use of a population of models for model comparison. In analogy, Plot C in **Figure 12** displays the distributions of D values (SPA-t-test). After applying a two-side t-test, we found that the improvement of SPA over t-test is significant (P= 0). For this distribution, 22.7% D values is positive, indicating that based on a random splitting of the data t-test will have a 22.7% chance to perform better than SPA. However, based on a large scale comparison, the overall performance of SPA is statistically better than t-test.

To conclude, in this section we have compared the performances of the original variable set and variable sets selected using t-test and SPA. We found evidences to support the use of the proposed model population analysis approach for statistical model comparison of different classification models.

5. Conclusions

A model population analysis approach for statistical model comparison is developed in this work. From our case studies, we have found strong evidences that support the use of model population analysis for the comparison of different variable sets or different modeling methods in both regression and classification. P values resulting from the proposed method in combination with the sign of the mean of D values clearly shows whether two models have the same performance or which model is significantly better.

6. Acknowledgements

This work was financially supported by the National Nature Foundation Committee of P.R. China (Grants No. 20875104 and No. 21075138) and the Graduate degree thesis Innovation Foundation of Central South University (CX2010B057).

7. References

[1] H. Akaike, A new look at the statistical model identification, IEEE Transactions on Automatic Control, 19 (1974) 716.
[2] G.E. Schwarz, Estimating the dimension of a model, Annals of Statistics, 6 (1978) 461.
[3] S. Wold, Cross-validatory estimation of the number of components in factor and principal component analysis, Technometrics, 20 (1978) 397.
[4] Q.-S. Xu, Y.-Z. Liang, Monte Carlo cross validation, Chemometr. Intell. Lab., 56 (2001) 1.
[5] P. Filzmoser, B. Liebmann, K. Varmuza, Repeated double cross validation, J Chemometr, 23 (2009) 160.

[6] J. Shao, Linear Model Selection by Cross-Validation, J Am. Stat. Assoc., 88 (1993) 486.
[7] M. Stone, Cross-validatory choice and assessment of statistical predictions, J. R. Stat. Soc. B, 36 (1974) 111.
[8] H.-D. Li, Y.-Z. Liang, Q.-S. Xu, D.-S. Cao, Model population analysis for variable selection, J. Chemometr., 24 (2009) 418.
[9] B. Efron, G. Gong, A Leisurely Look at the Bootstrap, the Jackknife, and Cross-Validation, Am. Stat., 37 (1983) 36.
[10] B. Efron, R. Tibshirani, An introduction to the bootstrap, Chapman&Hall, (1993).
[11] S. Wold, M. Sjöström, L. Eriksson, PLS-regression: a basic tool of chemometrics, Chemometr. Intell. Lab., 58 (2001) 109.
[12] H.-D. Li, Y.-Z. Liang, Q.-S. Xu, Support vector machines and its applications in chemistry, Chemometr. Intell. Lab., 95 (2009) 188
[13] D.S. Cao, Y.Z. Liang, Q.S. Xu, H.D. Li, X. Chen, A New Strategy of Outlier Detection for QSAR/QSPR, J. Comput. Chem., 31 (2010) 592.
[14] H.-D. Li, M.-M. Zeng, B.-B. Tan, Y.-Z. Liang, Q.-S. Xu, D.-S. Cao, Recipe for revealing informative metabolites based on model population analysis, Metabolomics, 6 (2010) 353.
[15] Q. Wang, H.-D. Li, Q.-S. Xu, Y.-Z. Liang, Noise incorporated subwindow permutation analysis for informative gene selection using support vector machines, Analyst, 136 (2011) 1456.
[16] H.-D. Li, Y.-Z. Liang, Q.-S. Xu, D.-S. Cao, B.-B. Tan, B.-C. Deng, C.-C. Lin, Recipe for Uncovering Predictive Genes using Support Vector Machines based on Model Population Analysis, IEEE/ACM T Comput Bi, 8 (2011) 1633.
[17] A.I. Bandos, H.E. Rockette, D. Gur, A permutation test sensitive to differences in areas for comparing ROC curves from a paired design, Statistics in Medicine, 24 (2005) 2873.
[18] Y. Ai-Jun, S. Xin-Yuan, Bayesian variable selection for disease classification using gene expression data, Bioinformatics, 26 (2009) 215.
[19] A. Vehtari, J. Lampinen, Bayesian model assessment and comparison using cross-validation predictive densities, Neural Computation, 14 (2002) 2439.
[20] T. Chen, E. Martin, Bayesian linear regression and variable selection for spectroscopic calibration, Anal. Chim. Acta, 631 (2009) 13.
[21] L. Breiman, Bagging Predictors, Mach. Learn., 24 (1996) 123.
[22] Y. Freund, R. Schapire, Experiments with a new boosting algorithm, Machine Learning: Proceedings of the Thirteenth International Conference, (1996) 148.
[23] L. Breiman, Random Forests, Mach. Learn., 45 (2001) 5.
[24] K. Ting, J. Wells, S. Tan, S. Teng, G. Webb, Feature-subspace aggregating: ensembles for stable andÂ unstable learners, Mach. Learn., 82 (2010) 375.
[25] H.B. Mann, D.R. Whitney, On a test of whether one of two random variables is stochastically larger than the other, Ann. Math. Statist., 18 (1947) 50.
[26] V. Centner, D.-L. Massart, O.E. de Noord, S. de Jong, B.M. Vandeginste, C. Sterna, Elimination of Uninformative Variables for Multivariate Calibration, Anal. Chem., 68 (1996) 3851.
[27] W. Cai, Y. Li, X. Shao, A variable selection method based on uninformative variable elimination for multivariate calibration of near-infrared spectra, Chemometr. Intell. Lab., 90 (2008) 188.

[28] H.-D. Li, Y.-Z. Liang, Q.-S. Xu, D.-S. Cao, Key wavelengths screening using competitive adaptive reweighted sampling method for multivariate calibration, Anal. Chim. Acta, 648 (2009) 77.

[29] J.-H. Jiang, R.J. Berry, H.W. Siesler, Y. Ozaki, Wavelength Interval Selection in Multicomponent Spectral Analysis by Moving Window Partial Least-Squares Regression with Applications to Mid-Infrared and Near-Infrared Spectroscopic Data, Anal. Chem., 74 (2002) 3555.

[30] C. Reynes, S. de Souza, R. Sabatier, G. Figueres, B. Vidal, Selection of discriminant wavelength intervals in NIR spectrometry with genetic algorithms, J. Chemometr., 20 (2006) 136.

[31] Q.-J. Han, H.-L. Wu, C.-B. Cai, L. Xu, R.-Q. Yu, An ensemble of Monte Carlo uninformative variable elimination for wavelength selection, Anal. Chim. Acta, 612 (2008) 121.

[32] B.-B. Tan, Y.-Z. Liang, L.-Z. Yi, H.-D. Li, Z.-G. Zhou, X.-Y. Ji, J.-H. Deng, Identification of free fatty acids profiling of type 2 diabetes mellitus and exploring possible biomarkers by GC–MS coupled with chemometrics, Metabolomics, 6 (2009) 219.

[33] W. Fan, H.-D. Li, Y. Shan, H.-Y. Lv, H.-X. Zhang, Y.-Z. Liang, Classification of vinegar samples based on near infrared spectroscopy combined with wavelength selection, Analytical Methods, 3 (2011) 1872.

[34] Source codes of CARS-PLS for variable selection: http://code.google.com/p/carspls/

[35] Source codes of CARS-PLSLDA for variable selection: http://code.google.com/p/cars2009/

[36] H.-D. Li, Y.-Z. Liang, Q.-S. Xu, Uncover the path from PCR to PLS via elastic component regression, Chemometr. Intell. Lab., 104 (2010) 341.

[37] M. Stone, R.J. Brooks, Continuum Regression: Cross-Validated Sequentially Constructed Prediction Embracing Ordinary Least Squares, Partial Least Squares and Principal Components Regression, J. R. Statist. Soc. B, 52 (1990) 237.

[38] A. Bjorkstrom, R. Sundberg, A generalized view on continuum regression, Scand. J. Statist., 26 (1999) 17.

[39] B.M. Wise, N.L. Ricker, Identification of finite impulse response models with continuum regression, J. Chemometr., 7 (1993) 1.

[40] J.H. Kalivas, Cyclic subspace regression with analysis of the hat matrix, Chemometr. Intell. Lab., 45 (1999) 215.

[41] Source codes of Elastic Component Regression: http://code.google.com/p/ecr/

[42] M. Barker, W. Rayens, Partial least squares for discrimination, J. Chemometr., 17 (2003) 166.

[43] J.A. Westerhuis, H.C.J. Hoefsloot, S. Smit, D.J. Vis, A.K. Smilde, E.J.J. van Velzen, J.P.M. van Duijnhoven, F.A. van Dorsten, Assessment of PLSDA cross validation, Metabolomics, 4 (2008) 81.

[44] L.-Z. Yi, J. He, Y.-Z. Liang, D.-L. Yuan, F.-T. Chau, Plasma fatty acid metabolic profiling and biomarkers of type 2 diabetes mellitus based on GC/MS and PLS-LDA, FEBS Letters, 580 (2006) 6837.

[45] J. Trygg, E. Holmes, T.r. Lundstedt, Chemometrics in Metabonomics, Journal of Proteome Research, 6 (2006) 469.

[46] Source codes of Subwindow Permutation Analysis: http://code.google.com/p/spa2010/

2

Critical Aspects of Supervised Pattern Recognition Methods for Interpreting Compositional Data

A. Gustavo González

Department of Analytical Chemistry, University of Seville, Seville Spain

1. Introduction

A lot of multivariate data sets of interest to scientists are called compositional or "closed" data sets, and consists essentially of relative proportions. A recent search on the web by entering "chemical compositional data", led to more than 2,730,000 results within different fields and disciplines, but specially, agricultural and food sciences (August 2011 using Google searcher). The driving causes for the composition of foods lie on four factors (González, 2007): Genetic factor (genetic control and manipulation of original specimens), Environmental factor (soil, climate and symbiotic and parasite organisms), Agricultural factor (cultures, crop, irrigation, fertilizers and harvest practices) and Processing factor (post-harvest manipulation, preservation, additives, conversion to another food preparation and finished product). But the influences of these factors are hidden behind the analytical measurements and only can be inferred and uncover by using suitable chemometric procedures.

Chemometrics is a term originally coined by Svante Wold and could be defined as "The art of extracting chemically relevant information from data produced in chemical experiments" (Wold, 1995). Besides, chemometrics can be also defined as the application of mathematical, statistical, graphical or symbolic methods to maximize the chemical information which can be extracted from data (Rock, 1985). Within the jargon of chemometrics some other terms are very common; among them, multivariate analysis and pattern recognition are often used. Chemometricians use the term Multivariate Analysis in reference to the different approaches (mathematical, statistical, graphical...) when considering samples featured by multiple descriptors simultaneously. Pattern recognition is a branch of the Artificial Intelligence that seeks to identify similarities and regularities present in the data in order to attain natural classification and grouping. When applied to chemical compositional data, pattern recognition methods can be seen as multivariate analysis applied to chemical measurements to find classification rules for discrimination issues. Depending on our knowledge about the category or class membership of the data set, two approaches can be applied: Supervised or unsupervised learning (pattern recognition).

Supervised learning methods develop rules for the classification of unknown samples on the basis of a group of samples with known categories (known set). Unsupervised learning methods instead do not assume any known set and the goal is to find clusters of objects which may be assigned to classes. There is hardly any quantitative analytical method that does not make use of chemometrics. Even if one confines the scope to supervised learning pattern recognition, these chemometric techniques are increasingly being applied to of compositional data for classification and authentication purposes. The discrimination of the geographical origin, the assignation to Denominations of Origin, the classification of varieties and cultivars are typical issues in agriculture and food science.

There are analytical techniques such as the based on sensors arrays (electronic nose and electronic tongue) that cannot be imaginable without the help of chemometrics. Thus, one could celebrate the triumph of chemometrics...so what is wrong? (Pretsch & Wilkin, 2006). The answer could be supported by a very well-known quotation attributed to the british politician D'Israeli (Defernez & Kemsley, 1997) but modified by us as follows: "There are three kinds of lies: Lies, damned lies and chemometrics". Here we have changed the original word "statistics" into "chemometrics" in order to point out the suspicion towards some chemometric techniques even within the scientific community. There is no doubt that unwarranted reliance on poorly understood chemometric methods is responsible for such as suspicion.

By the way, chemometric techniques are applied using statistical/chemometric software packages that work as black boxes for final users. Sometimes the blindly use of "intelligent problem solvers" or similar wizards with a lot of hidden options as default may lead to misleading results. Accordingly, it should be advisable to use software packages with full control on parameters and options, and obviously, this software should be used by a chemometric *connaisseur*.

There are special statistical methods intended for closed data such as compositional ones (Aitchison, 2003; Egozcue et al., 2003; Varmuza & Filzmoser, 2009), but a detailed description of these methods may be outside the scope of this chapter.

2. About the data set

Supervised learning techniques are used either for developing classification rules which accurately predict the classification of unknown patterns or samples (Kryger, 1981) or for finding calibration relationships between one set of measurements which are easy or cheap to acquire, and other measurements which are expensive or labour intensive, in order to predict these later (Naes et al., 2004). The simplest calibration problem consists of predicting a single response (y-variable) from a known predictor (x-variable) and can be solved by using ordinary linear regression (OLR). When fitting a single response from several predictive variables, multiple linear regression (MLR) may be used; but for the sake of avoiding multicollinearity drawbacks, some other procedures such as principal component regression (PCR) or partial least squares regression (PLS) are a good choice. If faced to several response variables, multivariate partial least squares (PLS2) techniques have to be used. These procedures are common in multivariate calibration (analyte concentrations from NIR data) or linear free energy relationships (pKa values from molecular descriptors). However, in some instances like quantitative structure-activity relationships (QSAR), non linear strategies are needed, such as quadratic partial least squares regression (QPLS), or regression procedures based on artificial neural networks or support vector machines. All these calibration procedures have to be suitably validated by using validation procedures based on the knowledge of class memberships of the objects, in a similar way as discussed below in this chapter that is devoted to supervised learning for classification.

Let us assume that a known set of samples is available, where the category or class membership of every sample is *a priori* known. Then a suitable planning of the data-acquisition process is needed. At this point, chemical experience, *savoir faire* and intuition are invaluable in order to decide which measurements should be made on the samples and which variables of these measurements are most likely to contain class information.

In the case of compositional data, a lot of analytical techniques can be chosen. Analytical procedures based on these techniques are then selected to be applied to the samples. Selected analytical methods have to be fully validated and with an estimation of their uncertainty (González & Herrador, 2007) and carried out in Quality Assurance conditions (equipment within specifications, qualified staff, and documentation written as Standard Operational Procedures...). Measurements should be carried out at least duplicate and according to a given experimental design to ensure randomization and avoid systematic trends.

Difficulties arise when the concentration of an element is below the detection limit (DL). It is often standard practice to report these data simply as '<DL' values. Such 'censoring' of data, however, can complicate all subsequent statistical analyses. The best method to use generally depends on the amount of data below the detection limit, the size of the data set, and the probability distribution of the measurements. When the number of '< DL' observations is small, replacing them with a constant is generally satisfactory (Clarke, 1998). The values that are commonly used to replace the '< DL' values are 0, DL, or DL/2. Distributional methods such as the marginal maximum likelihood estimation (Chung, 1993) or more robust techniques (Helsel, 1990) are often required when a large number of '< DL' observations are present.

After all measurements are done we can built the corresponding data table or data matrix. A sample, object or pattern is described by a set of "p" variables, features or descriptors. So, all descriptors of one pattern form a 'pattern vector' and accordingly, a given pattern "i" can be seen as a vector $\vec{x}_i$ whose components are $x_{i1}, x_{i2}, \ldots x_{ij}, \ldots x_{ip}$ in the vectorial space defined by the features. In matricial form, pattern vectors are row vectors. If we have n patterns, we can build a data matrix $X_{n \times p}$ by assembling the different row pattern vectors. A change in perspective is also possible: A given feature "j" can be seen as a column vector $\vec{x}_j$ with components $x_{1j}, x_{2j}, \ldots x_{ij}, \ldots x_{nj}$ in the vectorial space defined by the patterns. We can also construct the data matrix by assembling the different feature column vectors. Accordingly, the data matrix can be considered as describing the patterns in terms of features or *vice versa*. This lead to two main classes of chemometric techniques called Q- and R-modes respectively. R-mode techniques are concerned with the relationships amongst the features of the experiment and examine the interplay between the columns of the data matrix. A starting point for R-mode procedures is the covariance matrix of mean centered variables $C = X^T X$ whose elements are given by $c_{jk} = \frac{1}{n-1}\sum_{l=1}^{n}\left(x_{lj} - \bar{x}_j\right)\left(x_{lk} - \bar{x}_k\right)$ where $\bar{x}_j$ and $\bar{x}_k$ are the mean of the observations on the jth

and k*th* feature. If working with autoscaled data, the sample correlation matrix and the covariance matrix are identical. The element r_{jk} of correlation matrix R represents the cosine between each pair of column vectors and is given by $r_{jk} = \frac{c_{jk}}{\sqrt{c_{jj}c_{kk}}}$. The diagonal elements of R are always unity. The alternative viewpoint considers the relationships between patterns, the Q-mode technique. This way normally starts with a matrix of distances between the objects in the n-dimensional pattern space to study the clustering of samples. Typical metric measurements are Euclidean, Minkowski, Manhattan, Hamming, Tanimoto and Mahalanobis distances (Varmuza, 1980).

3. Inspect data matrix

Once the data matrix has been built, it should be fully examined in order to ensure the suitable application of Supervised Learning methodology. A typical undesirable issue is the existence of missing data. Holes in the data matrix must be avoided; however some measurements may not have been recorded or are impossible to obtain experimentally due to insufficient sample amounts or due to high costs. Besides, data can be missing due to various malfunctions of the instruments, or responses can be outside the instrument range.

As stated above, most chemometric techniques of data analysis do not allow for data gaps and thereof different methods have been applied for handling missing values in data matrix. Aside from the extreme situations of casewise deletion or mean substitution, the use of iterative algorithms (IA) is a promising tool. Each iteration consists of two steps. The first step performs estimation of model parameters just as if there were no missing data. The second step finds the conditional expectation of the missing elements given the observed data and current estimated parameters. The detailed procedure depends on the particular application. The typical IA used in Principal Component Analysis can be summarized as (Walczak & Massart, 2001):

1. Fill in missing elements with their initial estimates (expected values, calculated as the mean of the corresponding row's and column's means)
2. Perform singular value decomposition of the complete data set
3. Reconstruct X with the predefined number of factors
4. Replace the missing elements with the predicted values and go to step 2 until convergence.

Replacement of missing values or censored data with any value is always risky since this can substantially change the correlation in the data. It is possible to deal with both missing values and outliers simultaneously (Stanimirova et al., 2007). An excellent revision dealing with zeros and missing values in compositional data sets using non-parametric imputation has been performed by Martin-Fernández et al. (2003).

On the other hand, a number of chemometric procedures are based on the assumption of normality of features. Accordingly, features should be assessed for normality. The well-known Kolmogorov-Smirnov, Shapiro-Wilks and Lilliefors tests are often used (González,

2007), although the data about the skewness and kurtosis of the distribution are also of interest in order to consider parametric or non parametric descriptive statistics. Some simple presentations such as the Box-and-whisker plots help in the visual identification of outliers and other unusual characteristics of the data set. The box-and-whisker plot assorted with a numerical scale is a graphical representation of the five-number summary of the data set (Miller & Miller, 2005) where it is described by its extremes, its lower and upper quartiles and the median and gives at a glance the spread and the symmetry of the data set. Box-and-whisker plots may reveal suspicious patterns that should be tested for outliers. Abnormal data can road chemometric techniques leading to misleading conclusions, especially when outliers are present in the training set. Univariate outlier tests such as Dean and Dixon (1951) or Grubbs (1969) assays are not suitable. Instead, multivariate criteria for outlier detection are more advisable. The techniques based on the Mahalanobis distance (Gemperline & Boyer, 1995), and the hat matrix leverage (Hoaglin & Welsch, 1978) have been often used for decades. Hat matrix $H = X\left(X^T X\right)^{-1} X^T$ has diagonal values h_{ii} called leverage values. Patterns having leverage values higher than 2p/n are commonly considered outliers. However these methods are unreliable for multivariate outlier detection. Numerous methods for outlier detection have been based on the singular value decomposition or Principal Component Analysis (PCA) (Jollife, 2002). Soft Independent Modelling of Class Analogy (SIMCA) has been also applied to outlier detection (Mertens et al. , 1994). Once outliers have been deleted, researchers usually remove them from the data set, but outliers could be corrected before applying the definite mathematical procedures by using robust algorithms (Daszykowski et al., 2007). Robust methods give better results, specially some improved algorithms such as resampling by the half-means (RHM) and smallest half-volume (SHV) (Egan & Morgan , 1998).

Within the field of food authentication, the wrong conclusions are, however, mostly due to data sets that do not keep all aspects of the food characterisation (partial or skewed data set) or do merge data measured with different techniques (Aparicio & Aparicio-Ruiz, 2002). A classical example is the assessment of the geographical origin of Italian virgin olive oils by Artificial Neural Networks (ANN). The differences between oils were mainly due to the use of different chromatographic columns (packed columns against capillary columns) when quantifying the free fatty acid (FFA) profile of the oils from Southern and Northern Italy respectively (Zupan et al. , 1994). The neural network thus mostly learned to recognise the chromatographic columns.

4. Data pre-treatment

Data transformation (scaling) can be applied either for statistical dictates to optimise the analysis or based on chemical reasons. Raw compositional data are expressed in concentration units that can differ by orders of magnitude (e.g., percentage, ppm or ppb), the features with the largest absolute values are likely to dominate and influence the rule development and the classification process. Thus, for statistical needs, transformation of raw data can be applied to uniformize feature values. Autoscaling and column range scaling are the most common transformation (Sharaf et al., 1986). In the autoscaling or Z-transformation, raw data are transformed according to

$$x_{ij}^{'} = \frac{x_{ij} - \overline{x}_j}{s_j} \text{ with } s_j = \sqrt{\frac{\sum_{i=1}^{n}\left(x_{ij} - \overline{x}_j\right)^2}{n-1}} \tag{1}$$

and can be seen as a parametric scaling by column standardisation leading to $\overline{x}_j^{'} = 0$ and $s_j^{'} = 1$.

Column range scaling or minimax scaling involves the following transformation

$$x_{ij}^{'} = \frac{x_{ij} - \min_j\left(x_{ij}\right)}{\max_j\left(x_{ij}\right) - \min_j\left(x_{ij}\right)} \tag{2}$$

Now, the transformed data verify $0 \le x_{ij}^{'} \le 1$. Range scaling can be considered as an interpolation of data within the interval (0,1). It is a non-parametric scaling but sensitive to outliers. When the data contain outliers and the preprocessing is necessary, one should consider robust way of data preprocessing in the context of robust Soft Independent Modelling of Class Analogy (SIMCA) method (Daszykowski et al., 2007).

The second kind of transformations are done for chemical reasons and comprise the called "constant-row sum" and "normalization variable" (Johnson & Ehrlich, 2002). Dealing with compositional data, concentrations can vary widely due to dilution away from a source. In the case of contaminated sediment investigations, for example, concentrations may decrease exponentially away from the effluent pipe. However, if the relative proportions of individual analytes remain relatively constant, then we would infer a single source scenario coupled with dilution far away from the source. Thus, a transformation is needed to normalize concentration/dilution effects. Commonly this is done using a transformation to a fractional ratio or percent, where each concentration value is divided by the total concentration of the sample: Row profile or constant row-sum transformation because the sum of analyte concentrations in each sample (across rows) sums unity or 100%:

$$x_{ij}^{'} = \frac{x_{ij}}{\sum_{j=1}^{p} x_{ij}} \tag{3}$$

leading to $\sum_{j=1}^{p} x_{ij}^{'} = 1 \text{ (or } 100\%)$. An alternative is to normalize the data with respect to a single species or compound set as reference congener, the normalization variable. This transformation involves setting the value of the normalization feature to unity, and the values of all other features to some proportion of 1.0, such that their ratios with respect to the normalization feature remain the same in the original metric.

When n > p, the rank of data matrix is p (if all features are independent). Thus, autoscaling and minimax transformations do not change data dimensionality because these treatments do not induce any bound between features. Row profiles instead build a relationship between features scores (the constant-row sum) that decreases the data dimensionality by 1 and the rank of data matrix is then p-1. Accordingly, the patterns fall on a hypersurface in the feature space and it is advisable to remove one feature to avoid problems involving matrix inversion when the rank of the matrix is less than p.

As a final remark it should be realized that when using some Supervised Learning techniques like SIMCA, the scaling of the data set is carried out only over the samples belonging to the same class (separate scaling). This is due because the own fundamentals of the methodology and has a beneficial effect on the classification (Derde et al., 1982).

5. Feature selection and extraction

Irrelevant features are very expensive ones, because they contribute to the chemical information with noise only; but even more expensive may be simply wrong features. Accordingly, they should be eliminated in order to circumvent disturb in classification. In almost all chemical applications of pattern recognition the number of original raw features is too large and a reduction of the dimensionality is necessary. Features which are essential for classification purposes are often called intrinsic features. Thus, a common practise to avoid redundant information consists of computing the correlation matrix of features R. Pair of most correlated features can be either combined or one of them is deleted. Researchers should be aware that the number of independent descriptors or features, p, must be much smaller than that of patterns, n. Otherwise, we can build a classification rule that even separates randomly selected classes of the training set (Varmuza, 1980). This assumes that the set of features is linearly independent (actually, it is the basis of the vectorial space) and the number of features is the dimensionality of the vectorial space. Accordingly, the true dimensionality should be evaluated for instance from an eigenanalysis (PCA) of the data matrix and extract the proper number of factors which correspond to the true dimensionality (d) of the space. Most efficient criteria for extracting the proper number of underlying factors are based on the Malinowski indicator function (Malinowski, 2002) and the Wold's Cross-Validation procedure (Wold, 1978). For most classification methods, a ratio n/d > 3 is advised and > 10, desirable. However, PCA-based methods like SIMCA or Partial Least Squares Discriminant Analysis (PLS-DA) can be applied without problem when p >> n. However, even in these instances there are suitable methods for selecting a subset of features and to build a final model based on it.

Therefore, when it is advisable the feature selection, weighing methods determine the importance of the scaled features for a certain classification problem. Consider a pattern vector $\vec{x}_i \equiv (x_{i1}, x_{i2}, \ldots x_{ip})$. Assuming that the data matrix X can be partitioned into a number Q of classes, let $\vec{x}_i^{(C)}$ a pattern vector belonging to class C. The averaged patterns $\vec{m}$ and $\vec{m}^{(C)}$ represent the general mean vector and the C-class mean, according to:

$$\vec{m} = \frac{\sum_{i=1}^{n} \vec{x}_i}{n} \quad \text{and} \quad \vec{m}^{(C)} = \frac{\sum_{i=1}^{n(C)} \vec{x}_i^{(C)}}{n} \qquad (4)$$

When they are applied to a selected j feature we have

$$m_j = \frac{\sum_{i=1}^{n} x_{ij}}{n} \quad \text{and} \quad m_j^{(C)} = \frac{\sum_{i=1}^{n(C)} x_{ij}^{(C)}}{n(C)} \tag{5}$$

Where n(C) is the number of patterns of class C.

Accordingly, we can explore the inter-class scatter matrix as well as the intra-class scatter matrix. The total scatter matrix can be obtained as

$$T = \sum_{i=1}^{n} (\vec{x}_i - \vec{m})(\vec{x}_i - \vec{m})^T \tag{6}$$

and its elements as

$$T_{jk} = \sum_{i=1}^{n} (x_{ij} - m_j)(x_{ik} - m_k) \tag{7}$$

The within classes scatter matrix, together with its element is given by

$$\begin{aligned} W &= \sum_{C=1}^{Q} \sum_{i=1}^{n(C)} (\vec{x}_i^{(C)} - \vec{m}^{(C)})(\vec{x}_i^{(C)} - \vec{m}^{(C)})^T \\ W_{jk} &= \sum_{C=1}^{Q} \sum_{i=1}^{n(C)} (x_{ij}^{(C)} - m_j^{(C)})(x_{ik}^{(C)} - m_k^{(C)}) \end{aligned} \tag{8}$$

And the between classes matrix and element,

$$\begin{aligned} B &= \sum_{C=1}^{Q} n(C)(\vec{m}^{(C)} - \vec{m})(\vec{m}^{(C)} - \vec{m})^T \\ B_{jk} &= \sum_{C=1}^{Q} n(C)(m_j^{(C)} - m_j)(m_k^{(C)} - m_k) \end{aligned} \tag{9}$$

For a case involving two classes 1 and 2 and one feature j we have the following:

$$\begin{aligned} T_{jj} &= \sum_{i=1}^{p} (x_{ij} - m_j)^2 \\ W_{jj} &= \sum_{i=1}^{n(1)} (x_{ij}^{(1)} - m_j^{(1)})^2 + \sum_{i=1}^{n(2)} (x_{ij}^{(2)} - m_j^{(2)})^2 \\ B_{jj} &= n(1)(m_j^{(1)} - m_j)^2 + n(2)(m_j^{(2)} - m_j)^2 \end{aligned} \tag{10}$$

Weighting features in Supervised Learning techniques can be then extracted from its relative importance in discriminating classes pairwise. The largest weight corresponds to the most important feature. The most common weighting factors are:

- Variance weights (VW) (Kowalski & Bender, 1972): $VW_j = \frac{B_{jj}}{W_{jj}}$

- Fisher weights (FW) (Duda et al., 2000): $FW_j = \dfrac{(m_j^{(1)} - m_j^{(2)})^2}{W_{jj}}$

- Coomans weights (g) (Coomans et al., 1978):

$$g_j = \frac{\left|m_j^{(1)} - m_j^{(2)}\right|}{s_j^{(1)} + s_j^{(2)}} \text{ with } s_j^{(C)} = \sqrt{\frac{\sum_{i=1}^{n(C)} (x_{ij}^{(C)} - m_j^{(C)})^2}{n(C)}}$$

A multi-group criterion is the called Wilks' λ or McCabe U statistics (McCabe, 1975). This is a general statistic used as a measure for testing the difference among group centroids. All classes are assumed to be homogeneous variance-covariance matrices and the statistic is defined as

$$\lambda = \frac{\det W}{\det T} = \frac{SSW}{SST} = \frac{SSW}{SSB + SSW} \tag{11}$$

Where SSW, SSB and SST refer to the sum of squares corresponding to the scatter matrices W, B and T, respectively, as defined above. Remembering that the ratio $\eta = \sqrt{\dfrac{BSS}{WSS}}$ is the coefficient of canonical correlation, $\eta = \sqrt{1-\lambda}$, and hence when $\eta \to 1$ for intrinsic features, $\lambda \to 0$ and more significant are the centroid difference. Before calculation of the statistic, data should be autoscaled. This later criterion as well as the largest values of Rao's distance or Mahalanobis distance is generally used in Stepwise Discriminant Analysis (Coomans et al., 1979). Certain Supervised Learning techniques enable feature selection according to its own philosophy. Thus, for instance, SIMCA test the intrinsic features according the values of two indices called discriminating power and modelling power (Kvalheim & Karstang, 1992). Using ANNs for variable selection is attractive since one can globally adapt the variables selector together with the classifier by using the called "pruning" facilities. Pruning is a heuristic method to feature selection by building networks that do not use those variables as inputs. Thus, various combinations of input features can be added and removed, building new networks for each (Maier et al., 1998).

Genetic algorithms are also very useful for feature selection in fast methods such as PLS (Leardi & Lupiañez, 1998).

6. Development of the decision rule

In order to focus the commonly used Supervised Learning techniques of pattern recognition we have selected the following methods: K-Nearest Neighbours (KNN) (Silverman & Jones, 1989), Linear Discriminant Analysis (LDA) (Coomans et al., 1979), Canonical Variate Analysis (CVA) (Cole & Phelps, 1979), Soft Independent Modelling of Class Analogy (SIMCA) (Wold, 1976), Unequal dispersed classes (UNEQ) (Derde & Massart, 1986), PLS-DA (Stahle & Wold, 1987), Procrustes Discriminant Analysis (PDA) (González-Arjona et al., 2001), and methods based on ANN such as Multi-Layer Perceptrons (MLP) (Zupan & Gasteiger, 1993; Bishop, 2000), Supervised Kohonen Networks (Melssen et al., 2006),

Kohonen Class-Modelling (KCM) (Marini et al., 2005), and Probabilistic Neural Networks (PNN) (Streit & Luginbuhl, 1994). Recently, new special classification techniques arose. A procedure called Classification And Influence Matrix Analysis (CAIMAN) has been introduced by Todeschini *et al* (2007). The method is based on the leverage matrix and models each class by means of the class dispersion matrix and calculates the leverage of each sample with respect to each class model space. Since about two decades another new classification (and regression) revolutionary technique based on statistical learning theory and kernel latent variables has been proposed: Support Vector Machines (SVM) (Vapnik, 1998; Abe, 2005; Burges, 1998). The purpose of SVM is separate the classes in a vectorial space independently on the probabilistic distribution of pattern vectors in the data set (Berrueta et al., 2007). This separation is performed with the particular hyperplane which maximizes a quantity called margin. The margin is the distance from a hyperplane separating the classes to the nearest point in the data set (Pardo & Sberveglieri, 2005). The training pattern vectors closest to the separation boundary are called *support vectors.* When dealing with a non linear boundary, the kernel method is applied. The key idea of kernel method is a transformation of the original vectorial space (input space) to a high dimensional Hilbert space (feature space), in which the classes can be separated linearly. The main advantages of SVM against its most direct concurrent method, ANN, are the easy avoiding of overfitting by using a penalty parameter and the finding of a deterministic global minimum against the non deterministic local minimum attained with ANN.

Some of the mentioned methods are equivalent. Let us consider some couples: CVA and LDA and PLS-DA and PDA. CVA attempts to find linear combinations of variables from each set that exhibit maximum correlation. These may be referred to as canonical variates, and data can be displayed as scatterplot of one against the other. The problem of maximizing the correlation can be formulated as an eigenanalysis problem with the largest eigenvalue providing the maximized correlation and the eigenvectors giving the canonical variates. Loadings of original features in the canonical variates and cumulative proportions of eigenvalues are interpreted, partly by analogy with PCA. Note that if one set of features are dummy variables giving group indicators, and then CVA is mathematically identical to LDA (González-Arjona et al., 2006). PLS-DA finds latent variables in the feature space which have a maximum covariance with the y variable. PDA may be considered equivalent to PLS-DA. The only difference is that in PDA, eigenvectors are obtained from the covariance matrix $Z^T Z$ instead of $X^T X$, with $Z = Y^T X$ where Y is the membership target matrix constructed with ones and zeros: For a three classes problem, sample labels are 001, 010 and 100. Accordingly, we can consider CVA equivalent to LDA and PLS-DA equivalent to PDA.

Researchers should be aware of apply the proper methods according to the nature and goals of the chemical problem. As Daszykowski and Walczak pointed out in his excellent survey (Daszykowski & Walczak, 2006), in many applications, unsupervised methods such as PCA are used for classification purposes instead of the supervised approach. If the data set is well structured, then PCA-scores plot can reveal grouping of patterns with different origin, although the lack of these groups in the PCA space does not necessarily mean that there is no statistical difference between these samples. PCA by definition maximizes data variance, but the main variance cannot be necessarily associated with the studied effect (for instance, sample origin). Evidently, PCA can be used for exploration, compression and visualization of data trends, but it cannot be used as Supervised Learning classification method.

On the other hands, according to the nature of the chemical problem, some supervised techniques perform better than others, because its own fundamentals and scope. In order to consider the different possibilities, four paradigms can be envisaged:

1. *Parametric/non-parametric techniques*: This first distinction can be made between techniques that take account of the information on the population distribution. Non parametric techniques such as KNN, ANN, CAIMAN and SVM make no assumption on the population distribution while parametric methods (LDA, SIMCA, UNEQ, PLS-DA) are based on the information of the distribution functions. LDA and UNEQ are based on the assumption that the population distributions are multivariate normally distributed. SIMCA is a parametric method that constructs a PCA model for each class separately and it assumes that the residuals are normally distributed. PLS-DA is also a parametric technique because the prediction of class memberships is performed by means of model that can be formulated as a regression equation of Y matrix (class membership codes) against X matrix (González-Arjona et al., 1999).
2. *Discriminating (hard)/Class-Modelling (soft) techniques*: Pure classification, discriminating or hard classification techniques are said to apply for the first level of Pattern Recognition, where objects are classified into either of a number of defined classes (Albano et al., 1978). These methods operate dividing the hyperspace in as many regions as the number of classes so that, if a sample falls in the region of space corresponding to a particular category, it is classified as belonging to that category. These kinds of methods include LDA, KNN, PLS-DA, MLP, PNN and SVM. On the other hands, Class-Modelling techniques build frontiers between each class and the rest of the universe. The decision rule for a given class is a class box that envelopes the position of the class in the pattern space. So, three kinds of classification are possible: (i) an object is assigned to a category if it is situated inside the boundaries of only a class box, (ii) an object can be inside the boundaries (overlapping region) of more than one class box, or (iii) an object is considered to be an outlier for that class if it falls outside the class box. These are the features to be covered by methods designed for the so called second level of Pattern Recognition: The first level plus the possibility of outliers and multicategory objects. Thus, typical class modelling techniques are SIMCA and UNEQ as well as some modified kind of ANN as KCM. CAIMAN method is developed in different options: D-CAIMAN is a discriminating classification method and M-CAIMAN is a class modelling one.
3. *Deterministic/Probabilistic techniques*: A deterministic method classifies an object in one and only one of the training classes and the degree of reliability of this decision is not measured. Probabilistic methods provide an estimate of the reliability of the classification decision. KNN, MLP, SVM and CAIMAN are deterministic. Other techniques, including some kind of ANN are probabilistic (e.g., PNN where a Bayesian decision is implemented).
4. *Linear/Non-Linear separation boundaries*: Here our attention is focused on the mathematical form of the decision boundary. Typical non-linear classification techniques are based on ANN and SVM, specially devoted to apply for classification problems of non-linear nature. It is remarkable that CAIMAN method seems not to suffer of nonlinear class separability problems.

7. Validation of the decision rule

A very important issue is the improper model validation. This pitfall even appears in very simple cases, such as the fitting of a series of data points by using a polynomial function. If we use a parsimonic fitting where the number of points is higher than the number of polynomial coefficients, the fitting train the generalities of the data set. Overparametrized fitting where the number of points becomes equal to the number of polynimial coefficients, trains idiosyncrasies and leads to overtraining or overfitting. Thus, a complex fitting function may fit the noise, not just the signal. Overfitting is a Damocles' sword that gravitates over any attempt to model the classification rule. We are interested to an intermediate behaviour: A model which is powerful enough to represent the underlying structure of the data (generalities), but not so powerful that it faithfully models the noise (idiosyncrasies) associated to data. This balance is known as the bias-variance tradeoff . The bias-variance tradeoff is most likely to become a problem when we have relatively few data points. In the opposite case, there is no danger of overfitting, as the noise associated with any single data point plays an immaterial role in the overall fit.

If we transfer the problem of fitting a polynomial to data into the use of another functions, such as the discriminant functions of canonical variates issued from LDA, the number of discriminant functions will be p (the number of features) or Q-1 (Q is the number of classes), whichever is smaller. As a rule of thumb (Defernez & Kemsley, 1997), the onset of overfitting should be strongly suspected when the dimensionality $d > \frac{n-Q}{3}$. One of the simplest and most widely used means of preventing overfitting is to split the known data set into two sets: the training set and the validation, evaluation, prediction or test set.

Commonly, the known set is generally randomly divided into the training and validation sets, containing about P% and 100-P% samples of every class. Typical values are 75-25% or even 50-50% for training and validation sets. The classification performance is computed in average. Thus, the randomly generation of training and validation sets is repeated a number of times, 10 times for instance. Once the classification rule is developed, some workers consider as validation parameters the recalling efficiency (rate of training samples correctly classified by the rule) and, specially, the prediction ability (rate of evaluation samples correctly classified by the rule).

An alternative to the generation of training and validation sets are the cross-validation and the bootstrapping method (Efron and Gong, 1983). In the called k-fold cross validation, the know set is split into k subsets of approximately equal size. Then the training is performed k times, each time leaving out one of k the subsets, but using only the omitted subset to predict its class membership. From all predictions, the percentage of hits gives an averaged predictive ability. A very common and simple case of cross-validation is the leave-one-out method: At any given time, only a pattern is considered and tested and the remaining patterns form the training set. Training and prediction is repeated until each pattern was treated as test once. This later procedure is easily confused with jacknifing because both techniques involve omitting each pattern in turn, but cross-validation is used just for validation purposes and jacknife is applied in order to estimate the bias of a statistic.

In bootstrapping, we repeatedly analyze subsamples, instead of subsets of the known set. Each subsample is a random sample with replacement from the full sample (known set). Bootstrapping seems to perform better than cross-validation in many instances (Efron, 1983).

However, the performance rate obtained for validating the decision rule could be misleading because they do not consider the number of false positive and false negative for each class. These two concepts provide a deep knowledge of the classes' space. Accordingly, it seems to be more advisable the use of terms sensitivity (*SENS*) and specificity (*SPEC*) (González-Arjona et al., 2006) for validating the decision rule. The *SENS* of a class corresponds to the rate of evaluation objects belonging to the class that are correctly classified, and the *SPEC* of a class corresponds to the rate of evaluation objects not belonging to the class that are correctly considered as belonging to the other classes. This may be explained in terms of the first and second kind of risks associated with prediction. The first kind of errors (α) corresponds to the probability of erroneously reject a member of the class as a non-member (rate of false negative, FN). The second kind of errors (β) corresponds to the probability of erroneously classify a non-member of the class as a member (rate of false positive, FP). Accordingly, for a given class A, and setting n_A as the number of members of class A, $\overline{n}_A$ as the number of non-members of class A, $\langle n_A \rangle$ as the number of members of class A correctly classified as "belonging to class A" and $\langle \overline{n}_A \rangle$ as the number of non-members of class A classified as "not belonging to class A", we have (Yang et al., 2005):

$$\begin{array}{ll} TP = \langle n_A \rangle & FP = \overline{n}_A - \langle \overline{n}_A \rangle \\ TN = \langle \overline{n}_A \rangle & FN = n_A - \langle n_A \rangle \end{array} \tag{12}$$

TP and TN being the number of True Positive and True Negative members of the considered class. Accordingly,

$$\begin{array}{l} SENS = \dfrac{\langle n_A \rangle}{n_A} = 1 - \alpha = 1 - \dfrac{\mathrm{FN}}{n_A} = \dfrac{TP}{TP + FN} \\ SPEC = \dfrac{\langle \overline{n}_A \rangle}{\overline{n}_A} = 1 - \beta = 1 - \dfrac{\mathrm{FP}}{\overline{n}_A} = \dfrac{TN}{TN + FP} \end{array} \tag{13}$$

It is clear that values close to unity for both parameters indicates a successfully validation performance.

With these parameters it can be built the called *confusion matrix* for class A:

$$^{C}M_A = \begin{bmatrix} TN & FP \\ FN & TP \end{bmatrix} \tag{14}$$

As it has been outlined, the common validation procedure consists of dividing the known set into two subsets, namely training and validation set. However, the validation procedure has to be considered with more caution in case of some kinds of ANN such as MLP because they suffer a special overfitting damage. The MLP consists of formal neurons and connection (weights) between them. As it is well known, neurons in MLP are commonly arranged in three layers: an input layer, one hidden layer (sometimes plus a bias neuron) and an output layer. The number of hidden nodes in a MLP indicates the complexity of the relationship in a way very similar to the fitting of a polynomial to a data set. Too many connections have the risk of a network specialization in training noise and poor prediction ability. Accordingly, a first action should be minimizing the number of neurons of the hidden layer. Some authors (Andrea & Kalayeh, 1991) have proposed the parameter ρ which plays a major role in determining the best architecture:

$$\rho = \frac{\text{Number of data points in the training set}}{\text{Sum of the number of connections in the network}} \tag{15}$$

In order to avoid overfitting it is recommended that $1 < \rho < 2.2$.

Besides, the overfitting problem can be minimized by monitoring the performance of the network during training by using an extra verification set different from training set. This verification set is needed in order to stop the training process before the ANN learns idiosyncrasies present in the training data that leads to overfitting (González, 2007).

8. Concluding remarks

The selection of the supervised learning technique depends on the nature of the particular problem. If we have a data set composed only by a given number of classes and the rule is going to be used on test samples that we know they may belong to one of the former established classes only, then we can select a discriminating technique such as LDA, PLS-DA, SVM or some kind of discriminating ANN (MLP or PNN). Otherwise, class modelling techniques such as SIMCA, UNEQ or KCM are useful. Class modelling tools offer at least two main advantages: To identify samples which do not fall in any of the examined categories (and therefore can be either simply outlying observations or members of a new class not considered in the known set) and to take into account samples that can simultaneously belong to more than one class (multiclass patterns).

If the idiosyncrasy of the problem suggests that the boundaries could be of non-linear nature, then the use of SVM or ANN is the best choice.

In cases where the number of features in higher than the number of samples (p > n), a previous or simultaneous step dealing with feature selection is needed when non-PCA based techniques are used (KNN, LDA, ANN, UNEQ). PCA-based methods such as SIMCA and PLS-DA can be applied without need of feature selection. This characteristic is very interesting beyond of compositional analysis, when samples are characterized by a spectrum, like in spectrometric methods (FT-IR, FT-Raman, NMR...). A different behaviour of these two methods against the number of FP and FN has been noticed (Dahlberg et al., 1997). SIMCA is focused on class specificities, and hence it detects strangers with high accuracy (only when the model set does not contain outliers. Otherwise, robust SIMCA model can be used), but sometimes fails to recognize its own members if the class is not homogeneous enough or the training set is not large enough. PLS-DA, on the contrary, deals with an implicitly closed universe (since the Y variables have a constant sum) so that it ignores the possibility of strangers. However, this has the advantage to make the method more robust to class inhomogeneities, since what matters most in class differences.

In compositional data, as pointed out Berrueta et al. (2007), the main problem is class overlap, but with a suitable feature selection and adequate sample size, good classification performances can be achieved. In general, non-linear methods such as ANN or SVM are rarely needed and most classification problems can be solved using linear techniques (LDA, CVA, PLS_DA).

Sometimes, several different types of techniques can be applied to the same data set. Classification methods are numerous and then the main problem is to select the most suitable one, especially dealing with quantitative criteria like prediction ability or misclassification percentage. In order to carry out the comparison adequately, the McNemar's test is a good choice (Roggo et al., 2003). Two classification procedures A and B are trained and the same validation set is used. Null hypothesis is that both techniques lead to the same misclassification rate. McNemar's test is based on a χ^2 test with one degree of freedom if the number of samples is higher than 20. The way to obtain the McNemar's statistic is as follows:

$$\text{McNemar's value} = \frac{\left(|n_{01} - n_{10}| - 1\right)^2}{n_{01} + n_{10}} \tag{16}$$

with

n_{00}: number of samples misclassified by both methods A and B

n_{01}: number of samples misclassified by method A but not by B

n_{10}: number of samples misclassified by method B but not by A

n_{11}: number of samples misclassified by neither method A nor B

$n_{val} = n_{00} + n_{01} + n_{10} + n_{11} =$ number of patterns in the validation set

The critical value for a 5% significance level is 3.84. In order to get insight about this procedure, the paper of Roggo et al (2006) is very promising.

Finally, a last consideration about problems with the data set representativeness. As it has been claimed in a published report a LDA was applied to differentiate 12 classes of oils on the basis of the chromatographic data, where some classes contained two or three members only (and besides, the model was not validated). There is no need of being an expertise chemometrician to be aware of two or three samples are insufficient to draw any relevant conclusion about the class to which they belong. There are more sources of possible data variance than the number of samples used to estimate class variability (Daszykowski & Walczak, 2006). The requirements of a sufficient number of samples for every class could be envisaged according to a class modelling technique to extract the class dimensionality and consider, for instance, a number of members within three to ten times this dimensionality.

Aside from this representativity context, it should be point out that when the aim is to classify food products or to build a classification rule to check the authentic origin of samples, they have to be collected very carefully according to a well established sampling plan. Often not enough care is taken about it, and thus is it hardly possible to obtain accurate classification models.

9. References

Abe, S. (2005). Support vector machines for pattern classification. Springer, ISBN:1852339299, London, UK

Aitchison, J. (2003). The statistical analysis of compositional data. The Blackburn Press, ISBN:1930665784, London, UK

Albano, C.; Dunn III, W.; Edlund, U.; Johansson, E.; Norden, B.; Sjöström, M. & Wold, S. (1978). Four levels of Pattern Recognition. Analytica Chimica Acta. Vol. 103, pp. 429-443. ISSN:0003-2670

Andrea, T.A.; Kalayeh, H. (1991). Applications of neural networks in quantitative structure-activity relationships of dihydrofolate reductase inhibitors. Journal of Medicinal Chemistry. Vol. 34, pp. 2824-2836. ISSN: 0022-2623

Aparicio, R. & Aparicio-Ruíz, R. (2002). Chemometrics as an aid in authentication, In: Oils and Fats Authentication, M. Jee (Ed.), 156-180, Blackwell Publishing and CRC Press, ISBN:1841273309, Oxford, UK and FL, USA

Bishop, C.M. (2000). Neural Networks for Pattern Recognition, Oxford University Press, ISBN:0198538642, NY, USA

Berrueta, L.A.; Alonso-Salces, R.M. & Héberger, K. (2007). Supervised pattern recognition in food analysis. Journal of Chromatography A. Vol. 1158, pp. 196-214. ISSN:0021-9673

Burges, C.J.C. (1998). A tutorial on support vector machines for pattern recognition. Data Mining and Knowledge Discovery. Vol. 2, pp. 121-167. ISSN:1384-5810

Chung, C.F. (1993). Estimation of covariance matrix from geochemical data with observations below detection limits. Mathematical Geology. Vol. 25, pp. 851-865. ISSN:1573-8868

Clarke, J.U. (1998). Evaluation of censored data methods to allow statistical comparisons among very small samples with below detection limits observations. Environmental Science & Technology. Vol. 32, pp. 177-183. ISSN:1520-5851

Cole, R.A. & Phelps, K. (1979). Use of canonical variate analysis in the differentiation of swede cultivars by gas-liquid chromatography of volatile hydrolysis products. Journal of the Science of Food and Agriculture. Vol. 30, pp. 669-676. ISSN:1097-0010

Coomans, D.; Broeckaert, I.; Fonckheer, M; Massart, D.L. & Blocks, P. (1978). The application of linear discriminant analysis in the diagnosis of thyroid diseases. Analytica Chimica Acta. Vol. 103, pp. 409-415. ISSN:0003-2670

Coomans, D.; Massart, D.L. & Kaufman, L. (1979) Optimization by statistical linear discriminant analysis in analytical chemistry. Analytica Chimica Acta. Vol. 112, pp. 97-122. ISSN:0003-2670

Dahlberg, D.B.; Lee, S.M.; Wenger, S.J. & Vargo, J.A. (1997). Classification of vegetable oils by FT-IR. Applied Spectroscopy. Vol. 51, pp. 1118-1124. ISSN:0003-7028

Daszykowski, M. & Walczak, B. (2006). Use and abuse of chemometrics in chromatography. Trends in Analytical Chemistry. Vol. 25, pp. 1081-1096. ISSN:0165-9936

Daszykowski, M.; Kaczmarek, K.; Stanimirova, I.; Vander Heyden, Y. & Walczak, B. (2007). Robust SIMCA-bounding influence of outliers. Chemometrics and Intelligent Laboratory Systems. Vol. 87, pp. 95-103. ISSN:0169-7439

Dean, R.B. & Dixon, W.J. (1951). Simplified statistics for small number of observations. Analytical Chemistry. Vol. 23, pp. 636-638. ISSN:0003-2700

Defernez, M. & Kemsley, E.K. (1997). The use and misuse of chemometrics for treating classification problems. Trends in Analytical Chemistry. Vol. 16, pp. 216-221. ISSN:0165-9936

Derde, M.P.; Coomans, D. & Massart, D.L. (1982). Effect of scaling on class modelling with the SIMCA method. Analytica Chimica Acta. Vol. 141, pp. 187-192. ISSN:0003-2670

Derde, M.P. & Massart, D.L. (1986). UNEQ: A class modelling supervised pattern recognition technique. Microchimica Acta. Vol. 2, pp. 139-152. ISSN:0026-3672

Duda, R.O.; Hart, P.E. & Stork, D.G. (2000). Pattern classification. 2nd edition. Wiley, ISBN:0471056693, NY, USA

Efron, B. (1983). Estimating the error rate of a prediction rule: Improvement on cross validation. Journal of the American Statistical Association. Vol. 78, pp. 316-331. ISSN:0162-1459

Efron, B. & Gong, G. (1983). A leisurely look at the bootstrap, the jacknife and cross validation. The American Statiscian. Vol. 37, pp. 36-48. ISSN:0003-1305

Egan, W.J. & Morgan, S.L. (1998). Outlier detection in multivariate analytical chemical data. Analytical Chemistry. Vol. 70, pp. 2372-2379. ISSN:0003-2700

Egozcue, J.J.; Pawlowsky-Glahn, V.; Mateu-Figueros, G.; Barcelo-Vidal, C. (2003). Isometric logratio transformation for compositional data analysis. Mathematical Geology. Vol. 35, pp. 279-300. ISSN:1573-8868

Gemperline, P.J. & Boyer, N.R. (1995). Classification of near-infrared spectra using wavelength distances: Comparisons to the Mahalanobis distance and Residual Variance methods. Analytical Chemistry. Vol.67, pp. 160-166. ISSN:0003-2700

González, A.G. (2007). Use and misuse of supervised pattern recognition methods for interpreting compositional data. Journal of Chromatograpy A. Vol. 1158, pp. 215-225. ISSN:0021-9673

González, A.G. & Herrador, M.A. (2007). A practical guide to analytical method validation, including measurement uncertainty and accuracy profiles. Trends in Analytical Chemistry. Vol. 26, pp. 227-237. ISSN:0165-9936

González-Arjona, D.; López-Pérez, G. & González, A.G. (1999). Performing Procrustes discriminant analysis with HOLMES. Talanta. Vol. 49, pp. 189-197. ISSN:0039-9140

González-Arjona, D.; López-Pérez, G. & González, A.G. (2001). Holmes, a program for performing Procrustes Transformations. Chemometrics and Intelligent Laboratory Systems. Vol. 57, pp. 133-137. ISSN:0169-7439

González-Arjona, D.; López-Pérez, G. & González, A.G. (2006). Supervised pattern recognition procedures for discrimination of whiskeys from Gas chromatography/Mass spectrometry congener analysis. Journal of Agricultural and Food Chemistry. Vol. 54, pp. 1982-1989. ISSN:0021-8561

Grubbs, F. (1969). Procedures for detecting outlying observations in samples. Technometrics. Vol. 11, pp. 1-21. ISSN:0040-1706

Helsel, D.R. (1990).Less than obvious: Statistical treatment of data below the detection limit. Environmental Science & Technology. Vol. 24, pp. 1766-1774. ISSN: 1520-5851

Hoaglin, D.C. & Welsch, R.E. (1978). The hat matrix in regression and ANOVA. The American Statiscian. Vol. 32, pp. 17-22. ISSN:0003-1305

Holger, R.M.; Dandy, G.C. & Burch, M.D. (1998). Use of artificial neural networks for modelling cyanobacteria Anabaena spp. In the river Murray, South Australia. Ecological Modelling. Vol. 105, pp. 257-272. ISSN:0304-3800

Jollife, I.T. (2002). Principal Component Analysis. 2nd edition, Springer, ISBN:0387954422, NY, USA

Johnson, G.W. & Ehrlich, R. (2002). State of the Art report on multivariate chemometric methods in Environmental Forensics. Environmental Forensics. Vol. 3, pp. 59-79. ISSN:1527-5930

Kryger, L. (1981). Interpretation of analytical chemical information by pattern recognition methods-a survey. Talanta. Vol. 28, pp. 871-887. ISSN:0039-9140

Kowalski, B.R. & Bender, C.F. (1972). Pattern recognition. A powerful approach to interpreting chemical data. Journal of the American Chemical Society. Vol. 94, pp. 5632-5639. ISSN:0002-7863

Kvalheim, O.M. & Karstang, T.V. (1992). SIMCA-Classification by means of disjoint cross validated principal component models, In: Multivariate Pattern Recognition in Chemometrics, illustrated by case studies, R.G. Brereton (Ed.), 209-245, Elsevier, ISBN:0444897844, Amsterdam, Netherland

Leardi, R. & Lupiañez, A. (1998). Genetic algorithms applied to feature selection in PLS regression: how and when to use them. Chemometrics and Intelligent Laboratory Systems. Vol. 41, pp. 195-207. ISSN:0169-7439

Malinowski, E.R. (2002). Factor Analysis in Chemistry. Wiley, ISBN:0471134791, NY, USA

Marini, F.; Zupan, J. & Magrí, A.L. (2005). Class modelling using Kohonen artificial neural networks. Analytica Chimica Acta. Vol.544, pp. 306-314. ISSN:0003-2670

Martín-Fernández, J.A.; Barceló-Vidal, C. & Pawlowsky-Glahn, V. (2003). Dealing with zeros and missing values in compositional data sets using nonparametric imputation. Mathematical Geology. Vol. 35, pp. 253-278. ISSN:1573-8868

McCabe, G.P. (1975). Computations for variable selection in discriminant analysis. Technometrics. Vol. 17, pp. 103-109. ISSN:0040-1706

Melssen, W.; Wehrens, R. & Buydens, L. (2006). Supervised Kohonen networks for classification problems. Chemometrics and Intelligent Laboratory Systems. Vol. 83, pp. 99-113. ISSN:0169-7439

Mertens, B.; Thompson, M. & Fearn, T. (1994). Principal component outlier detection and SIMCA: a synthesis. Analyst. Vol. 119, pp. 2777-2784. ISSN:0003-2654

Miller, J.N. & Miller, J.C. (2005). Statistics and Chemometrics for Analytical Chemistry. 4th edition. Prentice-Hall, Pearson. ISBN:0131291920. Harlow, UK

Naes, T.; Isaksson, T.; Fearn, T.; Davies, T. (2004). A user-friendly guide to multivariate calibration and classification. NIR Publications, ISBN;0952866625, Chichester, UK

Pardo, M. & Sberveglieri, G. (2005). Classification of electronic nose data with support vector machines. Sensors and Actuators. Vol. 107, pp. 730-737. ISSN:0925-4005

Pretsch, E. & Wilkins, C.L. (2006). Use and abuse of Chemometrics. Trends in Analytical Chemistry. Vol. 25, p. 1045. ISSN:0165-9936

Rock, B.A. (1985). An introduction to Chemometrics, 130th Meeting of the ACS Rubber Division. October 1985. Available from http://home.neo.rr.com/catbar/chemo/int_chem.html

Roggo, Y.; Duponchel, L. & Huvenne, J.P. (2003). Comparison of supervised pattern recognition methods with McNemar's statistical test: Application to qualitative analysis of sugar beet by near-infrared spectroscopy. Analytica Chimica Acta. Vol. 477, pp. 187-200. ISSN:0003-2670

Sharaf, M.A.; Illman, D.A. & Kowalski, B.R. (1986). Chemometrics. Wiley, ISBN:0471831069, NY, USA

Silverman, B.W. & Jones, M.C. (1989). E. Fix and J.L. Hodges (1951): An important contribution to non parametric discriminant analysis and density estimation. International Statistical Review. Vol. 57, pp. 233-247. ISSN:0306-7734

So, S.S. & Richards, W.G. (1992). Application of Neural Networks: Quantitative structure activity relationships of the derivatives of 2,4-diamino-5-(substituted-benzyl) pyrimidines as DHFR inhibitors. Journal of Medicinal Chemistry. Vol. 35, pp. 3201-3207. ISSN:0022-2623

Stahle, L. & Wold, S. (1987). Partial least squares analysis with cross validation for the two class problem: A monte-Carlo study. Journal of Chemometrics. Vol. 1, pp. 185-196. ISSN:1099-128X

Stanimirova, I.; Daszykowski, M. & Walczak, B. (2007). Dealing with missing values and outliers in principal component analysis. Talanta. Vol. 72, pp. 172-178. ISSN:0039-9140

Streit, R.L. & Luginbuhl, T.E. (1994). Maximun likelihood training of probabilistic neural networks. IEEE Transactions on Neural Networks. Vol. 5, pp. 764-783. ISSN:1045-9227

Todeschini, R.; Ballabio, D.; Consonni, V.; Mauri, A. & Pavan, M. (2007). CAIMAN (Classification And Influence Matrix Analysis): A new approach to the classification based on leverage-scale functions. Chemometrics and Intelligent Laboratory Systems. Vol. 87, pp. 3-17. ISSN:0169-7439

Vapnik, V.N. (1998). Statistical learning theory. Wiley, ISBN:0471030031, NY, USA

Varmuza, K. (1980). Pattern recognition in chemistry. Springer, ISBN:0387102736, Berlin, Germany

Varmuza, K. & Filzmoser, P. (2009). Introduction to multivariate statistical analysis in chemometrics, CRC Press, Taylor & Francis Group, ISBN:14005975, Boca Ratón, FL, USA

Walczak, B. & Massart, D.L. (2001). Dealing with missing data: Part I and Part II. Chemometrics and Intelligent Laboratory System. Vol. 58, pp. 15-27 and pp. 29-42. ISSN:0169-7439

Wold, S. (1976). Pattern recognition by means of disjoint principal component models. Pattern Recognition. Vol. 8, pp. 127-139. ISSN:0031-3203

Wold, S. (1978). Cross validatory estimation of the number of components in factor and principal components models. Technometrics. Vol. 20, pp. 397-405. ISSN:0040-1706

Wold, S. (1995). Chemometrics, what do we mean with it, and what do we want from it? Chemometrics and Intelligent Laboratory Systems. Vol. 30, pp. 109-115. ISSN:0169-7439

Yang, Z.; Lu, W.;Harrison, R.G.; Eftestol, T.; Steen,P.A. (2005). A probabilistic neural network as the predictive classifier of out-of-hospital defibrillation outcomes. Resuscitation. Vol. 64, pp. 31-36. ISSN:0300-9572

Zupan, J. & Gasteiger, J. (1993). Neural Networks for chemists. VCH, ISBN:1560817917, Weinheim, Germany

Zupan, J.; Novic, M.; Li, X. & Gasteiger, J. (1994). Classification of multicomponent analytical data of olive oils using different neural networks. Analytica Chimica Acta. Vol. 292, pp. 219-234. ISSN:0003-2670

3

Analysis of Chemical Processes, Determination of the Reaction Mechanism and Fitting of Equilibrium and Rate Constants

Marcel Maeder and Peter King
Department of Chemistry, University of Newcastle, Australia
Jplus Consulting Ltd, Perth, Australia

1. Introduction

This chapter is intended to demonstrate some recent approaches to the quantitative determination of chemical processes based on the quantitative analysis of experimental spectrophotometric measurements. In this chapter we will discuss kinetic processes, equilibrium processes and also processes that include a combination of kinetic and equilibrium steps.

We also emphasise the advantage of 'global' multivariate (multiwavelength) data analysis which has the advantage of allowing the robust determination of more complex mechanisms than single wavelength analysis and also has the benefit of yielding the spectra of all the participating species.

Rather than dwell on the mathematical derivation of the complex numerical algorithms and a repetition of the fundamentals of non-linear regression methods and least squares fitting which are available from a wide variety of sources (Martell and Motekaitis 1988; Polster and Lachmann 1989; Gans 1992; Press, Vetterling et al. 1995; Maeder and Neuhold 2007), we aim to show the experimentalist how to obtain the results they are interested, using purpose designed global analysis software and a variety of worked examples. We will be using ReactLab, a suite of versatile and powerful reaction modelling and analysis tools developed by the authors. Other academic and commercial applications exist for multivariate and related types of analysis and the reader is encouraged to explore these for comparative purposes. All offer different features and benefits but will not be discussed here.

2. Spectrophotometry, the ideal technique for process analysis

Any spectroscopic technique is ideal for the analysis of chemical processes as there is no interference in the underlying chemistry by the measurement technique. This is in sharp contrast to say chromatographic analysis or other separation methods which are totally unsuitable for the analysis of dynamic equilibrium systems. Such methods are also of very limited use for kinetic studies which often are too fast on the chromatographic time scale of typically tens of minutes to hours (except where reactions are first quenched and the intermediates stabilised). In contrast most forms of spectroscopy provide a completely non-invasive snapshot of a sample's composition at a single instant.

Amongst the different spectroscopies routinely available to the chemist, light absorption spectrophotometry in the UV-Visible (UV/Vis) is most common for several reasons: instruments are relatively inexpensive and accurate, they provide stable referenced signals as they are usually split or double beam instruments, there is a simple relationship between concentration and the measured absorbance signal (Beer-Lambert's law) and many compounds absorb somewhere in the accessible wavelength region. As a consequence there is a considerable amount of software available for the analysis of spectrophotometric data. This is the case both for kinetic and equilibrium investigations. NMR spectroscopy is a powerful method for structural investigations but it is less commonly used for quantitative analytical purposes. A theoretically very powerful alternative to UV/Vis absorption spectroscopy is FT-IR spectroscopy. The richness of IR spectra is very attractive as there is much more information contained in an IR spectrum compared with a relatively structureless UV/Vis spectrum. The main disadvantage is the lack of long term stability as FT-IR instruments are single beam instruments. Other difficulties include solvent absorption and the lack of non-absorbing cell materials, particularly for aqueous solutions. However, attenuated total reflection or ATR is a promising novel measurement technique in the IR. Near-IR spectroscopy is very similar to UV/Vis spectroscopy and is covered by the present discussions.

Of course fluorescence detection is a very sensitive and important tool particularly in kinetics studies and can yield important mechanistic information where intermediates do not possess chromophores and are therefore colourless or are studied at very low concentrations. In the main fluorescence studies are carried out at a single emission wavelength or adopting the total fluorescence approach (using cut-off filters), so there is no wavelength discrimination in the data. Whilst this type of measurement can be analysed by the methods described below and is essentially equivalent to analysing single wavelength absorption data. We will in the following discussion concentrate on the general case of processing multiwavelength measurements

3. The experiment, structure of the data

For kinetic investigations the absorption of the reacting solution is measured as a function of reaction time; for equilibrium investigations the absorption is recorded as a function of the reagent addition or another independent variable such as pH. Absorption readings can of course be taken at a single wavelength but with modern instrumentations it is routine and advantageous to record complete spectra vs. time or reagent addition. This is particularly prevalent with the use of photodiode array (PDA) based spectrophotometers and online detectors.

In the case of kinetics, depending on the rate of a chemical reaction the mixing of the reagents that undergo the reaction has to be done fast using a stopped-flow instrument or it can be done manually for slower reactions in the cuvette of a standard UV-Vis spectrometer with suitably triggered spectral data acquisition. A series of spectra are collected at time intervals following the mixing event to cover the reaction time of interest. The measured spectra change as the reaction proceeds from reagents to products (Wilkins 1991; Espenson 1995).

For equilibrium investigations the spectra of a series of pre-mixed and equilibrated solutions have to be recorded (Martell and Motekaitis 1988; Polster and Lachmann 1989). This is most commonly done as a titration where small amounts of a reagent are added stepwise to the solution under investigation. Titrations can be done in the cuvette, requiring internal stirring after each addition, prior to the absorption measurement, or the solutions can be mixed externally with transfer of the equilibrated solutions into the cuvette performed manually or using a flow cell and automatic pumping. In an alternative configuration optical probes can be coupled to the optical path in some spectrometers and placed into the solution contained in an external titration vessel (Norman and Maeder 2006). Often the pHs of the equilibrated titration solutions are recorded together with the absorption spectra where protonation equilibria are a feature of the mechanism.

For both kinetic and equilibrium investigations the measurement data can be arranged in a data matrix **D** which contains row-wise the recorded spectra as a function of time or reagent addition. The number of columns of **D** is the number of wavelengths, *nlam*, over which the spectra are taken. For single wavelength data the matrix reduces to a single column (vector). The number of rows, *nspectra,* corresponds to the number of spectra recorded during the process (one at each time interval for kinetics or reagent addition for an equilibrium titration). The dimensions of **D** thus are *nspectra*×*nlam*. For spectra taken on a mechanical scanning instrument, the number of wavelengths can be 1 to typically some 10 or 20 but for diode array instruments it can easily be in excess of 1000 depending on the solid state detector pixel resolution (typically these provide a resolution progression of 256, 512 and 1024 pixels). The number of spectra taken is typically much larger on a stopped-flow instrument equipped with a fast diode array detector with a typical minimum spectrum acquisition time of the order of a millisecond. Frequently a logarithmic time base is an option which enables both fast and slower events to be resolved in a single kinetic experiment. A graphical representation of a date matrix **D** is given in Figure 1.

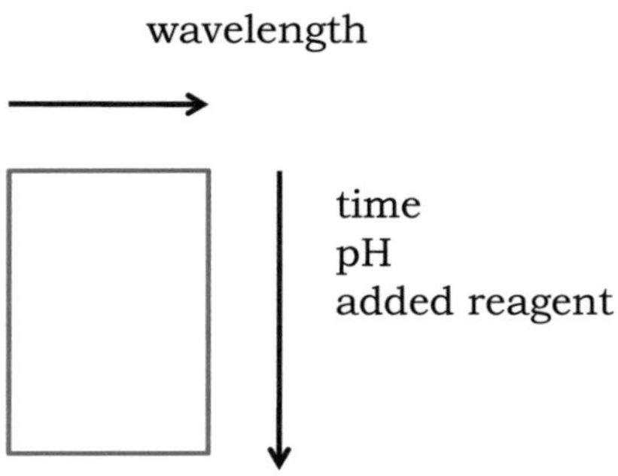

Fig. 1. Graphical representation of a data matrix **D**, the spectra are arranged as the rows.

For both kinetic and equilibrium investigations, we obtain a series of spectra each of which represent the solution at one particular moment during the process. The spectra are taken as a function of time or reagent addition.

4. Information to be gained from the measurements

The purpose of collecting this type of data is to determine the chemical reaction mechanism that describes the underlying process in terms of identifiable steps together with the associated key parameters; the rates and/or equilibrium constants which define the interconversions and stabilities of the various species. This may initially be a purely academic exercise to characterise a novel chemical reaction for publication purposes but ultimately defines the behaviour of the participating species for any future research into this or related chemistry as well as being the foundation for commercially important applications e.g. drug binding interactions in pharmaceutical development or reaction optimisation in industrial processes.

The objective is therefore to find the chemical model which best fits the data and validate and refine this model with subsequent experiments under other conditions. The clear benefit of multi-wavelength measurements is that the model must satisfy (fit) the data at all measurement wavelengths simultaneously and this significantly helps the accurate determination of multiple parameters and also allows determination of the individual spectra of the participating species.

5. Beer-Lambert's law

Before we can start the possible ways of extracting the useful parameters from the measured data set, the rate constants in the case of kinetics, the equilibrium constants in the case of equilibria, we need to further investigate the structure of the data matrix **D**. According to Beer-Lambert's law for multicomponent systems, the total absorption at any particular wavelength is the sum over all individual contributions of all absorbing species at this wavelength. It is best to write this as an equation:

$$D(i,j) = \sum_{k=1}^{ncomp} C(i,k) \times A(k,j) \tag{1}$$

where:
D(i,j): absorption of the i-th solution at wavelength j
C(i,k); concentration of the k-th component in the i-th solution
A(k,j): molar absorptivity of the k-th component at the j-th wavelength
ncomp: number of components in the system under investigation.

Thus equation (1) represents a system of $i \times j$ equations with many unknowns, i.e. all elements of **C** ($nspectra \times ncomp$) and all elements of **A** ($ncomp \times nlam$).

It is extremely useful to realise that the structure of Beer-Lambert's law allows the writing of Equation (1) in a very elegant matrix notion, Equation (2) and Figure 2

$$\mathbf{D} = \mathbf{C} \times \mathbf{A} \tag{2}$$

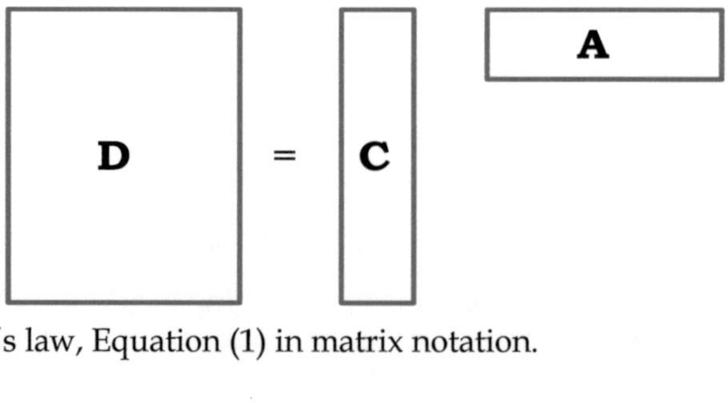

Fig. 2. Beer-Lambert's law, Equation (1) in matrix notation.

The matrix **D**(*nspectra*×*nlam*) is the product of a matrix of concentrations **C**(*nspectra*×*ncomp*) and a matrix **A**(*ncomp*×*nlam*). **C** contains as columns the concentration profiles of the reacting components and the matrix **A** contains, as rows, their molar absorption spectra.

Equations (1) and (2) and Figure 2 represent the ideal case of perfect absorption readings without any experimental noise. This of course is not realistic and both equations have to be augmented by an error term, R(i,j) which is the difference between the ideal value and its measured counterpart, equation (3) and Figure 3.

$$D(i,j) = \sum_{k=1}^{ncomp} C(i,k) \times A(k,j) + R(i,j)$$

$$\mathbf{D} = \mathbf{C} \times \mathbf{A} + \mathbf{R} \tag{3}$$

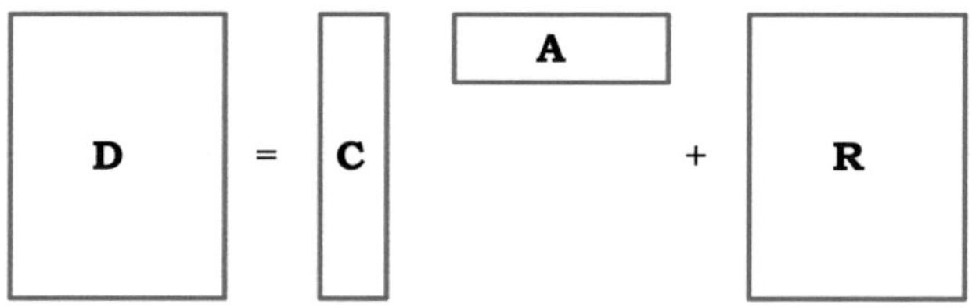

Fig. 3. Beer-Lambert's law including the residuals.

The goal of the fitting is to determine that set of matrices **C** and **A** for which the sum over all the squares of the residuals, *ssq*, is minimal,

$$ssq = \sum_{i=1}^{nspectra} \sum_{j=1}^{nlam} R(i,j) \tag{4}$$

At first sight this looks like a very daunting task. However, as we will see, it is manageable.

Ideally the final square sum achieved should be numerically equal to the sum of the squares of the Gaussian noise in the measurement - usually instrumental in origin. At this point the fit cannot be further improved, though this is not a guarantee that the model is the correct one.

6. The chemical model

The first and central step of any data analysis is the computation of the matrix **C** of concentration profiles based on the proposed chemical model and the associated parameters such as, but not exclusively, the rate and or equilibrium constants. Initially these key parameters may be only rough estimates of the true values.

So far the explanations are valid for kinetic and equilibrium studies. The difference between these two investigation lies in the different computations required for the calculation of the concentration profiles in the matrix **C**.

7. Kinetics

The chemical model for a kinetic investigation is a set of reaction equations which describe the process under investigation. Consider as an example the basic enzymatic reaction scheme

$$E + S \underset{k_{-1}}{\overset{k_{+1}}{\rightleftharpoons}} ES \xrightarrow{k_2} E + P \tag{5}$$

An enzyme *E* reacts rapidly and reversibly with the substrate *S* to form an enzyme substrate complex *ES*. This is followed by the 1st order chemical conversion of the substrate and release of product. The free enzyme is then available to undergo another catalytic cycle.

Before proceeding to a ReactLab based mechanistic analysis it is informative to briefly outline the classical approach to the quantitative analysis of this and similar basic enzyme mechanisms. The reader is referred to the many kinetics textbooks available for a more detailed description of these methods. The scheme in equation (5) was proposed by Michaelis and Menten in 1913 to aid in the interpretation of kinetic behaviour of enzyme-substrate reactions (Menten and Michaelis 1913). This model of the catalytic process was the basis for an analysis of measured initial rates (v) as a function of initial substrate concentration in order to determine the constants K_M (The Michaelis constant) and V_{max} that characterise the reaction. At low [S], v increases linearly, but as [S] increases the rise in v slows and ultimately reaches a limiting value V_{max}.

Analysis was based on the derived Michaelis Menten formula:

$$v = \frac{[E_0]\,[S]\,k_{cat}}{K_M + [S]} \tag{6}$$

Where $V_{max} = k_{cat}[E]_0$, and K_M is equal to the substrate concentration at which $v = ½\,V_{max}$. The key to this derivation is that the enzyme substrate complex *ES* is in dynamic equilibrium with free *E* and *S* and the catalytic step proceeds with a first order rate constant k_{cat}. This 'turnover number' k_{cat} is represented by k_2 in the scheme in equation (5).

It can be shown that under conditions where $k_2 << k_{-1}$ then K_M is in fact equal to the equilibrium dissociation constant K_1,

$$K_1 = \frac{[ES]}{[E]\,[S]} = \frac{k_{+1}}{k_{-1}} \tag{7}$$

Importantly, however, the parameter K_M is not always equivalent to this fundamental equilibrium constant K_1 when this constraint ($k_2 << k_{-1}$) doesn't apply.

Furthermore though the Michealis Menten scheme can be extended to cover more complex mechanisms with additional intermediates, the K_M and k_{cat} parameters now become even more complex combinations of individual rate and equilibrium constants. The k_{cat} and K_M parameters determined by these classical approaches are therefore not the fundamental constants defining the mechanism and significant effort is required to determine the true underlying equilibrium and microscopic rate constants.

In contrast direct analysis using ReactLab to fit the core mechanism to suitable data delivers the true underlying rate and equilibrium constants in a wholly generic way that can be applied without assumptions and also to more complex models.

This involves the modelling of the entire mechanism to deliver the matrix **C** comprising the concentration profiles of all the participating species. The reaction scheme in equation (5) defines a set of ordinary differential equations, ODE's, which need to be solved or integrated (Maeder and Neuhold 2007). Reaction schemes that only consist of first order reactions can be integrated analytically in which case the concentration can be calculated directly at any point using the resulting explicit function. Most other schemes, containing one or more second order reactions, require numerical integration. Numerical integration is usually done with variable step-size Runge-Kutta algorithms, unless the system is 'stiff' (comprising both very fast and slow reactions) for which special stiff solvers, such as Gear and Bulirsch-Stoer algorithms are available (Press, Vetterling et al. 1995).

Integration, explicit or numerical, requires the knowledge of the initial conditions, in the case of kinetics the initial concentrations of all interacting species. For the above example, equation

$$E + S \underset{k_{-1}}{\overset{k_{+1}}{\rightleftarrows}} ES \xrightarrow{k_2} E + P \tag{5},$$

and using the rate constants ($k_{+1}=10^3$ M^{-1} sec^{-1}, $k_{-1}=10^2$ sec^{-1}, $k_2=10^2$ sec^{-1}) and initial concentrations, ($[S]_0=1$ M, $[E]_0=10^{-4}$ M), the resulting concentration profiles generated by numerical integration and used to populate the columns of matrix **C** are shown in Figure 4.

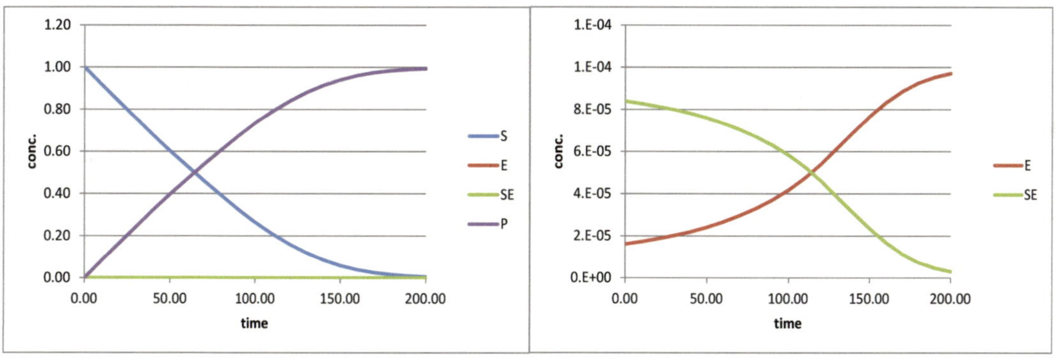

Fig. 4. Concentration profiles for the enzymatic reaction of equation (5); an expanded concentration axis is used in the right panel.

The transformation of the substrate into the product follows approximately zero-th order kinetics for most of the reaction whilst the substrate is in excess and all the enzyme sites are populated. Later in the reaction the substrate is exhausted and free enzyme released. The expanded plot in the right hand panel displays more clearly the small concentrations for the enzyme and the enzyme-substrate complex.

8. Equilibria

The chemical model for an equilibrium process is similar to the model of a kinetic process, only now there are exclusively equilibrium interactions, e.g.

$$A + H \xrightleftharpoons{K_{AH}} AH$$
$$AH + H \xrightleftharpoons{K_{AH_2}} AH_2 \tag{8}$$
$$B + H \xrightleftharpoons{K_{BH}} BH$$

The chemistry in this example comprises the protonation equilibria of the di-protic acid AH_2 and the mono-protic acid BH. The key difference now is that the steps are defined in terms of instantaneous stability or equilibrium constants, and the fast processes of the attainment of the equilibria are not observed.

Equilibrium investigations require a titration, which consists of the preparation of a series of solutions with different but known total component concentrations. In equilibrium studies we distinguish between components and species. Components are the building blocks; in the example (6) they are *A*, *B* and *H*; species are all the different molecules that are formed from the components during the titration, the example they are *A*, *AH*, AH_2, *B*, *BH*, *H* and *OH*. Note, the components are also species.

Instead of utilising numerical integration to compute the concentration profiles of the individual species as we did with kinetic time courses we instead use an iterative Newton-Raphson algorithm to determine the speciation based on the total component concentrations for each sample and the estimated equilibrium constants (Maeder and Neuhold 2007).

For a titration of 10ml of a solution with total component concentrations $[A]_{tot}$= 0.1M, $[B]_{tot}$=0.06 M and $[H]_{tot}$=0.4M with 5ml of 1.0M *NaOH* the concentration profiles of Figure 5 result. The protonation constants are $\log(K_{AH})=9$, $\log(K_{AH2})=3$ and $\log(K_{BH})=4$.

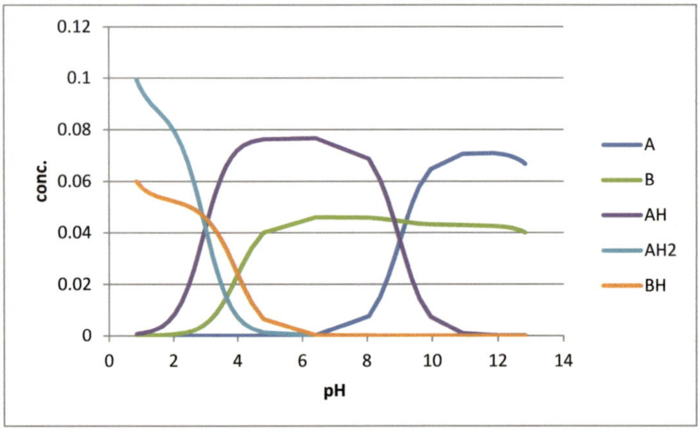

Fig. 5. Concentration profiles for the titration of an acidified solution of AH_2 and *BH* with *NaOH*.

8.1 Kinetics with coupled protonation equilibria

A significant recent development is the incorporation of instantaneous equilibria to kinetic analyses. Careful combination of numerical integration computations alongside the Newton-Raphson speciation calculations have made this possible (Maeder, Neuhold et al. 2002). This development has made the modelling of significantly more complex and realistic mechanisms possible. An example is the complex formation between ammonia and Ni^{2+} in aqueous solution as represented in equation (9).

$$Ni^{2+}+NH_3 \underset{k_{-ML}}{\overset{k_{ML}}{\rightleftarrows}} Ni(NH_3)^{2+} \tag{9}$$

$$Ni(NH_3)^{2+}+NH_3 \underset{k_{-ML_2}}{\overset{k_{ML_2}}{\rightleftarrows}} Ni(NH_3)_2^{2+}$$

$$NH_3+H^+ \overset{logK_{LH}}{\rightleftarrows} NH_4^+$$

Ni^{2+} is interacting with NH_3 to form the 1:1 and subsequently the 1:2 complexes. Importantly the ammonia is also involved in a protonation equilibrium. As a result the pH changes during the reaction and the rates of the complex formation reactions appear to change. The classical approach to this situation is to add buffers that approximately maintain constant pH and thus also the protonation equilibrium. Since buffers often interfere with the process under investigation the possibility of avoiding them is advantageous. This has only been made possible by this more sophisticated method of mechanistic analysis.

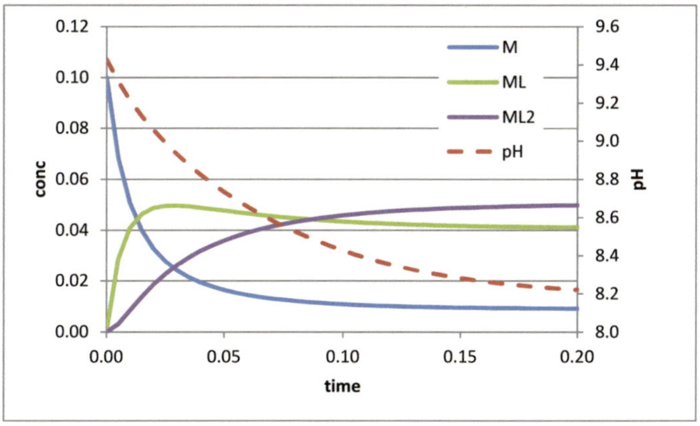

Fig. 6. The concentration profiles for the complex species in the reaction of Ni^{2+} and NH_3; also displayed is the pH of the reacting solution.

The concentration profiles for the complex species are shown in Figure 6. The patterns for a 'normal' consecutive reaction are distorted as the reaction slows down with the drop in pH from 9.5 to 8.2. The initial concentrations for this reaction are $[Ni^{2+}]_0$=0.1 M, $[NH_3]_0$=0.25 M and $[H^+]_0$=0.1 M.

9. Parameters

Any parameter that is used to calculate the matrix **C** of concentrations is potentially a parameter that can be fitted. Obvious parameters are the rate constants k and the equilibrium constants K; other less obvious parameters are the initial concentrations in kinetics or the total concentrations in equilibrium titrations. Concentration determinations are of course very common in equilibrium studies (quantitative titrations); for several reasons concentrations are not often fitted in kinetic studies. For first order reactions the concentrations are not defined at all unless there is additional spectroscopic information, i.e. molar absorptivities. For second order reactions they are in principle defined but only very poorly and thus kinetic determination is not a robust analytical technique in this case.

The parameters defining **C** are non-linear parameters and cannot be fitted explicitly, they need to be computed iteratively. Estimates are provided, a matrix **C** constructed and this is compared to the measurement according to the steps that follow below. Once this is complete it is possible to calculate shifts in these parameter estimates in a way that will improve the fit (i.e. reduce the square sum) when a new **C** is computed. This iterative improvement of the non-linear parameters is the basis of the non-linear regression algorithm at the heart of most fitting programs.

10. Calculation of the absorption spectra

The relationship between the matrix **C** and the measurement is based on equation (3). The matrix **A** contains the molar absorptivity for each species at each measured wavelength. All these molar absorptivities are unknown and thus also parameters to be determined. When spectra are collected the number of these parameters can be very large, but fortunately they are linear parameters and can be dealt with differently to the non-linear parameters discussed above.

Once the concentration profiles have been calculated, the matrix **A** of absorption spectra is computed. This is a linear least-squares calculation with an explicit solution

$$\mathbf{A} = \mathbf{C}^{+}\mathbf{D} \tag{10}$$

$\mathbf{C}^{+}$ is the pseudo-inverse of the concentration matrix **C**, it can be calculated as $(\mathbf{C}^{t}\mathbf{C})^{-1}\mathbf{C}^{t}$, or better using a numerically superior algorithm (Press, Vetterling et al. 1995).

11. Non-linear regression: fitting of the non-linear parameters

Fitting of the parameters requires the software to systematically vary all non-linear parameters, the rate and equilibrium constants as well as others such as initial concentrations, with the aim of minimising the sum of squares over all residuals, as defined in equation (4).

There are several algorithms for that task, the simplex algorithm which is relatively easy to program and features robust convergence with a high price of slow computation times particularly for the fitting of many parameters. Significantly faster is the Newton-Gauss algorithm; additionally it delivers error estimates for the parameters and with implementation of the Marquardt algorithm it is also very robust (Gans 1992; Maeder and Neuhold 2007).

As mentioned earlier non-linear regression is an iterative process and, provided the initial parameter estimates are not too poor and the model is not under-determined by the data, will converge to a unique minimum yielding the best fit parameters. With more complex models it is often necessary to fix certain parameters (either rate constants, equilibrium constants or complete spectra) particularly if they are known through independent investigations and most fitting applications will allow this type of constraint to be applied.

12. ReactLab analysis tools

ReactLab™ (Jplus Consulting Ltd) is a suite of software which is designed to carry out the fitting of reaction models to either kinetic or equilibrium multiwavelength (or single wavelength) data sets. All the core calculations described above are handled internally and the user simply provides the experimental measurements and a reaction scheme that is to be fitted to the data. A range of relevant supporting options are available as well as comprehensive graphical tools for visualising the data and the result of the analysis. To facilitate this all data, models and results are provided in pre-formatted Excel workbooks to allow post processing of results or customised plots to be added after the main analysis is complete.

13. Representing the chemical model

As has been discussed, at the root of the analysis is the generation of the species concentration matrix **C**.

Fitting a proposed reaction mechanism, or part of it, to the data therefore requires the determination of **C** from a reaction scheme preferably as would be written by a chemist. The ReactLab representation of the Ni^{2+}/NH_3 complexation mechanism of Equation (9) is shown in Figure 7. Forward arrows, >, are used to represent rate constants and the equal sign, =, represents an instantaneous equilibrium, e.g. a protonation equilibrium.

Reactants	Reaction Type	Products	Label	Parameters k / log K	±	Fit ☑
M+L	>	ML	k_ML	5.845E+02		☑
ML	>	M+L	k_-ML	1.114E+00		☐
ML+L	>	ML2	k_ML2	2.965E+02		☑
ML2	>	ML +L	k_-ML2	2.051E+00		☐
L+H	=	LH	logK_LH	9.250E+00		☐

Fig. 7. The ReactLab definition of the mechanism in Equation (7) with rate and equilibrium constants used to compute the concentration profiles of Figure 6.

To get from this scheme to our intermediate matrix **C** involves a number of key computational steps requiring firstly the dissection of the mechanism into its fundamental mathematical building blocks. Many analysis tools require the user to take this initial step manually, and therefore understand some fairly sophisticated underlying mathematical principles. Whilst this is no bad thing it does complicate the overall process and provide a significant barrier to trying and becoming familiar with this type of direct data fitting using numerical integration based algorithms. A significant advance has been the development of model editors and translators which carry out this process transparently. These can be found in a variety of data fitting applications including ReactLab.

14. Examples

In the following we will guide the reader through the steps required for the successful analysis of a number of example data sets. This section consists of two examples each from kinetics and equilibrium studies.

Example 1: Consecutive reaction scheme $\mathbf{A} \xrightarrow{k_1} \mathbf{B} \xrightarrow{k_2} \mathbf{C}$

The data set comprises a collection spectra measured as a function of time. These are arranged as rows of the 'Data' worksheet in Figure 8, the first spectrum in the cells D6:X6, the second in D7:X7, and so on. For each spectrum the measurement time is required and these times are collected in column C. The vector of wavelengths at which the spectra were acquired is stored in the row 5 above the spectra. The two inserted figures display the data as a function of time and of wavelength.

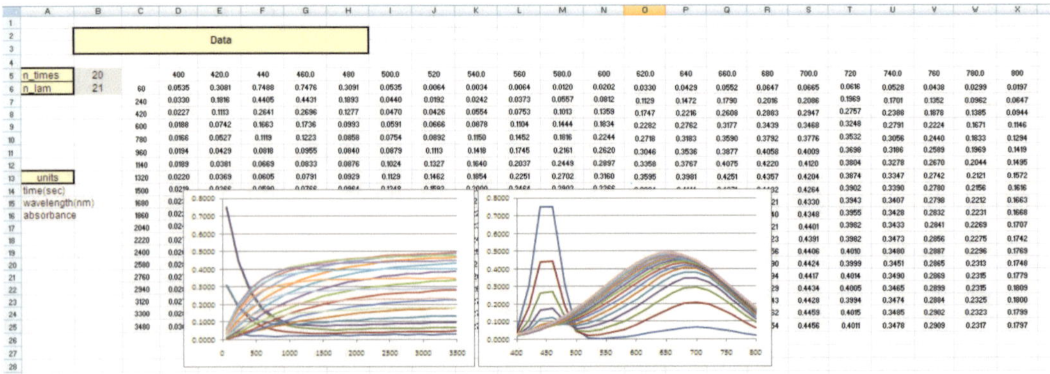

Fig. 8. The data arranged in an excel spreadsheet; the figure on the left displays the kinetic traces at all wavelengths, the figure on the right displays the measured spectra.

Prior to the fitting, the chemical reaction model on which the analysis will be based needs to be defined. As mentioned above ReactLab and other modern programs incorporate a model translator that allows the definition in a natural chemistry language and which subsequently translates automatically into internal coefficient information that allows the automatic construction of the mathematical expressions required by the numerical and speciation algorithms. Note for each reaction an initial guess for the rate constant has to be supplied. The ReactLab model is for this reaction is shown in Figure 9.

Reactants	Reaction Type	Products	Label	Parameters k / log K
A	>	B	k1	1.000E-02
B	>	C	k2	1.000E-04

Fig. 9. The definition of the chemical model for the consecutive reaction scheme $\mathbf{A} \xrightarrow{k_1} \mathbf{B} \xrightarrow{k_2} \mathbf{C}$

The 'compiler' recognises that there are 3 reacting species, *A*, *B*, *C*, and 2 rate constants. For the initial concentrations the appropriate values have to be supplied by the user. In the example $[A]_{init}$=0.001 M $[B]_{init}$ and $[C]_{init}$ are zero. Further the spectral status of each species needs to be defined, in the example all 3 species are 'colored' i.e. they do absorb in the wavelength range of the data, see Figure 10. The alternative 'non-absorbing' indicates that the species does not absorb in the measured range. Advanced packages including ReactLab also allow the implementation of 'known' spectra which need to be introduced elsewhere in the workbook.

n_species	3
n_par	2
n_aux_par	0

Species	A	B	C
init []	1.00E-03	0.00E+00	0.00E+00
Spectrum	colored	colored	colored

Fig. 10. For the consecutive reaction scheme there are 3 reaction species for which initial concentrations need to be given.

The program is now in a position to first calculate the concentration of all species as a function of time and subsequently their absorption spectra. The results for the present initial guesses for the rate constants are displayed in Figure 11.

Concentration profiles and spectra

Conc	A	B	C
60.000	0.001	0.000	0.000
240.000	0.000	0.001	0.000
420.000	0.000	0.001	0.000
600.000	0.000	0.001	0.000
780.000	0.000	0.001	0.000
960.000	0.000	0.001	0.000
1140.000	0.000	0.001	0.000
1320.000	0.000	0.001	0.000
1500.000	0.000	0.001	0.000
1680.000	0.000	0.001	0.000
1860.000	0.000	0.001	0.000
2040.000	0.000	0.001	0.000
2220.000	0.000	0.001	0.000
2400.000	0.000	0.001	0.000
2580.000	0.000	0.001	0.000
2760.000	0.000	0.001	0.000
2940.000	0.000	0.001	0.000
3120.000	0.000	0.001	0.000
3300.000	0.000	0.001	0.000
3480.000	0.000	0.001	0.000

Spectra	A	B	C
400.000	85.110	18.107	57.409
420.000	518.773	78.660	-94.665
440.000	1270.894	182.713	-379.762
460.000	1262.451	188.570	-282.028
480.000	501.917	95.929	202.108
500.000	54.337	45.918	564.432
520.000	-39.698	47.079	766.776
540.000	-61.797	62.775	939.915
560.000	-76.845	84.277	1097.238
580.000	-94.684	114.446	1219.188
600.000	-114.847	153.564	1286.297
620.000	-131.343	199.005	1298.891
640.000	-157.195	249.899	1234.535
660.000	-173.180	295.129	1119.170
680.000	-181.284	326.036	971.539
700.000	-181.735	332.577	830.737
720.000	-171.840	312.778	712.015
740.000	-148.511	269.552	622.870
760.000	-114.841	213.772	550.179
780.000	-90.220	157.249	485.204
800.000	-63.624	107.082	418.616

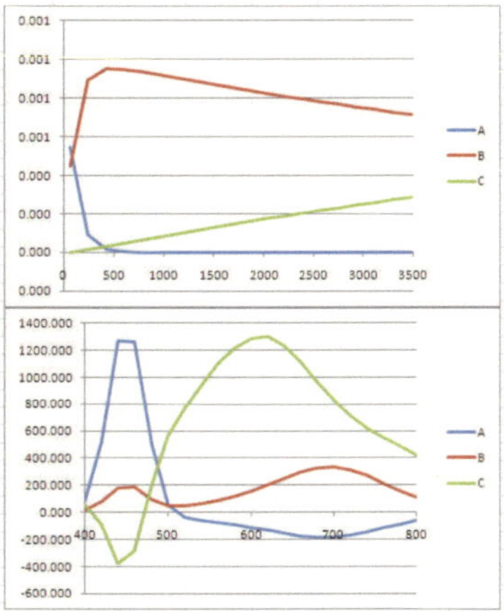

Fig. 11. The concentration profiles and absorption spectra, as calculated with initial guesses for the rate constants shown in Figure 9.

The concentration profiles indicate a fast first reaction *A>B* and a much slower subsequent reaction *B>C*. However, the calculated, partially negative absorption spectra clearly demonstrate that there is 'something wrong', that the initial guesses for the rate constants are obviously not correct. In this example the deviations are not too severe indicating the model itself is plausible.

Clicking the Fit button initiates the iterative improvement of the parameters and after a few iterations the 'perfect' results are evident. This of course is supportive of the validity of the model itself. If the scheme is wrong and cannot account for the detail in the data, a good fit will be unobtainable. The ReactLab GUI at the end of the fit is given in Figure 12.

On the other hand an over complex model has to be carefully avoided as any data can usually be fitted with enough parameters (including artefacts!). Occam's razor should be assiduously applied accepting the simplest model that fits the data as the most likely.

Of course it is also a risk that a model is underdetermined by the data. Put simply the information in the measurement is not sufficient to deliver a unique solution, and the program will not converge properly and usually oscillate delivering one of an infinite number of solutions (usually combinations of rates). Whilst this does not imply the model is

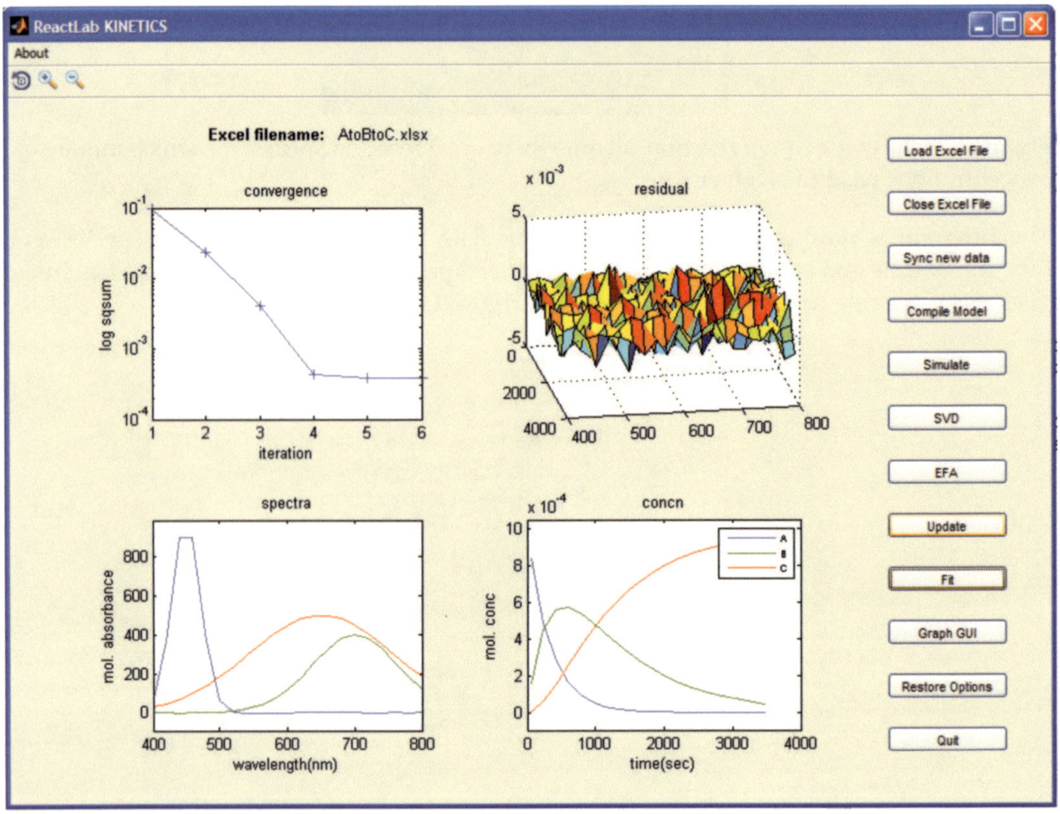

Fig. 12. The ReactLab GUI displaying the progress of *ssq*, the residuals, the absorption spectra and the concentration profiles after the fitting.

incorrect, further work will be required to determine and fix key parameters or spectra in order to resolve the problem.

Example 2: Kinetic analysis of the reaction of Ellmans reagent (*DTNB*) and thioglycerol *RS*.

This example illustrates the direct fitting of a simplified model followed by the correct and more complex model to a data set collected using a PDA on a stopped flow (data courtesy of TgK Scientific Ltd, UK).

Ellmans reagent, 5,5′-Dithio-bis(2-nitrobenzoic acid) or *DTNB* is a commercially available reagent for quantifying thiols both in pure and biological samples and measuring the number of thiol groups on proteins. The reaction yields a colored thiolate, *RS-TNB*, ion which absorbs at 412nm and can be used to quantify the original thiol concentration. In this particular case the reaction with thioglycerol, *RS-*, leads to a 2 step disulphide exchange reaction and is particularly suited for establishing the dead-time of stopped flow instruments (Paul, Kirschner et al. 1979). The reaction is represented in Figure 13. The model in the ReactLab definition is given in Figure 16.

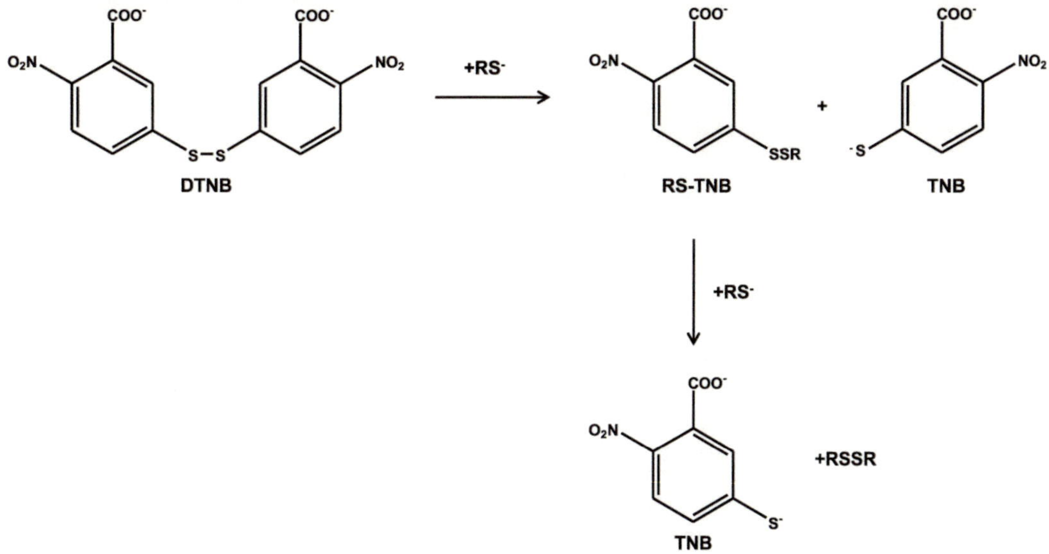

Fig. 13. The 2-step reaction of *DTNB* with a thiolate, RS^-.

A 3-D representation of the spectra measured at different time intervals for a total of 1.5 sec. on a stopped-flow instrument is shown in Figure 14.

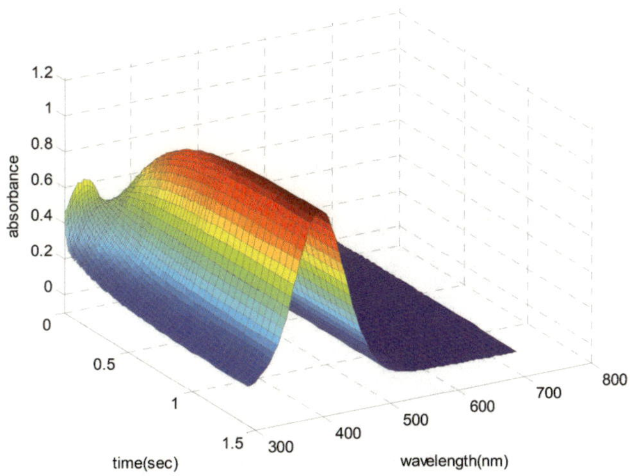

Fig. 14. Spectral changes resulting from the reactions of Ellmans reagent (*DTNB*) and thioglycerol (RS^-).

In the experiment a large excess of thioglycerol was used and thus the two second order reactions can be approximated with a two-step pseudo first order sequential mechanism. Thus, we first attempt to fit this biphasic reaction with a simple consecutive reaction scheme with three colored species $\mathbf{A} \xrightarrow{k_1} \mathbf{B} \xrightarrow{k_2} \mathbf{C}$ (Figure 15). The fitted rates are 75sec^{-1} and 3.9sec^{-1}.

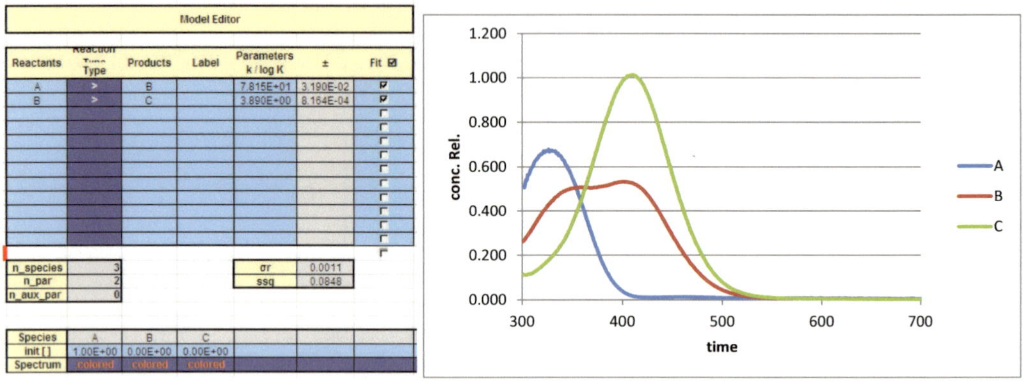

Fig. 15. A simple consecutive reaction model with best fit rates and spectra.

The problems with this fit are two-fold. First we know the reactions are second order and that the intermediate spectrum for species *B* is implausible with two peaks.

The complete model as defined in ReactLab, together with the fit at 412 nm, is displayed in Figure 16. Whilst the square sum is not significantly improved the spectra are now correct according to literature sources and the corresponding rates for the 2 steps are 1.57 $\times 10^4 \pm 6$ $M^{-1}sec^{-1}$ and 780 $\pm$ 2 $M^{-1}sec^{-1}$.

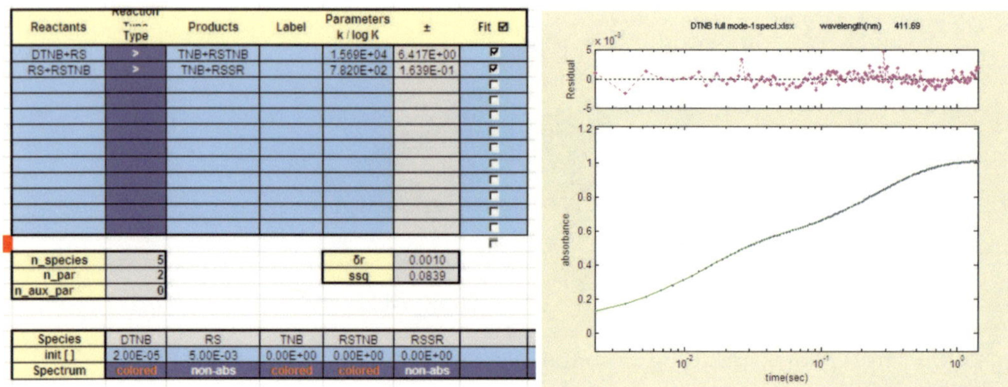

Fig. 16. Fitting the complete mechanism to the data yields the correct rates. The right panel displays the quality of the fit at 412 nm on a logarithmic time axis.

The concentration profiles and spectra resulting from a fit to the correct model are now as in Figure 17.

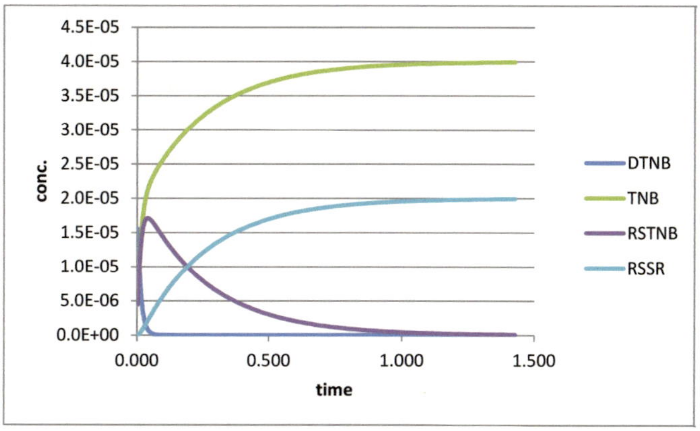

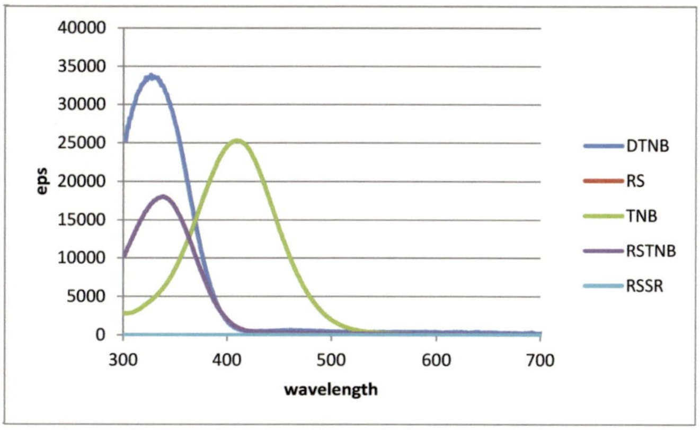

Fig. 17. Concentration profiles and molar absorption spectra for the analysis based on the complete reaction scheme.

This example does serve to demonstrate that good fits can be obtained with an incorrect or simplistic model and that some insight and care is required to establish the correct mechanism and obtain genuine parameters. What is certainly true is that the second model could only be fitted because of the numerical integration of the more complex second order mechanisms. This was a trivial change of model configuration in ReactLab and could not have been achieved using classical analysis approaches. Secondly the importance of dealing with whole spectra is highlighted in that the spectra resulting for the fit provide important insight into the underlying chemistry and must make sense in this respect. Single wavelength kinetic analysis has no such indirect reinforcement.

By way of a final comment on this example; we noted that the data was collected under pseudo first order conditions i.e. one reagent in excess. This ubiquitous approach was essential to enable the determination of second order rate constants using a first order fit by classical analysis using explicit functions (usually sums of exponentials). In the pseudo first order simplification a 2nd order rate constant is calculated from the observed pseudo first order rate constant.

Numerical integration methods eliminate the need for this constraint and therefore any requirement to work under pseudo first order conditions (or indeed the comparable constraint of keeping the reactant concentrations equal).

Example 3: Equilibrium investigation, concentration determination of a diprotic and a strong acid

Titrations can be used for the determination of equilibrium constants and insofar the analysis is very similar to a kinetic investigation. Titrations are also an important analytical tool for the determination of concentrations, in real terms this is probably the more common application.

Let us consider a titration of a solution of the diprotic acid AH_2 in the presence of an excess of a strong acid, e.g. *HCl*. The equilibrium model only includes the diprotic acid as the strong acid is always completely dissociated, see the ReactLab model in Fig. 18.

Reactants	Reaction Type	Products	Label	Parameters k / log K
A+H	=	AH		8.000E+00
AH+H	=	AH2		3.000E+00

Fig. 18. The model for a diprotic acid that undergoes two protonation equilibria.

The components are *A* and *H*, the species are *A*, *AH*, AH_2, *H*, *OH*. For the components the total concentrations have to be known in each solution during the titration. They are collected in the columns E and F of a spreadsheet. The measured spectra are collected from the cell N7 on, see Figure 19

	A	B	C	D	E	F	G	H	M	N	O	P	Q
1									Expand				
2		Data and Component concentrations											
3													
4													
5				Total component []						lam			
6	n_spectra	151	Vadd(ml)	Vtot(ml)	A	H				400.0	410.0	420.0	430.
7	n_lam	21	0.000	10.000	0.100	0.250				0.000	0.000	0.001	-0.0C
8			0.100	10.100	0.099	0.244				0.000	0.001	0.000	0.00
9			0.200	10.200	0.098	0.239				0.001	0.001	-0.001	0.00
10			0.300	10.300	0.097	0.234				-0.001	-0.001	0.001	0.00
11			0.400	10.400	0.096	0.229				-0.001	0.000	0.000	-0.0C
12			0.500	10.500	0.095	0.224				0.000	-0.001	0.000	0.00

Fig. 19. Data entry for a titration, the crucial total concentrations are stored in the columns E and F, the spectra to the right (incomplete in this Figure).

In the example 10ml of a solution containing *A* and *H* are titrated with 0.1 ml aliquots of base. The concentrations $[A]_{tot}$ and $[H]_{tot}$ are computed from the volumes and concentrations of the original solution in the 'beaker' and in the 'burette'. These concentrations are stored in the main sheet in the rows 37 and 38 as seen in Figure 20.

The definition of the total concentrations of *A* and *H* in the 'beaker' are defined in the cells C37, D37, the same component concentrations in the burette solution in the row below. Often these concentrations are known and then there is nothing to be added. In the example the component concentrations in the 'beaker' are to be fitted. This is achieved by defining

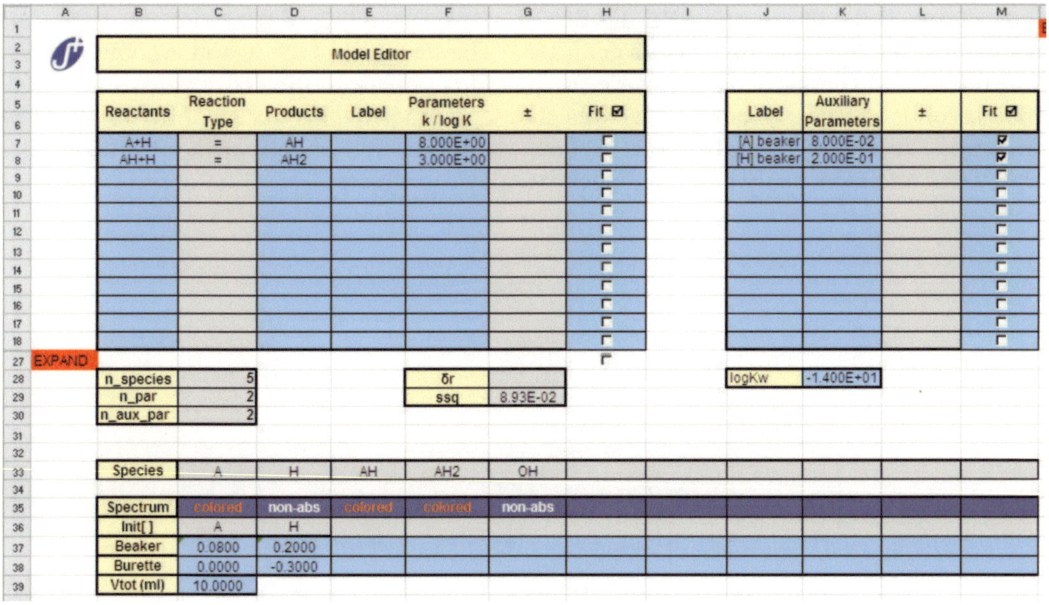

Model Editor

Reactants	Reaction Type	Products	Label	Parameters k / log K	±	Fit
A+H	=	AH		8.000E+00		☐
AH+H	=	AH2		3.000E+00		☐

Label	Auxiliary Parameters	±	Fit
[A] beaker	8.000E-02		☑
[H] beaker	2.000E-01		☑

n_species	5
n_par	2
n_aux_par	2
δr	
ssq	8.93E-02
logKw	-1.400E+01

Species	A	H	AH	AH2	OH
Spectrum	colored	non-abs	colored	colored	non-abs
Init[ ]	A	H			
Beaker	0.0800	0.2000			
Burette	0.0000	-0.3000			
Vtot (ml)	10.0000				

Fig. 20. Model entry and information on concentrations and spectral status.

them as auxiliary parameters in the cells K7, K8; the contents of cells C37, D37, are now references to the auxiliary parameters which are fitted.

Fitting results in values for the concentrations and their error estimates, Figure 21.

Label	Auxiliary Parameters	±	Fit ☑
[A] beaker	1.000E-01	2.126E-04	☑
[H] beaker	2.499E-01	2.152E-04	☑

Fig. 21. The result of the fitting of the concentrations, complete with error analysis.

Example 4: Equilibrium Interaction between Cu^{2+} and *PHE* (1,9-Bis(2-hydroxyphenyl)-2,5,8-triazanonane)

In this example we demonstrate the analysis of 'pH-metered' titrations, a mode of titration that is common in for the investigation of equilibria in aqueous solution. In this kind of titration the independent variable is the pH, rather than the added volume of reagent as has been the case in all previous examples. As before the measured spectra are the dependent variables. An important rationale for 'pH-metered' titrations is the fact that it is often difficult to completely exclude CO_2 during the whole titration. Unknown amounts of CO_2 result in the addition of unknown amounts of acid via formation of carbonate species. In 'pH-metered' titrations the effect of this impurity is minimal as long as none of the carbonate species interfere with the process under investigation; the effect in the 'default mode' can be much more pronounced. The price to pay for that advantage is more complex data acquisition as the pH has to be measured and recorded together with the spectra after each addition of reagent.

The example is the titration of *PHE*, 1,9-Bis(2-hydroxyphenyl)-2,5,8-triazanonane, with Cu^{2+} in aqueous solution.(Gampp, Haspra et al. 1984) The structure of the ligand is shown below. It forms several complexes: *ML*, where the ligand is penta-coordinated presumable via all three secondary amine groups as well as the deprotonated phenolates; and two partially protonated species *MLH* and *MLH2*, in these complexes one or both of the phenolates are protonated and most likely not or only very weakly coordinated.

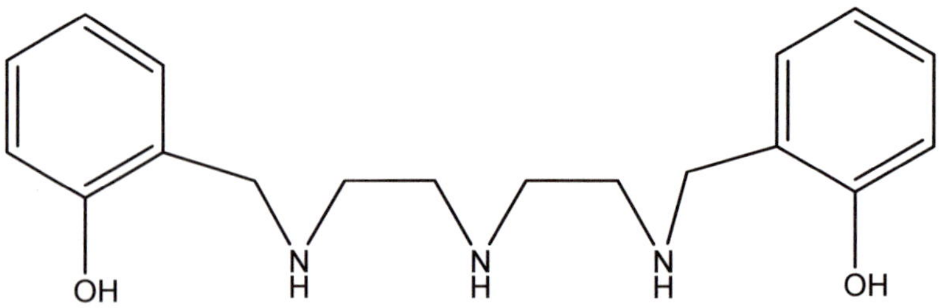

In this titration a solution of 7.23×10^{-4} M Cu^{2+} and 1.60×10^{-3} M *PHE* with an excess *HCl* were titrated with a total of approx. 750µL *NaOH* solution. After each addition of the base the pH and the spectrum were measured. The total concentrations of metal and ligand are entered for each sample in the **'Data'** worksheet. Note that the columns for the total concentration of the protons is left empty: the measured pH in column M is defining the free proton concentration which in turn is used to compute all species concentrations in conjunction with the total concentrations of in this case the metal ion and the ligand provided, see Figure 22.

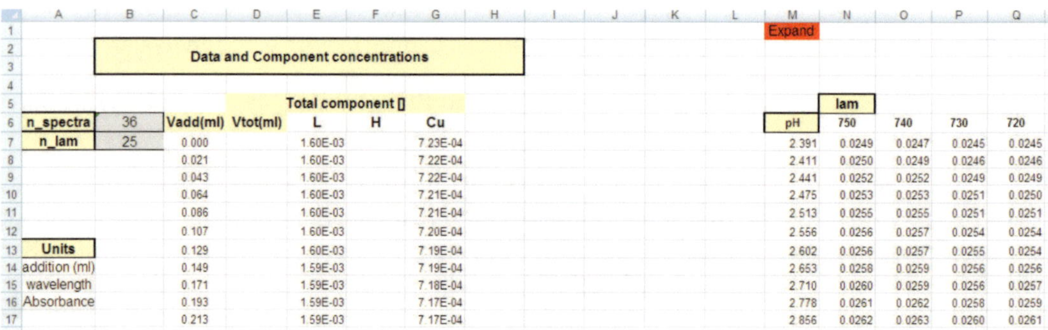

	A	B	C	D	E	F	G	H	I	J	K	L	M	N	O	P	Q
1													Expand				
2		Data and Component concentrations															
3																	
4																	
5					Total component []									lam			
6	n_spectra	36	Vadd(ml)	Vtot(ml)	L	H	Cu						pH	750	740	730	720
7	n_lam	25	0.000		1.60E-03		7.23E-04						2.391	0.0249	0.0247	0.0245	0.0245
8			0.021		1.60E-03		7.22E-04						2.411	0.0250	0.0249	0.0246	0.0246
9			0.043		1.60E-03		7.22E-04						2.441	0.0252	0.0252	0.0249	0.0249
10			0.064		1.60E-03		7.21E-04						2.475	0.0253	0.0253	0.0251	0.0250
11			0.086		1.60E-03		7.21E-04						2.513	0.0255	0.0255	0.0251	0.0251
12			0.107		1.60E-03		7.20E-04						2.556	0.0256	0.0257	0.0254	0.0254
13	Units		0.129		1.60E-03		7.19E-04						2.602	0.0256	0.0257	0.0255	0.0254
14	addition (ml)		0.149		1.59E-03		7.19E-04						2.653	0.0258	0.0259	0.0256	0.0256
15	wavelength		0.171		1.59E-03		7.18E-04						2.710	0.0260	0.0259	0.0256	0.0257
16	Absorbance		0.193		1.59E-03		7.17E-04						2.778	0.0261	0.0262	0.0258	0.0259
17			0.213		1.59E-03		7.17E-04						2.856	0.0262	0.0263	0.0260	0.0261

Fig. 22. Only the total concentrations of the metal and ligand are required, the column for the protons is left empty. Column M contains the pH values and the entry 'pH' in cell M6 to indicate a 'pH-metered' titration.

The measurement is displayed in Figure 23, each curve is the measured absorption at one particular wavelength.

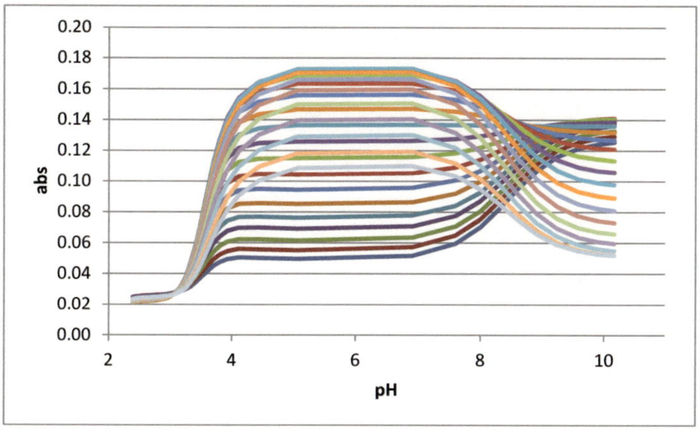

Fig. 23. The measurement, here displayed as a series of titration curves at the different wavelengths.

The ligand *PHE* has five protonation constants which have to be determined independently. The successive logK values are 10.06, 10.41, 9.09, 7.94 and 4.18. Note that in the model the protonations are defined as overall stabilities, see Figure 24.

The results of the analysis are summarised in Figure 24 and Figure 25. Again the protonation equilibria for the complex species are defined as formation constants, the logK values for the protonation equilibria $ML+H \rightleftarrows MLH$ and $MLH+H \rightleftarrows MLH_2$ are 8.42 and 3.92.

Reactants	Reaction Type	Products	Label	Parameters logK/logβ	±	Fit ☑
L+H	=	LH		11.060		☐
L+2H	=	LH2		21.470		☐
L+3H	=	LH3		30.560		☐
L+4H	=	LH4		38.500		☐
L+5H	=	LH5		42.680		☐
Cu+L	=	CuL		22.563	0.011	☑
Cu+L+H	=	CuLH		30.979	0.011	☑
Cu+L+2H	=	CuLH2		34.895	0.002	☑
						☐
						☐
						☐
						☐

n_species	12	σr	4.63E-04
n_par	8	ssq	1.76E-04
n_aux_par	0		

Fig. 24. The fitted equilibrium constants for the formation of the *ML*, *MLH* and *MLH2* complexes.

The concentration profiles are represented in two different modes, the left part has the measured pH as the x-axis and only the metal species are shown, the right part shows all species concentrations as a function of the added volume of base. This figure reveals that a substantial excess of acid has been added to the initial solution and the first 0.2 mL of base are used to neutralise this excess

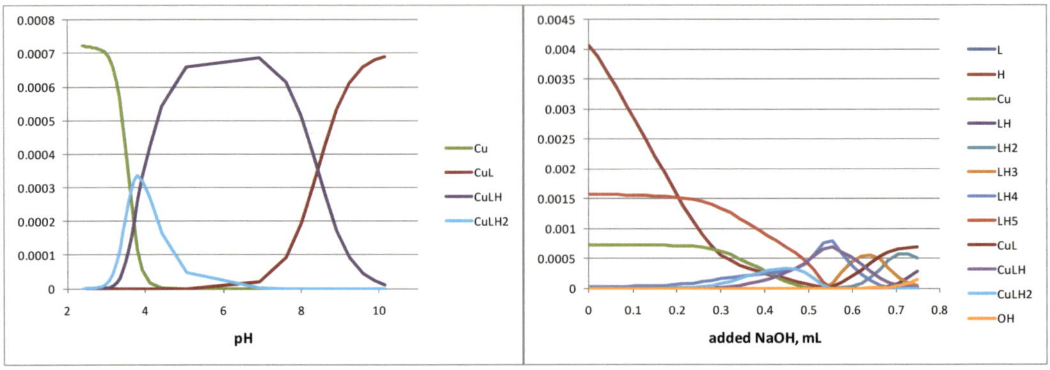

Fig. 25. The concentration profiles of the complexes as a function of pH and all species as a function of the volume of added base.

15. Conclusion

Data fitting is a well-established method that has been extensively used for the analysis of chemical processes since the beginnings of instrumental methods. Generally, increasing sophistication of instrumentation has inspired parallel developments in numerical methods for appropriate analysis of ever better and more plentiful data. This chapter concentrates on spectroscopic methods for the investigation of chemical processes; it details the structure of the data and the principles of model based data fitting. It is rounded of by a collection of typical and illustrative examples using a modern data analysis software package. The aim is to demonstrate the power and effortlessness of modern data analysis of typical data sets.

16. References

Espenson, J. H. (1995). *Chemical Kinetics and Reaction Mechanisms*. New York, McGraw-Hill.

Gampp, H., D. Haspra, et al. (1984). "Copper(II) Complexes with Linear Pentadentate Chelators." *Inorganic Chemistry* 23: 3724-4730.

Gans, P. (1992). Data Fitting in the Chemical Sciences, Wiley.

Maeder, M. and Y.-M. Neuhold (2007). *Practical Data Analysis in Chemistry*. Amsterdam, Elsevier.

Maeder, M., Y. M. Neuhold, et al. (2002). "Analysis of reactions in aqueous solution at non-constant pH: No more buffers?" *Phys. Chem. Chem. Phys.* 5: 2836-2841.

Martell, A. E. and R. J. Motekaitis (1988). *The Determination and Use of Stability Constants*. New York, VCH.

Menten, L. and M. I. Michaelis (1913). "Die Kinetik der Invertinwirkung." *Biochem Z* 49: 333-369.

Norman, S. and M. Maeder (2006). "Model-Based Analysis for Kinetic and Equilibrium Investigations." *Critical Reviews in Analytical Chemistry* 36: 199-209.

Paul, C., K. Kirschner, et al. (1979). "Calibration of Stopped-Flow Spectrophotometers Using a Two-Step Disulfide Exchenge Reaction." *Analytical Biochemistry* 101: 442-448.

Polster, J. and H. Lachmann (1989). *Spectrometric Titrations: Analysis of Chemical Equilibria*. Weinheim, VCH.

Press, W. H., W. T. Vetterling, et al. (1995). *Numerical Recipes in C*. Cambridge, Cambridge University Press.

Wilkins, R. G. (1991). *Kinetics and Mechanism of Reactions of Transition Metal Complexes*. Weinheim, VCH.

4

Exploratory Data Analysis with Latent Subspace Models

José Camacho
Department of Signal Theory, Telematics and Communication,
University of Granada, Granada
Spain

1. Introduction

Exploratory Data Analysis (EDA) has been employed for decades in many research fields, including social sciences, psychology, education, medicine, chemometrics and related fields (1) (2). EDA is both a data analysis philosophy and a set of tools (3). Nevertheless, while the philosophy has essentially remained the same, the tools are in constant evolution. The application of EDA to current problems is challenging due to the large scale of the data sets involved. For instance, genomics data sets can have up to a million of variables (5). There is a clear interest in developing EDA methods to manage these scales of data while taking advantage of *the basic importance of simply looking at data* (3).

In data sets with a large number of variables, collinear data and missing values, projection models based on latent structures, such as Principal Component Analysis (PCA) (6) (7) (1) and Partial Least Squares (PLS) (8) (9) (10), are valuable tools within EDA. Projection models and the set of tools used in combination simplify the analysis of complex data sets, pointing out to special observations (outliers), clusters of similar observations, groups of related variables, and crossed relationships between specific observations and variables. All this information is of paramount importance to improve data knowledge.

EDA based on projection models has been successfully applied in the area of chemometrics and industrial process analysis. In this chapter, several standard tools for EDA with projection models, namely score plots, loading plots and biplots, are revised and their limitations are elucidated. Two recently proposed tools are introduced to overcome these limitations. The first of them, named Missing-data methods for Exploratory Data Analysis or MEDA for short (11), is used to investigate the relationships between variables in projection subspaces. The second one is an extension of MEDA, named observation-based MEDA or *o*MEDA (33), to discover the relationships between observations and variables. The EDA approach based on PCA/PLS with scores and loading plots, MEDA and *o*MEDA is illustrated with several real examples from the chemometrics field.

This chapter is organized as follows. Section 2 briefly discusses the importance of subspace models and score plots to explore the data distribution. Section 3 is devoted to the investigation of the relationship among variables in a data set. Section 4 studies the relationship between observations and variables in latent subspaces. Section 5 presents a EDA case study of Quantitative Structure-Activity Relationship (QSAR) modelling and section 6 proposes some concluding remarks. Examples and Figures were computed using the MATLAB programming environment, with the PLS-Toolbox (32) and home-made software. A MATLAB toolbox with the tools employed in this chapter is available at http://wdb.ugr.es/ josecamacho/.

2. Patterns in the data distribution

The distribution of the observations in a data set contains relevant information for data understanding. For instance, in an industrial process, one outlier may represent an abnormal situation which affects the process variables to a large extent. Studying this observation with more detail, one may be able to identify if it is the result of a process upset or, very commonly, a sensor failure. Also, clusters of observations may represent different operation points. Outliers, clusters and trends in the data may be indicative of the degree of control in the process and of assignable sources of variation. The identification of these sources of variation may lead to the reduction of the variance in the process with the consequent reduction of costs.

The distribution of the observations can be visualized using scatter plots. For obvious reasons, scatter plots are limited to three dimensions at most, and typically to two dimensions. Therefore, the direct observation of the data distribution in data sets with several tens, hundreds or even thousands of variables is not possible. One can always construct scatter plots for selective pairs or thirds of variables, but this is an overwhelming and often misleading approach. Projection models overcome this problem. PCA and PLS can be used straightforwardly to visualize the distribution of the data in the latent subspace, considering only a few latent variables (LVs) which contain most of the variability of interest. Scatter plots of the scores corresponding to the LVs, the so-called score plots, are used for this purpose.

Score plots are well known and accepted in the chemometric field. Although simple to understand, score plots are paramount for EDA. The following example may be illustrative of this. In Figure 1, three simulated data sets of the same size (100×100) are compared. Data simulation was performed using the technique named Approximation of a DIstribution for a given COVariance matrix (15), or ADICOV for short. Using this technique, the same covariance structure was simulated for the three data sets but with different distributions: the first data set presents a multi-normal distribution in the latent subspace, the second one presents a severe outlier and the third one presents a pair of clusters. If the scatter plot of the observations in the plane spanned by the first two variables is depicted (first row of Figure 1), the data sets seem to be almost identical. Therefore, unless an extensive exploration is performed, the three data sets may be though to come from a similar data generation procedure. However, if a PCA model for each data set is fitted and the score plots corresponding to the first 2 PCs are shown (second row of Figure 1), differences among the three data sets are made apparent: in the second data set there is one outlier (right side of Figure 1(e)) and in the third data set there are two clusters of observations. As already discussed, the capability to find these details is paramount for data understanding, since outliers and clusters are very informative of the underlaying phenomena. Most of the times these details are also apparent in the original variables, but finding them may be a tedious work. Score plots after PCA modelling are perfectly suited to discover large deviations among the observations, avoiding the overwhelming task of visualizing each possible pair of original variables. Also, score plots in regression models such as PLS are paramount for model interpretation prior to prediction.

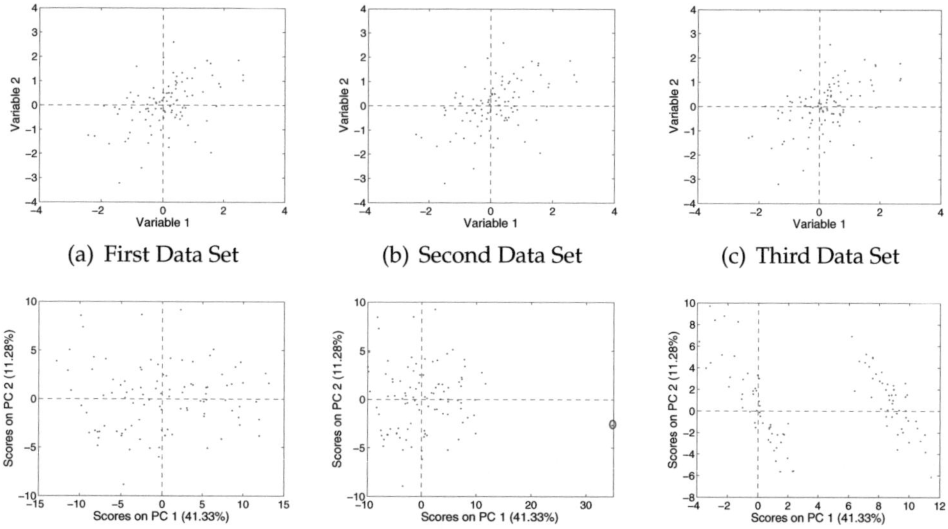

(a) First Data Set (b) Second Data Set (c) Third Data Set

(d) First Data Set (e) Second Data Set (f) Third Data Set

Fig. 1. Experiment with three simulated data sets of dimension 100×100. Data simulation was performed using the ADICOV technique (15). In the first row of figures, the scatter plots corresponding to the first two variables in the data sets are shown. In the second row of figures, the scatter plots (score plots) corresponding to the first two PCs in the data sets are shown.

3. Relationships among variables

PCA has been often employed to explore the relationships among variables in a data set (19; 20). Nevertheless, it is generally accepted that Factor Analysis (FA) is better suited than PCA to study these relationships (1; 7). This is because FA algorithms are designed to distinguish between shared and unique variability. The shared variability, the so-called communalities in the FA community, reflect the common factors–common variability–among observable variables. The unique variability is only present in one observable variable. The common factors make up the relationship structure in the data. PCA makes no distinction between shared and unique variability and therefore it is not suited to find the structure in the data.

When either PCA or FA are used for data understanding, a two step procedure is typically followed (1; 7). Firstly, the model is calibrated from the available data. Secondly, the model is rotated to obtain a so-called simple structure. The second step is aimed at obtaining loading vectors with as much loadings close to 0 as possible. That way, the loading vectors are easier to interpret. It is generally accepted that oblique transformations are preferred to the more simple orthogonal transformations (19; 20), although in many situations the results are similar (1).

The limitation of PCA to detect common factors and the application of rotation methods will be illustrated using the pipelines artificial examples (14). Data for each pipeline are simulated according to the following equalities:

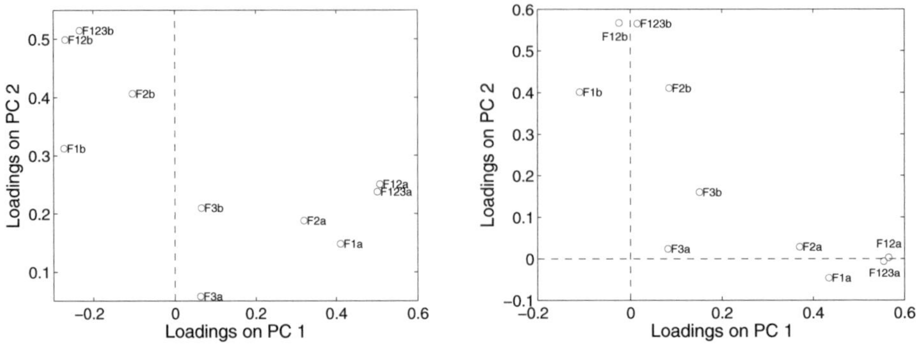

(a) PCA loadings (b) Varimax rotation

Fig. 2. Loading plots of the PCA model fitted from the data in the two pipelines example: (a) original loadings and (b) loadings after varimax rotation.

$$F12 = F1 + F2$$

$$F123 = F1 + F2 + F3$$

where F1, F2 and F3 represent liquid flows which are generated independently at random following a normal distribution of 0 mean and standard deviation 1. A 30% of measurement noise is generated for each of the five variables in a pipeline at random, following a normal distribution of 0 mean:

$$\tilde{x}_i = (x_i + \sqrt{0.3} \cdot n)/(\sqrt{1.3})$$

where $\tilde{x}_i$ is the contaminated variable, x_i the noise-free variable and n the noise generated. This simulation structure generates blocks of five variables with three common factors: the common factor F1, present in the observed variables F1, F12 and F123; the common factor F2, present in the observed variables F2, F12 and F123; and the common factor F3, present in F3 and F123. Data sets of any size can be obtained by combining the variables from different pipelines. In this present example, a data set with 100 observations from two pipelines for which data are independently generated is considered. Thus, the size of the data set is 100 $\times$ 10 and the variability is built from 6 common factors.

Figure 2 shows the loading plots of the PCA model of the data before and after rotation. Loading plots are interpreted so that close variables, provided they are far enough from the origin of coordinates, are considered to be correlated. This interpretation is not always correct. In Figure 2(a), the first component separates the variables corresponding to the two pipelines. The second component captures variance of most variables, specially of those in the second pipeline. The two PCs capture variability corresponding to most common factors in the data at the same time, which complicates the interpretation. As already discussed, PCA is focused on variance, without making the distinction between unique and shared variance. The result is that the common factors are not aligned with the PCs. Thus, one single component reflects several common factors and the same common factor may be reflected in several components. As a consequence, variables with high and similar loadings in the same subset of components do not necessarily need to be correlated, since they may present very different loadings in others components. Because of this, inspecting only a pair of components may lead to incorrect conclusions. A good interpretation would require inspecting and interrelating all pairs of components with relevant information, something which may be challenging in many situations. This problem affects the interpretation and it is the reason why FA is generally preferred to PCA.

Figure 2(b) shows the resulting loadings after applying one of the most used rotation methods: the varimax rotation. Now, the variables corresponding to each pipeline are grouped towards one of the loading vectors. This highlights the fact that there are two main and orthogonal sources of variability, each one representing the variability in a pipeline. Also, in the first component variables collected from pipeline 2 present low loadings whereas in the second component variables collected from pipeline 1 present low loadings. This is the result of applying the notion of simple structure, with most of the loadings rotated towards 0. The interpretation is simplified as a consequence of improving the alignment of components with common factors. This is especially useful in data sets with many variables.

Although FA and rotation methods may improve the interpretation, they still present severe limitations. The derivation of the structure in the data from a loading plot is not straightforward. On the other hand, the rotated model depends greatly on the normalization of the data and the number of PCs (1; 21). To avoid this, several alternative approaches to rotation have been suggested. The point in common of these approaches is that they find a trade-off between variance explained and model simplicity (1). Nevertheless, imposing a simple structure has also drawbacks. Reference (11) shows that, when simplicity is pursued, there is a potential risk of simplifying even the true relationships in the data set, missing part of the data structure. Thus, the indirect improvement of data interpretation by imposing a simple structure may also report misleading results in certain situations.

3.1 MEDA

MEDA is designed to find the true relationships in the data. Therefore, it is an alternative to rotation methods or in general to the simple structure approach. A main advantage of MEDA is that, unlike rotation or FA methods, it is applied over any projection subspace without actually modifying it. The benefit is twofold. Firstly, MEDA is straightforwardly applied in any subspace of interest: PCA (maximizing variance), PLS (maximizing correlations) and any other. On the contrary, FA methods are typically based on complicated algorithms, several of which have not been extended to regression. Secondly, MEDA is also useful for model interpretation, since common factors and components are easily interrelated. This is quite useful, for instance, in the selection of the number of components.

MEDA is based on the capability of missing values estimation of projection models (22–27). The MEDA approach is depicted in Figure 3. Firstly, a projection model is fitted from the calibration $N \times M$ matrix $\mathbf{X}$ (and optionally $\mathbf{Y}$). Then, for each variable m, matrix $\mathbf{X}_m$ is constructed, which is a $N \times M$ matrix full with zeros except in the m-th column where it contains the m-th column of matrix $\mathbf{X}$. Using $\mathbf{X}_m$ and the model, the scores are estimated with a missing data method. The known data regression (KDR) method (22; 25) is suggested at this point. From the scores, the original data is reconstructed and the estimation error computed. The variability of the estimation error is compared to that of the original data according to the following index of goodness of prediction:

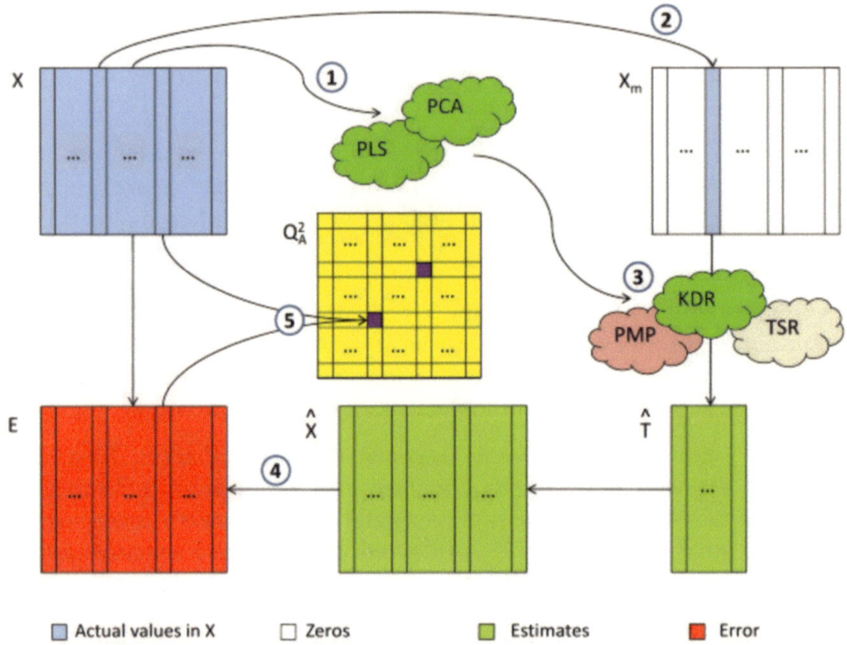

Fig. 3. MEDA technique: (1) model calibration, (2) introduction of missing data, (3) missing data imputation, (4) error computation, (5) computation of matrix $\mathbf{Q}_A^2$.

$$q^2_{A,(m,l)} = 1 - \frac{\|\hat{\mathbf{e}}_{A,(l)}\|^2}{\|\mathbf{x}_{(l)}\|^2}, \quad \forall l \neq m. \tag{1}$$

where $\hat{\mathbf{e}}_{A,(l)}$ corresponds to the estimation error for the l-th variable and $\mathbf{x}_{(l)}$ is its actual value. The closer the value of the index is to 1, the more related variables m and l are. After all the indices corresponding to each pair of variables are computed, matrix $\mathbf{Q}_A^2$ is formed so that $q^2_{A,(m,l)}$ is located at row m and column l. For interpretation, when the number of variables is large, matrix $\mathbf{Q}_A^2$ can be shown as a color map. Also, a threshold can be applied to $\mathbf{Q}_A^2$ so that elements over this threshold are set to 1 and elements below the threshold are set to 0.

The procedure depicted in Figure 3 is the original and more general MEDA algorithm. Nevertheless, provided KDR is the missing data estimation technique, matrix Q_A^2 can be computed from cross-product matrices following a more direct procedure. The value corresponding to the element in the i-th row and j-th column of matrix $\mathbf{Q}_A^2$ in MEDA is equal to:

$$q^2_{A,(m,l)} = \frac{2 \cdot S_{ml} \cdot S_{ml^A} - (S_{ml^A})^2}{S_{mm} \cdot S_{ll}}. \tag{2}$$

where S_{lm} stands for the cross-product of variables $\mathbf{x}_l$ and $\mathbf{x}_m$, i.e. $S_{lm} = \mathbf{x}_l^T \cdot \mathbf{x}_m$, and S_{lm^A}stands for the cross-product of variables $\mathbf{x}_l$ and $\mathbf{x}_m^A$, being $\mathbf{x}_m^A$ the projection of $\mathbf{x}_m$ in the model sub-space in coordinates of the original space. Thus, S_{lm} is the element in the l-th row

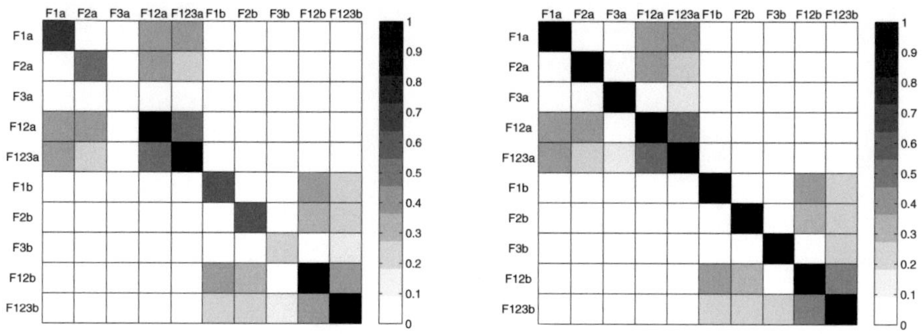

(a) 2 PCs (b) 6 PCs

Fig. 4. MEDA matrix from the PCA model fitted from the data in the two pipelines example.

and m-th column of the cross-product matrix $\mathbf{XX} = \mathbf{X}^T \cdot \mathbf{X}$ and S_{lm^A} corresponds to the element in the l-th row and m-th column of matrix $\mathbf{XX} \cdot \mathbf{P}_A \cdot \mathbf{P}_A^t$ in PCA and of matrix $\mathbf{XX} \cdot \mathbf{R}_A \cdot \mathbf{P}_A^t$ in PLS, with:

$$\mathbf{R}_A = \mathbf{W}_A \cdot (\mathbf{P}_A^T \cdot \mathbf{W}_A)^{-1}. \tag{3}$$

The relationship of the MEDA algorithm and cross-product matrices was firstly pointed out by Arteaga (28) and it can also be derived from the original MEDA paper (11). Equation (2) represents a direct and fast procedure to compute MEDA, similar in nature to the algorithms for model fitting from cross-product matrices, namely the eigendecomposition (ED) for PCA and the kernel algorithms (29) (30) (31) for PLS.

In Figure 4(a), the MEDA matrix corresponding to the 2 PCs PCA model of the example in the previous section, the two independent pipelines, is shown. The structure in the data is elucidated from this matrix. The separation between the two pipelines is shown in the fact that upper-right and lower-left quadrants are close to zero. The relationship among variables corresponding to factors F1 and F2 are also apparent in both pipelines. Since the variability corresponding to factors F3 is barely captured by the first 2 PCs, these are not reflected in the matrix. Nevertheless, if 6 PCs are selected, (Figure 4(b)) the complete structure in the data is clearly found.

MEDA improves the interpretation of both the data set and the model fitted without actually pursuing a simple structure. The result is that MEDA has better properties than rotation methods: it is more accurate and its performance is not deteriorated when the number of PCs is overestimated. Also, the output of MEDA does not depend on the normalization of the loadings, like rotated models do, an it is not limited to subspaces with two or three components at most. A comparison of MEDA with rotation methods is out of the scope of this chapter. Please refer to (11) for it and also for a more algorithmic description of MEDA.

3.2 Loading plots and MEDA

The limitations of loading plots and the application of MEDA were introduced with the pipelines artificial data set. This is further illustrated in this section with two examples provided with the PLS-toolbox (32): the Wine data set, which is used in the documentation of the cited software to show the capability of PCA for improving data understanding, and the

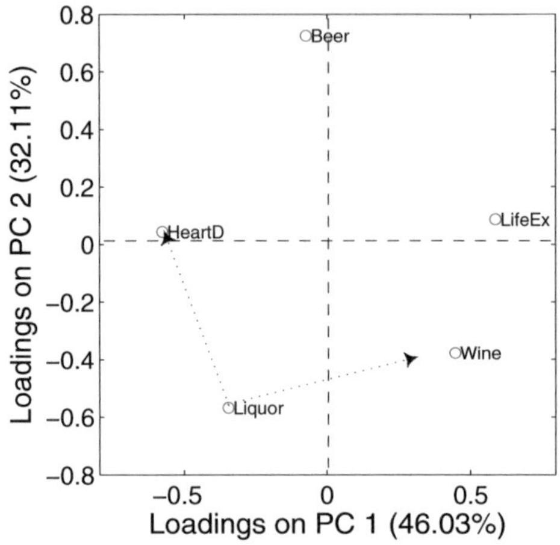

Fig. 5. Loading plot of the first 2 PCs from the Wine Data set provided with the PLS-toolbox (32).

PLSdata data set, which is used to introduce regression models, including PLS. The reading of the analysis and discussion of both data sets in (32) is recommended.

As suggested in (32), two PCs are selected for the PCA model of the Wine data set. The corresponding loading plot is shown in Figure 5. According to the reference, this plot shows that variables HeartD and LifeEx are negatively correlated, being this correlation captured in the first component. Also, "wine is somewhat positively correlated with life expectancy, likewise, liquor is somewhat positively correlated with heart disease". Finally, bear, wine and liquor form a triangle in the figure, which "suggests that countries tend to trade one of these vices for others, but the sum of the three tends to remain constant". Notice that although these conclusions are interesting, some of them are not so evidently shown by the plot. For instance, Liquor is almost as close to HeartD than to Wine. Is Liquor correlated to Wine as it is to HeartD?

MEDA can be used to improve the interpretation of loading plots. In Figure 6(a), the MEDA matrix for the first PC is shown. It confirms the negative correlation between HeartD and LifeEx, and the–lower–positive correlation between HeartD and Liqour and LifeEx and Wine. Notice that these three relationships are three different common factors. Nevertheless, they all manifest in the same component, making the interpretation with loading plots more complex. The MEDA matrix for the second PC in Figure 6(b) shows the relationship between the three types of drinks. The fact that the second PC captures this relationship was not clear in the loading plot. Furthermore, the MEDA matrix shows that Wine and Liquor are not correlated, answering to the question in the previous paragraph. Finally, this absence of correlation refutes that countries tend to trade wine for liquor or viceversa, although this effect may be true for bear.

In the PLSdata data set, the aim is to obtain a regression model that relates 20 temperatures measured in a Slurry-Fed Ceramic Melter (SFCM) with the level of molten glass. The x-block contains 20 variables which correspond to temperatures collected in two vertical thermowells. Variables 1 to 10 are taken from the bottom to the top in thermowell 1, and variables 11 to 20 from the bottom to the top in thermowell 2. The data set includes 300 training observations

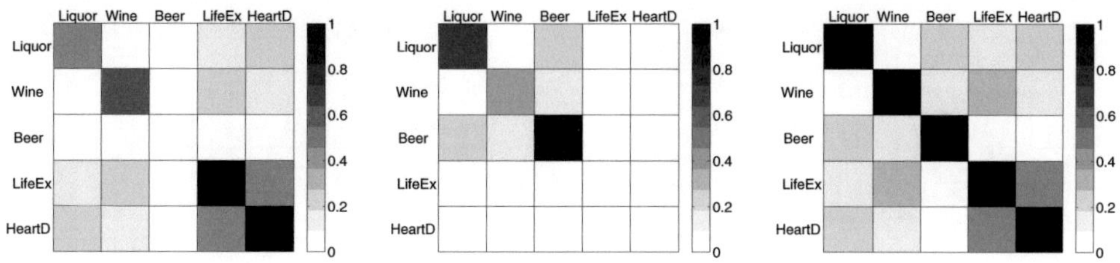

(a) First PC (b) Second PC (c) First 2 PCs

Fig. 6. MEDA matrices of the first PCs from the Wine data set provided with the PLS-toolbox (32).

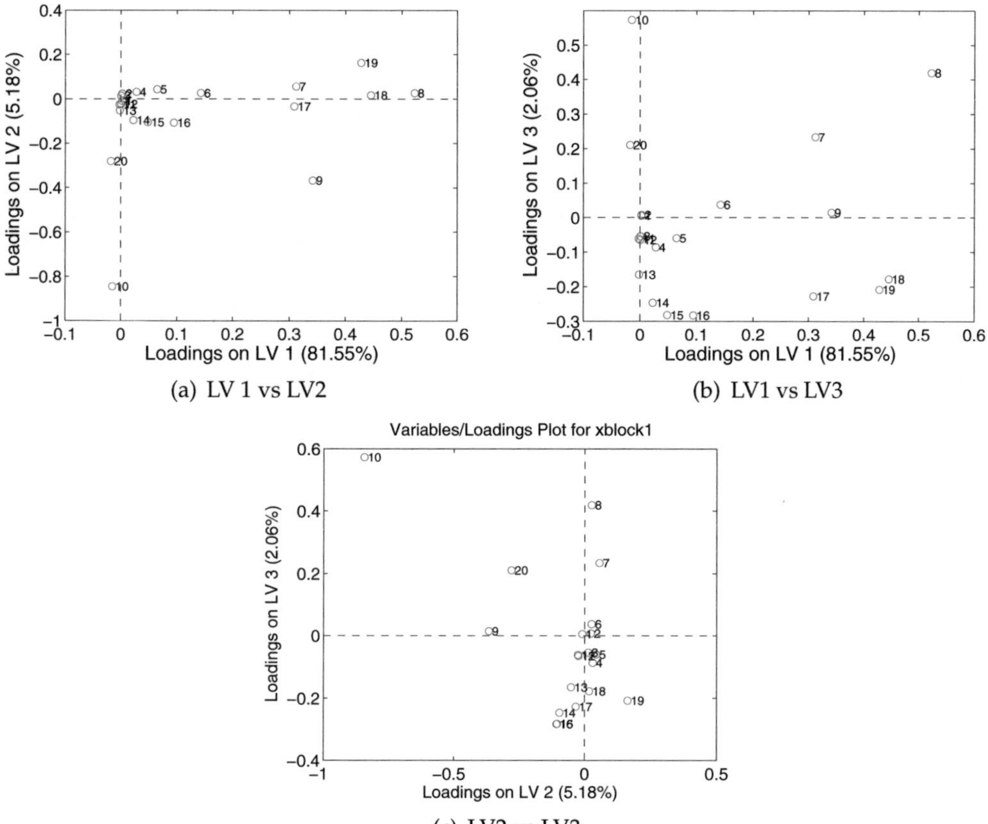

(a) LV 1 vs LV2 (b) LV1 vs LV3

(c) LV2 vs LV3

Fig. 7. Loading plots from the PLS model in the Slurry-Fed Ceramic Melter data set.

and 200 test observations. This same data set was used to illustrate MEDA with PCA in (11) with the temperatures and the level of molten glass together in the same block of data [1].

[1] There are some results of the analysis in (11) which are not coherent with those reported here, since the data sets used do not contain the same observations.

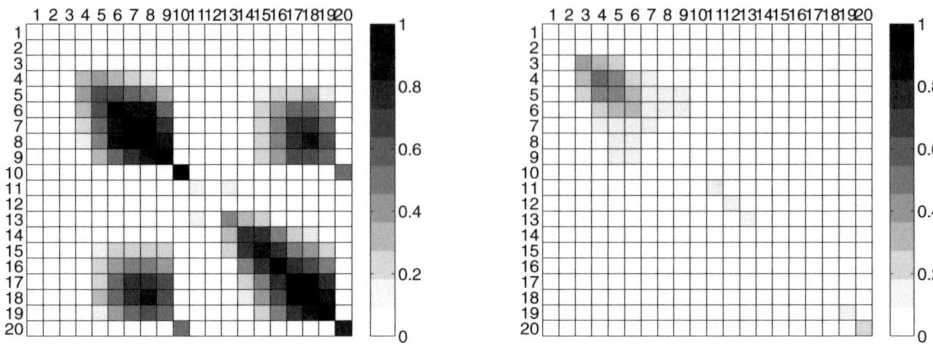

(a) First three LVs (b) Fourth LV

Fig. 8. MEDA matrices from the PLS model in the Slurry-Fed Ceramic Melter data set.

Following recommendations in (32), a 3 LVs PLS model from a mean-centered x-block was fitted. The loading plots corresponding to the three possible pairs of LVs are shown in Figure 7. The first LV captures the predictive variance in the temperatures, with higher loadings for higher indices of the temperatures in each thermowell. This holds exception made on temperatures 10 and 20, which present their predictive variance in the second and third LVs, being the variance in 10 between three and four times higher than that in 20. The third LVs also seems to discriminate between both thermowells. For instance, in Figure 7(c) most temperatures of the first thermowell are in the upper middle of the Figure, whereas most temperatures of the second thermowell are in the lower middle. Nevertheless, it should be noted that the third LV only captures 2% of the variability in the x-block and 1% of the variability in the y-block.

The information gained with the loading plots can be complemented with that in MEDA, which brings a much clearer picture of the structure in the data. The MEDA matrix for the PLS model with 3 LVs is shown in Figure 8(a). There is a clear auto-correlation effect in the temperatures, so that closer sensors are more correlated. This holds exception made on temperatures 10 and 20. Also, the corresponding temperatures in both thermowells are correlated, including 10 and 20. Finally, the temperatures at the bottom do not contain almost any predictive information of the level of molten glass. In (32), the predictive error by cross-validation is used to identify the number of LVs. Four LVs attain the minimum predictive error, but 3 LVs are selected since the fourth LV does not contribute much to the reduction of this error. In Figure 8(b), the contribution of this fourth LV is shown. It is capturing the predictive variability from the third to the fifth temperature sensors in the first thermowell, which are correlated. The corresponding variability in the second thermowell is already captured by the first 3 LVs. This is an example of the capability of MEDA for model interpretation, which can be very useful in the determination of the number of LVs. In this case and depending on the application of the model, the fourth LV may be added to the model in order to compensate the variance captured in both thermowells, even if the improvement in prediction performance is not high.

The information provided by MEDA can also be useful for variable selection. In this example, temperatures 1-2 and 11-12 do not contribute to a relevant degree to the regression model. As shown in Table 1, if those variables are not used in the model, its prediction performance

Variables	Complete model	[3 : 10 13 : 20]	[3 : 10]	[13 : 20]
LVs	3	3	3	3
X-block variance	88.79	89.84	96.36	96.93
Y-block variance	87.89	87.78	84.61	83.76
RMSEC	0.1035	0.1039	0.1166	0.1198
RMSECV	0.1098	0.1098	0.1253	0.1271
RMSEP	0.1396	0.1394	0.1522	0.1631

Table 1. Comparison of three PLS models in the Slurry-Fed Ceramic Melter data set. The variance in both blocks of data and the Root Mean Square Error of Calibration (RMSEC), Cross-validation (RMSECV) and Prediction (RMSEP) are compared.

remains the same. Also, considering the correlation among thermowells, one may be tempted to use only one of the thermowells for prediction, reducing the associated costs of maintaining two thermowells. If this is done, only 8 predictor variables are used and the prediction performance is reduced, but not to a large extent. Correlated variables in a prediction model help to better discriminate between true structure and noise. For instance, in this example, when only the sensors of one thermowell are used, the PLS model captures more x-block variance and less y-block variance. This is showing that more specific–noisy–variance in the x-block is being captured. Using both thermowells reduces this effect. Another example of variable selection with MEDA will be presented in Section 5.

3.3 correlation matrices and MEDA

There is a close similarity between MEDA and correlation matrices. To this regard, equation (2) simplifies the interpretation of the MEDA procedure. The MEDA index combines the original variance with the model subspace variance in S_{ml} and S_{ml^A}. Also, the denominator of the index in eq. (2) is the original variance. Thus, those pairs of variables where a high amount of the total variance of one of them can be recovered from the other are highlighted. This is convenient for data interpretation, since only factors of high variance are highlighted. On the other hand, it is easy to see that when the number of LVs, A, equals the rank of $\mathbf{X}$, then Q_A^2 is equal to the element-wise squared correlation matrix of $\mathbf{X}$, C^2 (11). This can be observed in the following element-wise equality:

$$q^2_{Rank(X),(m,l)} = \frac{S^2_{ml}}{S_{mm} \cdot S_{ll}} = c^2_{(m,l)}. \tag{4}$$

This equivalence shows that matrix Q_A^2 has a similar structure than the–element-wise squared–correlation matrix. To elaborate this similarity, a correlation matrix can be easily extended to the notion of latent subspace. The correlation matrix in the latent subspace, C_A, can de defined as the correlation matrix of the reconstruction of $\mathbf{X}$ with the first A LVs. Thus, $C_A = \mathbf{P}_A \cdot \mathbf{P}_A^t \cdot \mathbf{C} \cdot \mathbf{P}_A \cdot \mathbf{P}_A^t$ in PCA and $C_A = \mathbf{P}_A \cdot \mathbf{R}_A^t \cdot \mathbf{C} \cdot \mathbf{R}_A \cdot \mathbf{P}_A^t$ in PLS. If the elements of C_A are then squared, the element-wise squared correlation in the latent subspace, noted as C_A^2, is obtained. Strictly speaking, each element of C_A^2 is defined as:

$$c^2_{A,(m,l)} = \frac{S^2_{m^A l^A}}{S_{m^A m^A} \cdot S_{l^A l^A}}. \tag{5}$$

However, for the same reason explained before, if C_A^2 is aimed at data interpretation, the denominator should be original variance:

$$c^2_{A,(m,l)} = \frac{S^2_{m^A l^A}}{S_{mm} \cdot S_{ll}}. \tag{6}$$

If this equation is compared to equation (2), we can see that the main difference between MEDA and the–projected and element-wise squared–correlation matrix is the combination of original and projected variance in the numerator of the former. This combination is paramount for interpretation. Figure 9 illustrates this. The example of the pipelines is used again, but in this case ten pipelines and only 20 observations are considered, yielding a dimension of 20 × 50 in the data. Two data sets are simulated. In the first one, the pipelines are correlated. As a consequence, the data present three common factors represented by the three biggest eigenvalues in Figure 9(a). In the second one, each pipeline is independently generated, yielding a more distributed variance in the eigenvalues (Figure 9(b)). For matrices Q_A^2 and C_A^2 to infer the structure in the data, they should have large values in the elements which represent real structural information (common factors) and low values in the rest of the elements. Since in both data sets it is known a-priori which elements in the matrices represent actual common factors and which not, the mean values for the two groups of elements in matrices Q_A^2 and C_A^2 can be computed. The ratio of these means, computed by dividing the mean of the elements with common factors by the mean of the elements without common factors, is a measure of the discrimination capability between structure and noise of each matrix. The higher this index is, the better the discrimination capability is. This ratio is shown in Figures 9(c) and 9(d) for different numbers of PCs. Q_A^2 outperforms C_A^2 until all relevant eigenvalues are incorporated to the model. Also, Q_A^2 presents maximum discrimination capability for a reduced number of components. Notice that both alternative definitions of C_A^2 in equations (5) and (6) give exactly the same result, though equation (6) is preferred for visual interpretation.

4. Connection between observations and variables

The most relevant issue for data understanding is probably the connection between observations and variables. It is almost useless to detect certain details in the data distribution, such as outliers or clusters, if the set of variables related to these details are not identified. Traditionally, biplots (12) have been used for this purpose. In biplots, the scatter plots of loadings and scores are combined in a single plot. Apart from relevant considerations regarding the comparability of the axes in the plot, which is also important for any scatter plots, and of the scales in scores and loadings (18), biplots may be misleading just because of the loading plot included. In this point, a variant of MEDA, named observation-based MEDA or *o*MEDA, can be used to unveil the connection between observations and variables without the limitations of biplots.

4.1 *o*MEDA

*o*MEDA is a variant of MEDA to connect observations and variables. Basically, *o*MEDA is a MEDA algorithm applied over a combination of the original data and a dummy variable designed to cover the observations of interest. Take the following example: a number of subsets of observations $\{\mathbf{C}_1, ..., \mathbf{C}_N\}$ form different clusters in the scores plot which are located

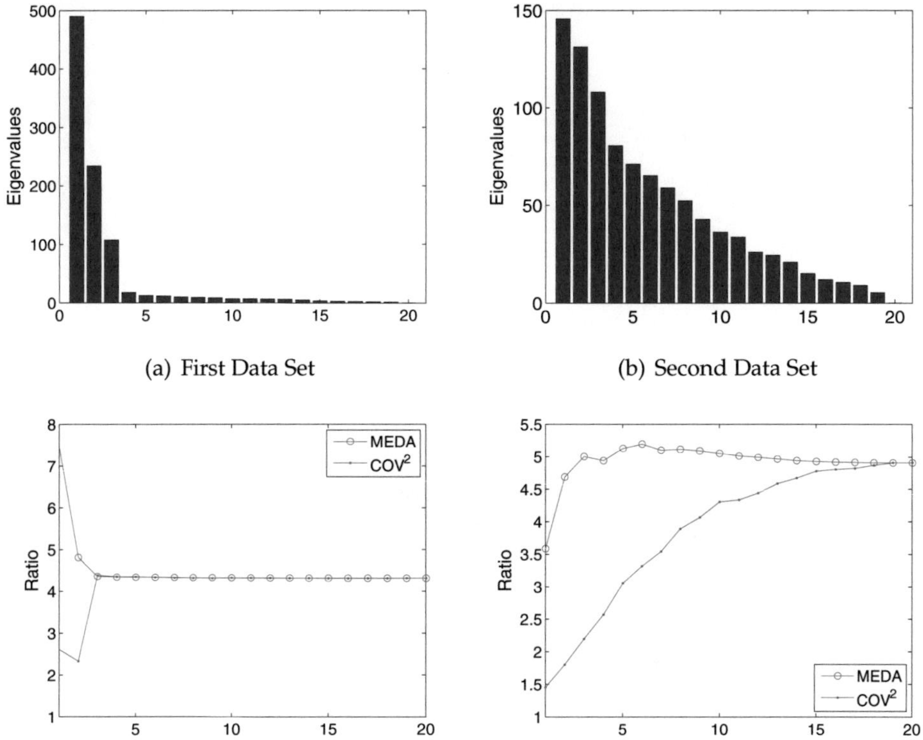

(a) First Data Set (b) Second Data Set

(c) First Data Set (d) Second Data Set

Fig. 9. Comparison of MEDA and the projected and element-wise squared covariance matrix in the identification of the data structure from the PCA model fitted from the data in the ten pipelines example: (a) and (b) show the eigenvalues when the pipelines are correlated and independent, respectively, and (c) and (d) show the ratio between the mean of the elements with common factors and the mean of the elements without common factors in the matrices.

far from the bulk of the data, $\mathbf{L}$. One may be interested in identifying, for instance, the variables related to the deviation of $\mathbf{C}_1$ from $\mathbf{L}$ without considering the rest of clusters. For that, a dummy variable $\mathbf{d}$ is created so that observations in $\mathbf{C}_1$ are set to 1, observations in $\mathbf{L}$ are set to -1, while the remaining observations are left to 0. Also, values other than 1 and -1 can be included in the dummy variable if desired. *o*MEDA is then performed using this dummy variable.

The *o*MEDA technique is illustrated in Figure 10. Firstly, the dummy variable is designed and combined with the data set. Then, a MEDA run is performed by predicting the original variables from the dummy variable. The result is a single vector, $\mathbf{d}_A^2$, of dimension $M \times 1$, being M the number of original variables. In practice, the *o*MEDA index is slightly different to that used in MEDA. Being $\mathbf{d}$ the dummy variable, designed to compare a set of observations with value 1 (or in general positive values) with another set with value -1 (or in general negative values), then the *o*MEDA index follows:

$$d_{A,(l)}^2 = \|\mathbf{x}_{(l)}^d\|^2 - \|\hat{\mathbf{e}}_{A,(l)}^d\|^2, \quad \forall l. \tag{7}$$

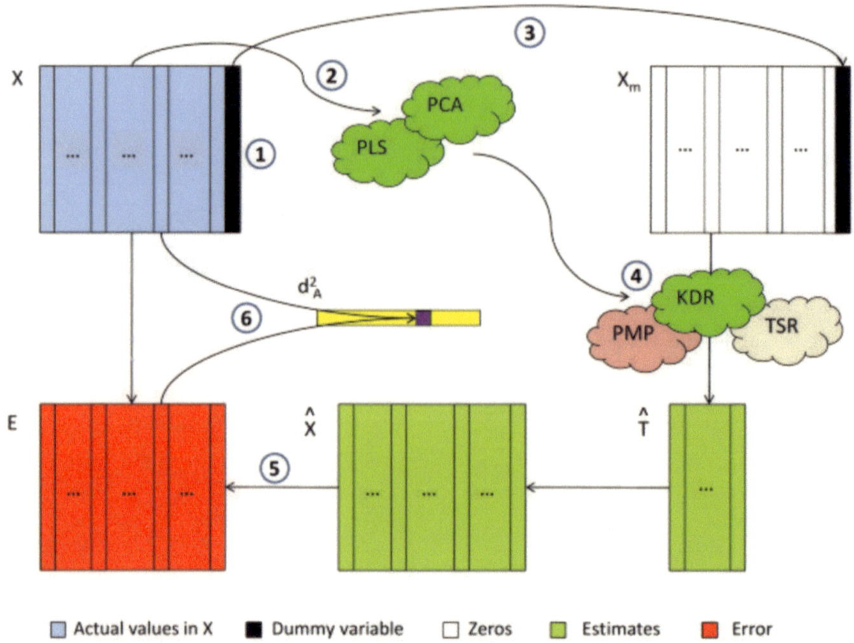

Fig. 10. *o*MEDA technique: (1) introduction of the dummy variable, (2) model calibration, (3) introduction of missing data, (4) missing data imputation, (5) error computation, (6) computation of vector $\mathbf{d}_A^2$.

where $\mathbf{x}_{(l)}^d$ represents the values of the l-th variable in the original observations different to 0 in $\mathbf{d}$ and $\hat{\mathbf{e}}_{A,(l)}^d$ is the corresponding estimation error. The main difference between the computation of index $d_{A,(l)}^2$ in *o*MEDA and that of MEDA is the absence of the denominator in the former. This modification is convenient to avoid high values in $d_{A,(l)}^2$ when the amount of variance of a variable in the reduced set of observations of interest is very low. Once $\mathbf{d}_A^2$ is computed for a given dummy variable, sign information can be added from the mean vectors of the two groups of observations considered (33).

In practice, in order to avoid any modification in the PCA or PLS subspace due to the introduction of the dummy variable, the *o*MEDA algorithm is slightly more complex than the procedure shown in Figure 10. For a description of this algorithm refer to (33). However, like in MEDA, the *o*MEDA vector can be computed in a more direct way by assuming KDR (26) is used as the missing data estimation procedure. If this holds, the *o*MEDA vector follows:

$$d_{A,(l)}^2 = 2 \cdot \mathbf{x}_{(l)}^t \cdot \mathbf{D} \cdot \mathbf{x}_{A,(l)} - \mathbf{x}_{A,(l)}^t \cdot \mathbf{D} \cdot \mathbf{x}_{A,(l)}, \tag{8}$$

where $\mathbf{x}_{(l)}$ represents the l-th variable in the–complete–set of original observations and $\mathbf{x}_{A,(l)}$ its projection in the latent subspace in coordinates of the original space and:

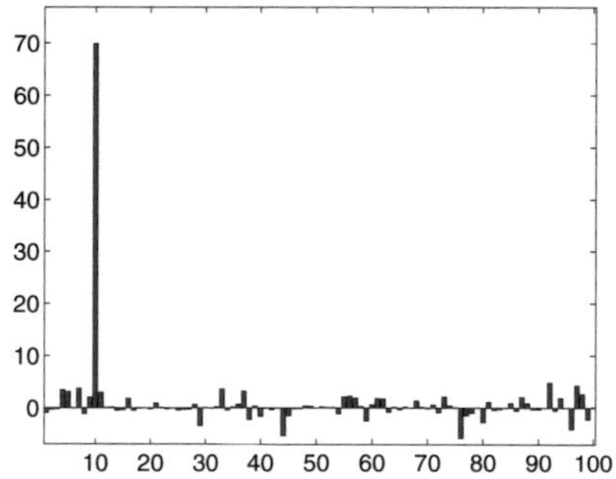

Fig. 11. *o*MEDA vector of two clusters of data from the 10 PCs PCA model of a simulated data set of dimension 100 × 100. This model captures 30% of the variability. Data present two clusters in variable 10.

$$\mathbf{D} = \frac{\mathbf{d} \cdot (\mathbf{d})^T}{\|\mathbf{d}\|^2}. \tag{9}$$

Finally, the equation can also be reexpressed as follows:

$$d^2_{A,(l)} = \frac{1}{N} \cdot (2 \cdot \Sigma^d_{(l)} - \Sigma^d_{A,(l)}) \cdot |\Sigma^d_{A,(l)}| \tag{10}$$

with $\Sigma^d_{(l)}$ and $\Sigma^d_{A,(l)}$ being the weighted sum of elements in $\mathbf{x}(l)$ and $\mathbf{x}_{A,(l)}$ according to the weights in $\mathbf{d}$, respectively. Equation (10) has two advantages. Firstly, it presents the *o*MEDA vector as a weighted sum of values, which is easier to understand. Secondly, it has the sign computation built in, due to the absolute value in the last element. Notice also that *o*MEDA inherits the combination of total and projected variance present in MEDA.

In Figure 11 an example of *o*MEDA is shown. For this, a 100 × 100 data set with two clusters of data was simulated. The distribution of the observations was designed so that both clusters had significantly different values only in variable 10 and then data was auto-scaled. The *o*MEDA vector clearly highlights variable 10 as the main difference between both clusters.

4.2 Biplots vs oMEDA

Let us return to the discussion regarding the relationship between the common factors and the components. As already commented, several common factors can be captured by the same component in a projection model. As a result, a group of variables may be located close in a loading plot without the need to be correlated. This is also true for the observations. Thus, two observations closely located in a score plot may be quite similar or quite different depending on their scores in the remaining LVs. However, score plots are typically employed to observe a general distribution of the observations. This exploration is more aimed at finding differences among observations rather than similarities. Because of this, the problem described for loading plots is not so relevant for the typical use of score plots. However, this is a problem when interpreting biplots. In biplots, deviations in the observations are related to deviations in the variables. Like loading plots, biplots may be useful to perform a fast view on the data, but any conclusion should be confirmed with another technique. *o*MEDA is perfectly suited for this.

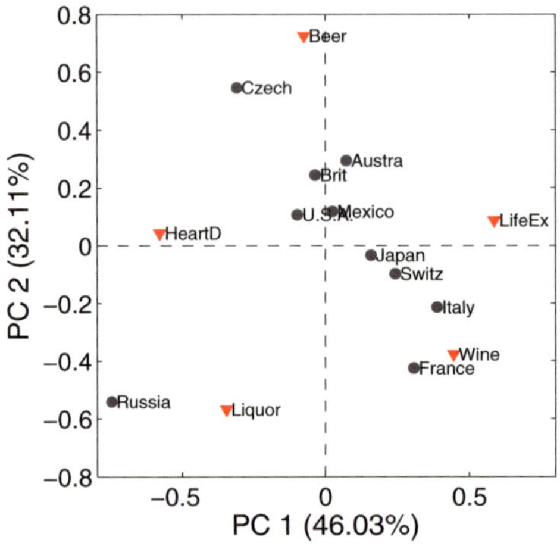

Fig. 12. Biplot of the first 2 PCs from the Wine Data set provided with the PLS-toolbox (32).

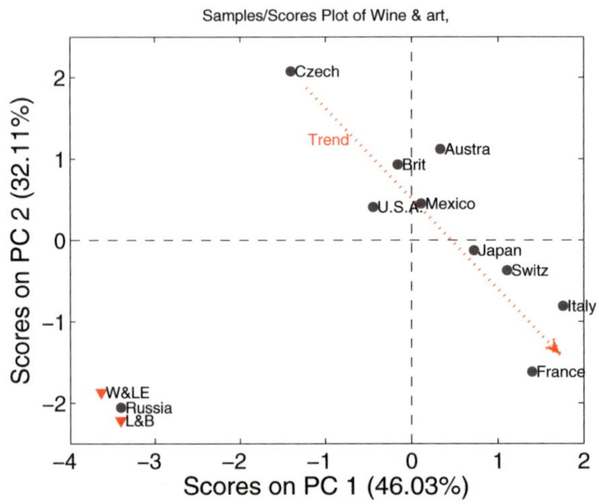

Fig. 13. Score plot of the first 2 PCs from the Wine Data set provided with the PLS-toolbox (32) and two artificial observations.

In Figure 12, the biplot of the Wine data set is presented. The distribution of the scores show that all countries but Russia follow a trend from the Czech Republic to France, while Russia is located far from this trend. The biplot may be useful to make hypothesis on the variables related to the special nature of Russia or to the trend in the rest of countries. Nevertheless, this hypothesis making is not straightforward. To illustrate this, in Figure 13 the scores are shown together with two artificial observations. The artificial observations were designed to lay close to Russia in the score plot of the first two PCs. In both cases, three of the five variables were left to their average value in the Wine data set and only two variables are set to approach Russia. Thus, observation W&LE only uses variables Wine and LifeEx to yield a point close to Russia in the score plot while the other variables are set to the average. Observation L&B only uses Liquor and Beer. With this example, it can be concluded that very little can be said about Russia only by looking at the biplot.

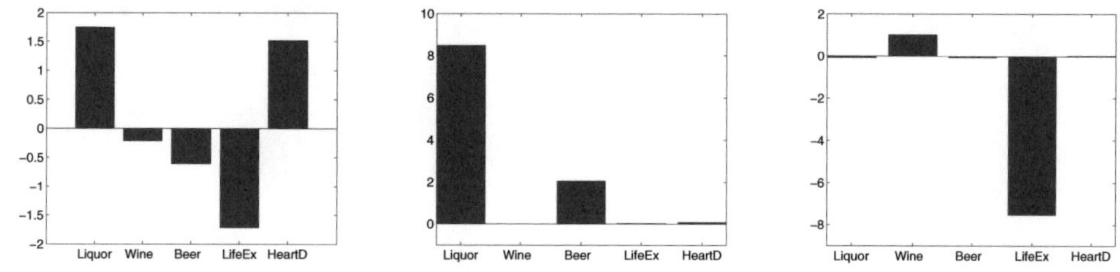

(a) Russia Vs The rest (b) L&B Vs The rest (c) W&LE Vs The rest

Fig. 14. oMEDA vectors of the first 2 PCs from the Wine Data set provided with the PLS-toolbox (32). Russia (a), L&B (b) and W&LE (c) compared to the rest of countries.

In Figure 14(a), the oMEDA vector to discover the differences between Russia and the rest of countries is shown. For this, a dummy variable is built where all observations except Russia are set to -1 and Russia is set to 1. oMEDA shows that Russia has in general less life expectancy and more heart disease and liquor consumption than the rest of countries. The same experiment is repeated for artificial observations L&B and W&LE in Figures 14(b) and 14(c). oMEDA clearly distinguishes among the three observations, while in the biplot they seem to be very similar.

To analyze the trend shown by all countries except Russia in Figure 12, the simplest approach is to compare the most separated observations, in this case France and the Czech Republic. The oMEDA vector is shown in Figure 15(a). In this case, the dummy variable is built so that France has value 1, the Czech Republic has value -1 and the rest of the countries have 0 value. Thus, positive values in the oMEDA vector identify variables with higher value in France than in the Czech Republic and negative values the opposite. oMEDA shows that the French consume more wine and less beer than Czech people. Also, according to the data, the former seem to be more healthy.

Comparing the two most separated observations may be misleading in certain situations. Another choice is to use the capability of oMEDA to unveil the variables related to any direction in a score plot. For instance, let us analyze the trend of the countries incorporating the information in all the countries. For this, different weights are considered in the dummy variable. We can think of these weights as-approximate–projections of the observations in the direction of interest. Following this approach, the weights listed in Table 2 are assigned, which approximate the projection of the countries in the imaginary line depicted by the arrow in Figure 13. Since Russia is not in the trend, it is left to 0. Using these weights, the resulting oMEDA vector is shown in Figure 15(b). In this case, the analysis of the complete set of observations in the trend resembles the conclusions in the analysis of the two most separated observations.

Country	**Weight**	**Country**	**Weight**
France	3	Mexico	-1
Italy	2	U.S.A.	-1
Switz	1	Austra	-1
Japan	1	Brit	-1
Russia	0	Czech	-3

Table 2. Weights used in the dummy variable for the oMEDA vector in Figure 15(b).

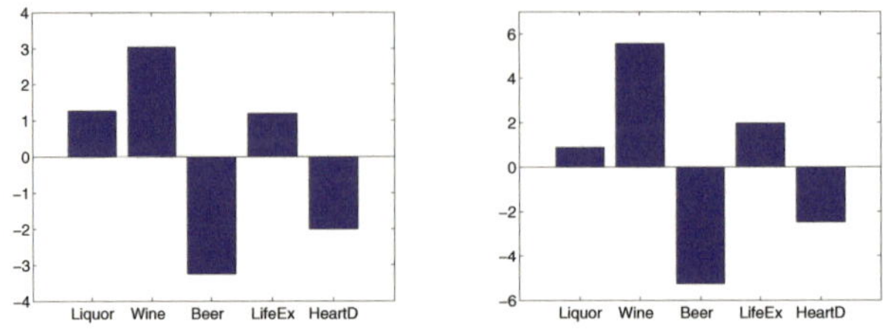

(a) France Vs Czech (b) All except Russia

Fig. 15. oMEDA vectors of the first 2 PCs from the Wine Data set provided with the PLS-toolbox (32). In (a), France and Czech Republic are compared. In (b), the trend shown in the score plot by all countries except Russia is analyzed.

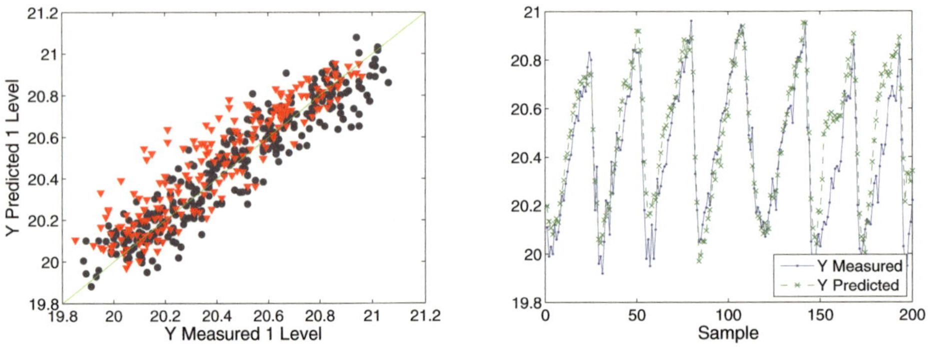

(a) Training & Test (b) Test

Fig. 16. Measured vs predicted values of molten glass level in the PLS model with 3 LVs fitted from the Slurry-Fed Ceramic Melter (SFCM) data set in (32).

Let us return to the PLSdata data set. A PLS model relating temperatures with the level of molten glass was previously fitted. As already discussed, the data set includes 300 training observations and 200 test observations. The measured and predicted values of both sets of observations are compared in Figure 16(a). The predicted values in the test observations (inverted triangles) tend to be higher than true values. This is also observed in Figure 16(b). The cause for this seems to be that the process has slightly moved from the operation point where training data was collected. oMEDA can be used to identify this change of operation point by simply comparing training and test observations in the model subspace. Thus, training (value 1) and test observations (value -1) are compared in the subspace spanned by the first 3 LVs of the PLS model fitted only from training data. The resulting oMEDA vector is shown in Figure 17. According to the result, considering the test observations have value -1 in the dummy variable, it can be concluded that the process has moved to a situation in which top temperatures are higher than during model calibration.

5. Case study: Selwood data set

In this section, an exploratory data analysis of the Selwood data set (34) is carried out. The data set was downloaded from http://michem.disat.unimib.it/chm/download/datasets.htm. It

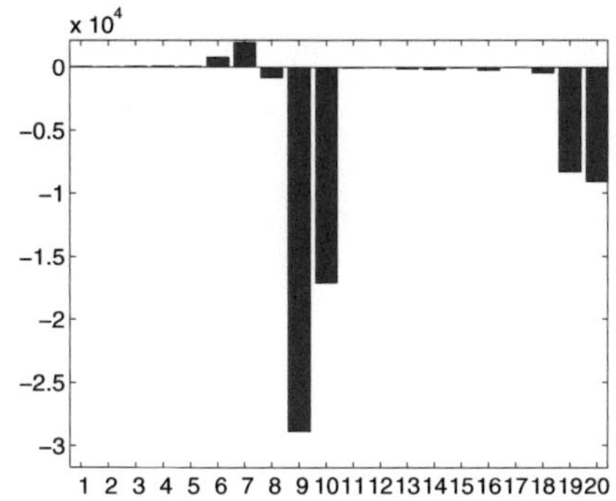

Fig. 17. oMEDA vector comparing training and test observations in the PLS model with 3 LVs fitted from the Slurry-Fed Ceramic Melter (SFCM) data set in (32).

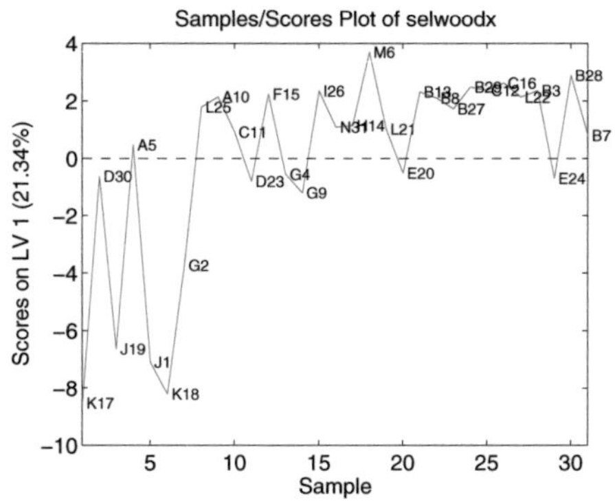

Fig. 18. Scores corresponding to the first LV in the PLS model with the complete set of descriptors of the Selwood dataset.

consists of 31 antifilarial antimycin A_1 analogues for which 53 physicochemical descriptors were calculated for Quantitative Structure-Activity Relationship (QSAR) modelling. The set of descriptors is listed in Table 3. These descriptors are used for predicting in vitro antifilarial activity (-LOGEC50). This data set has been employed for testing variables selection methods, for instance in (35; 36), in order to find a reduced number of descriptors with best prediction performance. Generally speaking, these variable selection methods are based on complex optimization algorithms which make use of heuristics to reduce the search space.

Indices	Descriptors
1:10	ATCH1 ATCH2 ATCH3 ATCH4 ATCH5 ATCH6 ATCH7 ATCH8 ATCH9 ATCH10
11:20	DIPV_X DIPV_Y DIPV_Z DIPMOM ESDL1 ESDL2 ESDL3 ESDL4 ESDL5 ESDL6
21:30	ESDL7 ESDL8 ESDL9 ESDL10 NSDL1 NSDL2 NSDL3 NSDL4 NSDL5 NSDL6
31:40	NSDL7 NSDL8 NSDL9 NSDL10 VDWVOL SURF_A MOFI_X MOFI_Y MOFI_Z PEAX_X
41:50	PEAX_Y PEAX_Z MOL_WT S8_1DX S8_1DY S8_1DZ S8_1CX S8_1CY S8_1CZ LOGP
51:53	M_PNT SUM_F SUM_R

Table 3. Physicochemical descriptors of the Selwood dataset.

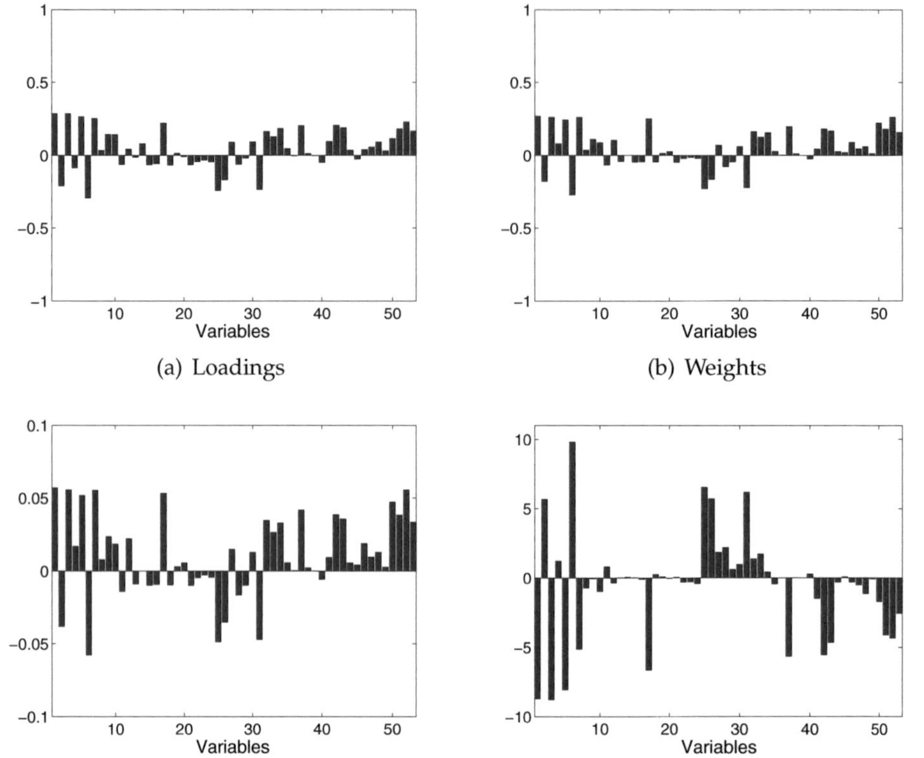

(a) Loadings (b) Weights

(c) Regression Coefficients (d) *o*Meda (K17, J1, J19 and K18 vs the rest)

Fig. 19. Several vectors corresponding to the first LV in the PLS model with the complete set of descriptors of the Selwood dataset.

First of all, a PLS model is calibrated between the complete set of descriptors and -LOGEC50. Leave-one-out cross-validation suggests one LV. The score plot corresponding to the 31 analogues in that LV are shown in Figure 18. Four of the compounds, namely K17, J1, J19 and K18, present an abnormally low score. This deviation is highly contributing to the variance in the first LV and the reason for it should be investigated. Two of these compounds, K17 and K18, were catalogued as outliers by (34), where the authors stated that "Chemically, these compounds are distinct from the bulk of the training set in that they have an n-alkyl side chain as opposed to a side chain of the phenoxy ether type". Since the four compounds present an abnormally low score in the first LV, typically the analyst may interpret the coefficients of that LV to try to explain this abnormality. In Figure 19, the loadings, weights and regression coefficients of the PLS model are presented together with the *o*MEDA vector. The latter identifies those variables related to the deviation of the four compounds from the rest. The *o*MEDA vector is similar, but with opposite sign, to the other vectors in several descriptors, but quite different in others. Therefore, the loadings, weights or coefficient vectors should not be used in this case for the investigation of the deviation, or otherwise one may arrive to incorrect conclusions. On the other hand, it may be worth to check whether the *o*MEDA vector is representative of the deviation in the four compounds. Performing *o*MEDA individually on each of the compounds confirm this fact (see Figure 20)

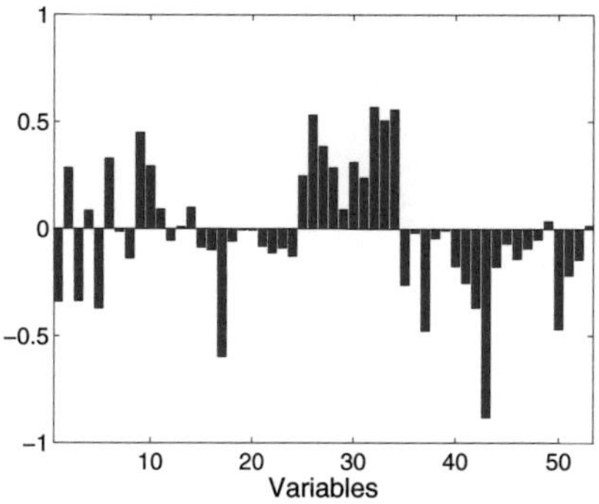

(a) K17

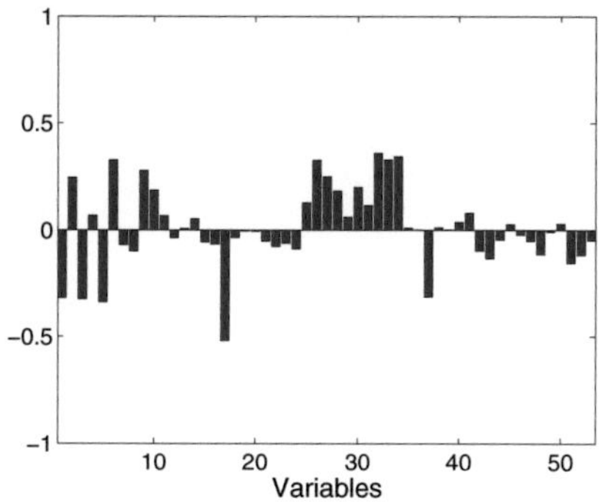

(b) J1

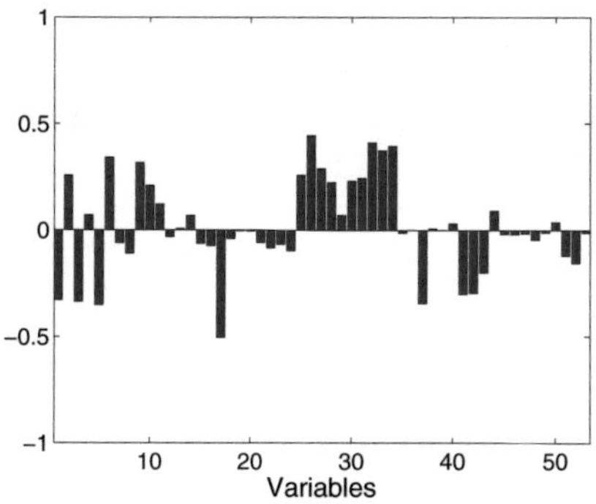

(c) J19

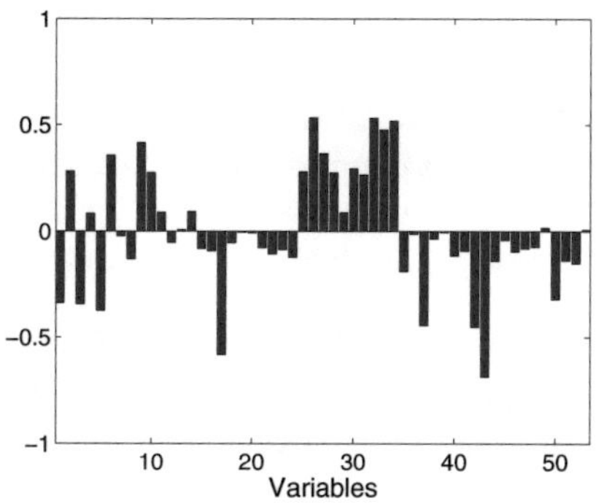

(d) K18

Fig. 20. *o*MEDA vectors corresponding to the first LV in the PLS model with the complete set of descriptors of the Selwood dataset. To compute each of the vectors, one of the four compounds K17, J1, J19 and K18 are set to 1 in the dummy variable and the other three are set to 0, while the rest of compounds in the data set are set to -1.

The subsequent step is to search for relevant descriptors (variable selection) For this, MEDA will be employed. In this point, there are two choices. The four compounds with low score in the first LV may be treated as outliers and separated from the rest of the data (34) or the complete data set may be modelled with a single QSAR model (35; 36). It should be noted that differences among observations in one model may not be found in a different model, so that the same observation may be an outlier or a normal observation depending on the model. Furthermore, as discussed in (35), the more general the QSAR model is, so that it models a wider set of compounds, the better. Therefore, the complete set of compounds will be considered in the remaining of the example. On the other hand, regarding the analysis tools used, there are different possibilities. MEDA may be applied over the PLS model relating the descriptors in the x-block and -LOGEC50 in the y-block. Alternatively, both blocks may be joined together in a single block of data and MEDA with PCA be applied. The second choice will be generally preferred to avoid over-fitting, but typically both approaches may lead to the same conclusions, like it happens in the present example.

The application of MEDA requires to select the number of PCs in the PCA model. Considering that the aim is to understand how the variability in -LOGEC50 is related to the descriptors in

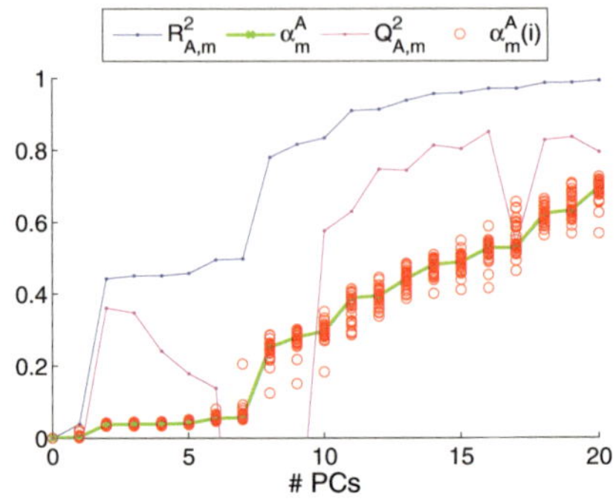

Fig. 21. Structural and Variance Information (SVI) plot of in vitro antifilarial activity (-LOGEC50). The data set considered combines -LOGEC50 with the complete set of descriptors of the Selwood dataset.

the data set, the Structural and Variance Information (SVI) plots are the adequate analysis tool (14). The SVI plots combine variance information with structural information to elucidate how a PCA model captures the variance of a single variable. The SVI plot of a variable v reveals how the following indices evolve with the addition of PCs in the PCA model:

- The R^2 statistic, which measures the variance of v.
- The Q^2 statistic, which measures the performance of the missing data imputation of v, or otherwise stated its prediction performance.
- The α statistic, which measures the portion of variance of v which is identified as unique variance, i.e. variance not shared with other variables.
- The stability of α, as an indicator of the stability of the model calibration.

Figure 21 shows the SVI plot of -LOGEC50 in the PCA model with the complete set of descriptors. The plot shows that the model remains quite stable until 5-6 PCs are included. This is seen in the closeness of the circles which represents the different instances of α computed on a leave-one-out cross-validation run. The main portion of variability in -LOGEC50 is captured in the second and eighth PCs. Nevertheless, is not until the tenth PC that the missing data imputation (Q^2) yields a high value. For more PCs, the captured variability is only slightly augmented. Since MEDA makes use of the missing data imputation of a PCA model, Q^2 is a relevant index. At the same time, from equation (2) is clear that MEDA is also influenced by captured variance. Thus, 10 PCs are selected. In any case, it should be noted that MEDA is quite robust to the overestimation in the number of PCs (11) and very similar MEDA matrices are obtained for 3 or more PCs in this example.

The MEDA matrix corresponding to the PCA model with 10 PCs from the data set which combines -LOGEC50 with the complete set of descriptors of the Selwood dataset is presented in Figure 22. For variable selection, the most relevant part of this matrix is the last column (or row), which corresponds to -LOGEC50. This vector is shown in Figure 23(a). Those descriptors with high value in this vector are the ones from which -LOGEC50 can be better predicted. Nevertheless, the selection of, say, the first n variables with higher value is not an adequate strategy because the relationship among the descriptors should also be considered. Let us select the descriptor with better prediction performance, in this case ATCH6, though

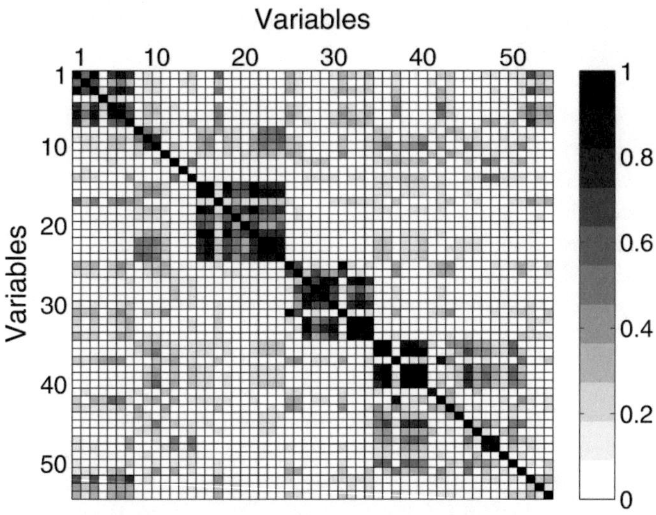

Fig. 22. MEDA matrix of the PCA model with 10 PCs from the data set which combines the in vitro antifilarial activity (-LOGEC50) with the complete set of descriptors of the Selwood dataset.

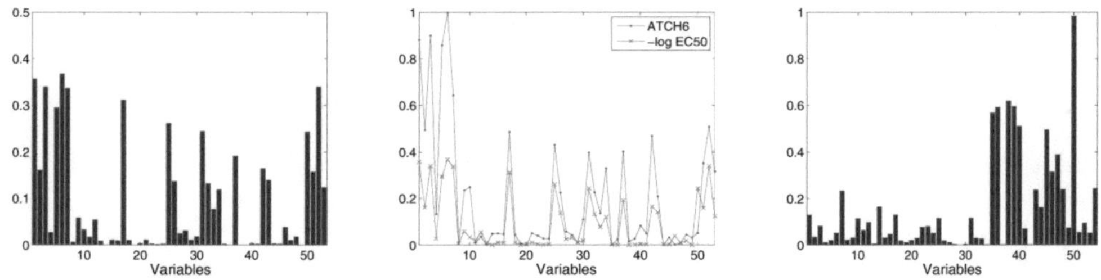

(a) (b) (c)

Fig. 23. MEDA vector corresponding to the in vitro antifilarial activity (-LOGEC50) (a) comparison between this vector and that corresponding to ATCH6 (b) and MEDA vector corresponding to LOGP (c) in the PCA model with 10 PCs from the data set which combines -LOGEC50 with the complete set of descriptors of the Selwood dataset.

ATCH1, ATCH3, ATCH7 or SUM_F have a very similar prediction performance. The least-squares regression model with ATCH6 as regressor attains a Q^2 equal to 0.30, more than the Q^2 attained by any number of LVs in the PLS model with the complete set of descriptors. If for instance ATCH6 and ATCH1 are used as regressors, $Q^2 = 0.26$ is obtained for least squares regression and $Q^2 = 0.31$ for PLS with 1 LV, which means an almost negligible improvement with the addition of ATCH1. The facts that the improvement is low and that the 1 LV PLS model outperforms the least squares model are caused by the correlation between ATCH6 and ATCH1, correlation clearly pointed out in the MEDA matrix (see the element at the sixth column and first row or the first column and sixth row) Clearly, both ATCH6 and ATCH1 are related to the same common factor in -LOGEC50. However, the variability in -LOGEC50 is the result of several sources of variability, which may be common factors with other descriptors.

Therefore, in order to introduce a new common factor in the model other than that in ATCH6, we need to find a descriptor related to -LOGEC50 but not to ATCH6. Also, the model may be improved by introducing a descriptor related to ATCH6 but not to -LOGEC50. For this, Figure 23(b) compares the columns in the MEDA matrix corresponding to ATCH6 and -LOGEC50. The comparison should not be performed in terms of direct differences between values. For instance, ATCH1 and ATCH6 are much more correlated than ATCH1 and -LOGEC50. It is the difference in shape which is informative. Thus, we find that -LOGEC50 present a high correlation with LOGP (variable 50) which is not found in ATCH6. Thus, LOGP presents a common factor with -LOGEC50 which is not present in ATCH6. Using LOGP and ATCH6 as regressors, the least squares model presents $Q^2 = 0.37$.

If an additional descriptor is to be added to the model, again it should present a different common factor with any of the variables in the model. The MEDA vector corresponding to LOGP is shown in Figure 23(c). This descriptor is related to a number of variables which are not related to -LOGEC50. This relationship represents a common factor in LOGP but not in -LOGEC50. The inclusion of a descriptor containing this common factor, for instance MOFI_Y (variable 38) may improve prediction because it may help to distinguish the portion of variability in LOGP which is useful to predict -LOGEC50 from the portion which is not. Using LOGP, ATCH6 and MOFI_Y as regressors yields $Q^2 = 0.56$, illustrating that the addition of a descriptor which is not related to the predicted variable may be useful for prediction.

In Figure 24, the two common factors described before, the one present in ATCH6 and -LOGEC50 and the one present in LOGP and MOFI_Y, are approximately highlighted in the MEDA matrix. If variables ATCH6 and MOFI_Y are replaced by others with the same common factors, the prediction performance of the model remains similar. However, LOGP is utmost for the model since is the only descriptor which relates the second common factor and -LOGEC50. These results are coherent with findings in the literature. Both (35) and (36) highlight the relevance of LOGP, and justify it with the results in several more publications. Furthermore, the top 10 models found in (35), presented in Table 4, follow the same patter of the solution found here. The models with three descriptors contain LOGP with one descriptor from the first and second common factors. The models with two descriptors contain LOGP and a variable with the second common factor.

Descriptors	Q^2
SUM_F (52) LOGP (50) MOFI_Y (38)	0.647
ESDL3 (17) LOGP (50) SURF_A (36)	0.645
SUM_F (52) LOGP (50) MOFI_Z (39)	0.644
LOGP (50) MOFI_Z (39)	0.534
ESDL3 (17) LOGP (50) MOFI_Y (38)	0.605
ESDL3 (17) LOGP (50) MOFI_Z (39)	0.601
LOGP (50) MOFI_Y (38)	0.524
LOGP (50) PEAX_X (40)	0.518
LOGP (50) SURF_A (36)	0.501
SUM_F (52) LOGP (50) PEAX_X (40)	0.599

Table 4. Top 10 models obtained after variable selection of the Selwood data set in (35)

Finally, in Figure 25 the plot of measured vs predicted values of -LOGEC50 in the model resulting from the exploration is shown. No outliers are identified, though the four

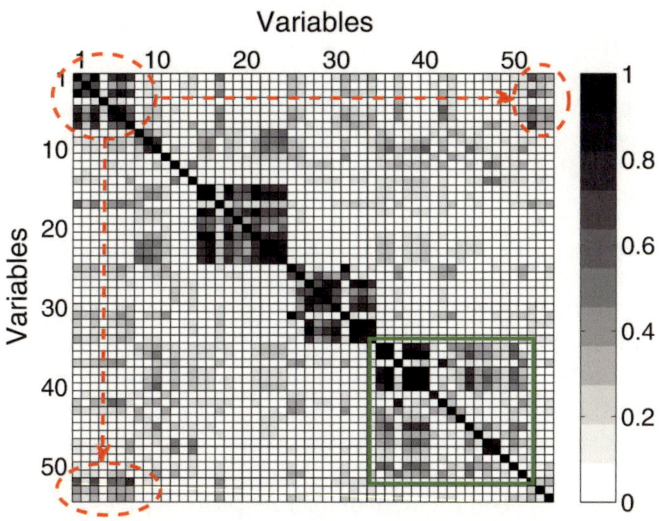

Fig. 24. MEDA matrix of the PCA model with 10 PCs from the data set which combines the in vitro antifilarial activity (-LOGEC50) with the complete set of descriptors of the Selwood dataset. Two common factors are highlighted. The first one is mainly found in descriptors 1 to 3, 5 to 7, 17, 52 and 53. The second one is mainly found in descriptors 35, 36, 38 to 40, 45, 47 and 50. Though the second common factor is not present in -LOGEC50, it is in LOGP.

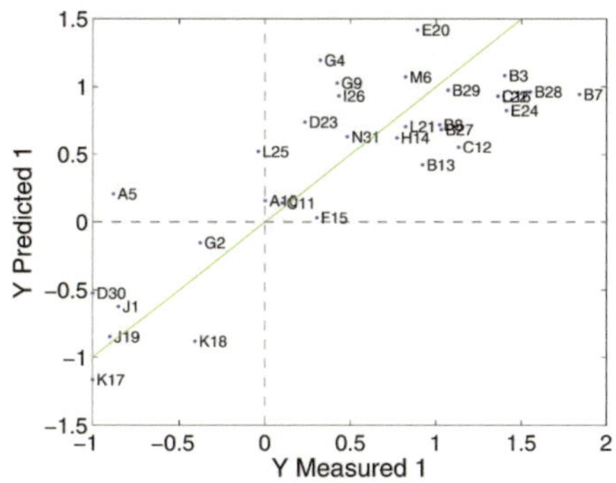

Fig. 25. Plot of measured vs predicted values of -LOGEC50 in the model with regressors LOGP, ATCH6 and MOFI_Y of the Selwood dataset.

compounds previously highlighted are found at the bottom left corner. This result support the non-convenience of isolating these compounds.

Notice that MEDA is not a variable selection technique per se and therefore other methods may be more powerful for this purpose. Nevertheless, being an exploratory method, the benefit of using MEDA for variable selection is that the general solution can be identified and understood, like in the present example. On the contrary, most variable selection approaches are based on highly complex algorithms which can only report a set of possible alternative solutions (e.g. Table 4).

6. Conclusion

In this chapter, new tools for exploratory data analysis are presented and combined with already well known techniques in the chemometrics field, such as projection models, score and loading plots. The shortcomings and potential pitfalls in the application of common tools are elucidated and illustrated with examples. Then, the new techniques are introduced to overcome these problems.

The Missing-data methods for Exploratory Data Analysis technique, MEDA for short, studies the relationships among variables. As it is discussed in the chapter, while chemometric models such as PCA and PLS are quite useful for data understanding, they have a main problem which complicates its interpretation: a single component captures several sources of variability or common factors and at the same time a single common factor is captured in several components. MEDA, like rotation methods or Factor Analysis (FA), is a tool for the identification of the common factors in subspace models, in order to elucidate the structure in the data. The output of MEDA is similar to a correlation matrix but with better properties associated. MEDA is the perfect complement of loading plots. It gives a different picture of the relationships among variables which is especially useful to find groups of related variables. Using a Quantitative Structure-Activity Relationship (QSAR) example, it was shown that the understanding of the relationships among variables in the data may lead to perform variable selection with similar performance of highly sophisticated algorithms, with the extra benefit that the global solution is not only found but also understood.

The second technique introduced in this chapter is a variant of MEDA, named observation-based MEDA or *o*MEDA. *o*MEDA was designed to identify the variables which differ between two groups of observations in a latent subspace, but it can be used for the more general problem of identifying the variables related to a given direction in the score plot. Thus, when a number of observations are located in a specific direction in the score plot, *o*MEDA gives the variables related to that distribution. *o*MEDA is the perfect complement of score plots and much more reliable than biplots. It can also be seen as an extension of contribution plots to groups of observations. It may be especially useful to check whether the distribution of a new set of observations agree with a calibration model.

Though MEDA and *o*MEDA are grounded on missing-data imputation methods and their original algorithms are complex to a certain extent, both tools can be computed with very simple equations. A MATLAB toolbox with the tools employed in this chapter, including MEDA, *o*MEDA, ADICOV and SVI plots, is available at http://wdb.ugr.es/ josecamacho/.

7. Acknowledgement

Research in this work is partially supported by the Spanish Ministry of Science and Technology through grant CEI BioTIC GENIL (CEB09-0010).

8. References

[1] Jolliffe I.T.. *Principal component analysis*. EEUU: Springer Verlag Inc. 2002.

[2] Han J., Kamber M.. *Data Mining: Concepts and Techniques*. agora.cs.illinois.edu: Morgan Kaufmann Publishers, Elsevier 2006.

[3] Keren Gideon, Lewis Charles. *A Handbook for data analysis in the behavioral sciences: statistical issues*. Hillsdale, NJ: Lawrence Erlbaum Associates 1993.

[4] Tukey John W. *Exploratory data analysis.* Addison-Wesley Series in Behavioral ScienceReading, MA: Addison-Wesley 1977.
[5] Teo Yik Y.. Exploratory data analysis in large-scale genetic studies *Biostatistics.* 2010;11:70-81.
[6] Pearson K.. On Lines and Planes of Closest Fit to Systems of Points in Space *Philosophical Magazine.* 1901;2:559–572.
[7] Jackson J.E.. *A User's Guide to Principal Components.* England: Wiley-Interscience 2003.
[8] Wold H., Lyttkens E.. Nonlinear iterative partial least squares (NIPALS) estimation procedures in *Bull. Intern. Statist. Inst. Proc., 37th session, London*:1–15 1969.
[9] Geladi P., Kowalski B.R.. Partial Least-Squares Regression: a tutorial *Analytica Chimica Acta.* 1986;185:1–17.
[10] Wold S., om M. Sj Eriksson L.. PLS-regression: a basic tool of chemometrics *Chemometrics and Intelligent Laboratory Systems.* 2001;58:109–130.
[11] Camacho J.. Missing-data theory in the context of exploratory data analysis *Chemometrics and Intelligent Laboratory Systems.* 2010;103:8–18.
[12] Gabriel K.R.. The biplot graphic display of matrices with application to principal component analysis *Biometrika.* 1971;58:453–467.
[13] Westerhuis J.A., Gurden S.P., Smilde A.K.. Generalized contribution plots in multivariate statistical process monitoring *Chemometrics and Intelligent Laboratory Systems.* 2000;51:95–114.
[14] Camacho J., Picó J., Ferrer A.. Data understanding with PCA: Structural and Variance Information plots *Chemometrics and Intelligent Laboratory Systems.* 2010;100:48–56.
[15] Camacho J., Padilla P., Díaz-Verdejo J., Smith K., Lovett D.. Least-squares approximation of a space distribution for a given covariance and latent sub-space *Chemometrics and Intelligent Laboratory Systems.* 2011;105:171–180.
[16] Kosanovich K.A., Dahl K.S., Piovoso M.J.. Improved Process Understanding Using Multiway Principal Component Analysis *Engineering Chemical Research.* 1996;35:138–146.
[17] Ferrer A.. Multivariate Statistical Process Control based on Principal Component Analysis (MSPC-PCA): Some Reflections and a Case Study in an Autobody Assembly Process *Quality Engineering.* 2007;19:311–325.
[18] Kjeldahl K., Bro R.. Some common misunderstandings in chemometrics *Journal of Chemometrics.* 2010;24:558–564.
[19] L. Fabrigar, D. Wegener, R. MacCallum, E. Strahan, Evaluating the use of exploratory factor analysis in psychological research, Psycological Methods 4 (3) (1999) 272–299.
[20] A. Costello, J. Osborne, Best practices in exploratory factor analysis: Four recommendations for getting the most from your analysis, Practical Assessment, Research & Evaluation 10 (7) (2005) 1–9.
[21] I. Jolliffe, Rotation of principal components: choice of normalization constraints, Journal of Applied Statistics 22 (1) (1995) 29–35.
[22] P. Nelson, P. Taylor, J. MacGregor, Missing data methods in pca and pls: score calculations with incomplete observations, Chemometrics and Intelligent Laboratory Systems 35 (1996) 45–65.
[23] D. Andrews, P. Wentzell, Applications of maximum likelihood principal component analysis: incomplete data sets and calibration transfer, Analytica Chimica Acta 350 (1997) 341Ű–352.
[24] B. Walczak, D. Massart, Dealing with missing data: Part i, Chemometrics and Intelligent Laboratoy Systems 58 (2001) 15–27.

[25] F. Arteaga, A. Ferrer, Dealing with missing data in mspc: several methods, different interpretations, some examples, Journal of Chemometrics 16 (2002) 408–418.

[26] F. Arteaga, A. Ferrer, Framework for regression-based missing data imputation methods in on-line mspc, Journal of Chemometrics 19 (2005) 439–447.

[27] M. Reis, P. Saraiva, Heteroscedastic latent variable modelling with applications to multivariate statistical process control, Chemometrics and Intelligent Laboratoy Systems 80 (2006) 57–66.

[28] F. Arteaga, Unpublished results.

[29] F. Lindgren, P. Geladi, S. Wold, The kernel algorithm for pls, Journal of Chemometrics 7 (1993) 45–59.

[30] S. de Jong, C. ter Braak, Comments on the pls kernel algorithm, Journal of Chemometrics 8 (1994) 169–174.

[31] B. Dayal, J. MacGregor, Improved pls algorithms, Journal of Chemometrics 11 (1997) 73–85.

[32] B. Wise, N. Gallagher, R. Bro, J. Shaver, W. Windig, R. Koch, PLSToolbox 3.5 for use with Matlab, Eigenvector Research Inc., 2005.

[33] Camacho J. Observation-based missing data methods for exploratory data analysis to unveil the connection between observations and variables in latent subspace models *Journal of Chemometrics* 25 (2011) 592 - 600.

[34] D.L. Selwood, D.J. Livingstone, J.C.W. Comley, A.B. O'Dowd, A.T. Hudson, P. Jackson, K.S. Jandu, V.S. Rose, J.N. Stables Structure-Activity Relationships of Antifiral Antimycin Analogues: A Multivariate Pattern Recognition Study, Journal of Medical Chemistry 33 (1990) 136–142.

[35] S.J. Cho, M.A. Hermsmeier, Genetic Algorithm Guided Selection: Variable Selection and Subset Selection, J. Chem. Inf. Comput. Sci. 42 (2002) 927–936.

[36] S.S. Liu, H.L. Liu, C.S. Yin, L.S. Wang, VSMP: A Novel Variable Selection and Modelling Method Based on the Prediction, J. Chem. Inf. Comput. Sci. 43 (2003) 964–969.

5

Experimental Optimization and Response Surfaces

Veli-Matti Tapani Taavitsainen
Helsinki Metropolia University of Applied Sciences
Finland

1. Introduction

Statistical design of experiments (DOE) is commonly seen as an essential part of chemometrics. However, it is often overlooked in chemometric practice. The general objective of DOE is to guarantee that the dependencies between experimental conditions and the outcome of the experiments (the responses) can be estimated reliably at minimal cost, i.e. with the minimal number of experiments. DOE can be divided into several subtopics, such as finding the most important variables from a large set of variables (screening designs), finding the effect of a mixture composition on the response variables (mixture designs), finding sources of error (variance component analysis) in a measurements system, finding optimal conditions in continuous processes (evolutionary operation, EVOP) or batch processes (response surface methodology, RSM), or designing experiments for optimal parameter estimation in mathematical models (optimal design).

Several good textbooks exist. Of the general DOE textbooks, i.e. the ones that are focused on any special field, perhaps (Box et. al., 2005), (Box & Draper, 2007) and (Montgomery, 1991) are the most widely used ones. Some of the DOE textbooks, e.g. (Bayne & Rubin, 1986), (Carlson & Carlson, 2005) and (Bruns et. al., 2006) focus on chemometric problems. Good textbooks covering other fields of applications include e.g. (Himmelblau, 1970) for chemical engineering, (Berthouex & Brown, 2002) and (Hanrahan, 2009) for environmental engineering, or (Haaland, 1989) for biotechnology. Many textbooks about linear models or quality technology also have good treatments of DOE, e.g. (Neter et. al., 1996), (Vardeman, 1994) and (Kolarik, 1995).

More extensive lists of DOE literature are given in many textbooks, see e.g. (Box & Draper, 2007) , or in the documentation of commercial DOE software packages, see e.g. (JMP, release 6)

This chapter focuses on common strategies of empirical optimization, i.e. optimization based on designed experiments and their results. The reader should be familiar with basic statistical concepts. However, for the reader's convenience, the key concepts needed in DOE will be reviewed. Mathematical prerequisites include basic knowledge of linear algebra, functions of several variables and elementary calculus. However, neither theory, nor the methodology is presented in a rigorous mathematical style; rather the style is relying on examples, common sense, and on pinpointing the key ideas.

The aim of this chapter is that the material could be used to guide chemists, chemical engineers and chemometricians in real applications requiring experimentation. Naturally, the examples presented have chemical/chemometric origin, but as with most statistical techniques, the field of possible applications is truly vast. The focus is on problems with quantitative variables and, correspondingly, on regression techniques. Qualitative (categorical) variables and analysis of variance (ANOVA) are merely mentioned.

Typical chemometric applications of RSM are such as optimization of chemical syntheses, optimization of chemical reactors or other unit operations of chemical processes, or optimization of chromatographic columns.

2. Optimization strategies

This section introduces the two most common empirical optimization strategies, the simplex method and the Box-Wilson strategy. The emphasis is on the latter, as it has a wider scope of applications. This section presents the basic idea; the techniques needed at different steps in following the given strategy are given in the subsequent sections.

2.1 The Nelder-Mead simplex strategy

The Nelder-Mead simplex algorithm was published already on 1965, and it has become a 'classic' (Nelder & Mead, 1965). Several variants and applications of it have been published since then. It is often also called the flexible polyhedron method. It should be noted that it has nothing to do with the so-called Dantzig's simplex method used in linear programming. It can be used both in mathematical and empirical optimization.

The algorithm is based on so-called simplices N-polytopes with $N+1$ vertices, where N is the number of (design) variables. For example, a simplex in two dimensions is a triangle, and a simplex in three dimensions is a tetrahedron. The idea behind the method is simple: a simplex provided with the corresponding response values (or function values in mathematical optimization) gives a minimal set of points to fit perfectly an N-dimensional hyperplane in a ($N+1$)-dimensional space. For example for two variables and the responses, the space is a plane in 3-dimensional space. Such a hyperplane is the simplest linear approximation of the underlying nonlinear function, often called a response surface, or rather a response hypersurface. The idea is to reflect the vertex corresponding to the worst response value along the hyperplane with respect to the opposing edge. The algorithm has special rules for cases in which the response at a reflected point doesn't give improvement, or if an additional expanded reflection gives improvement. These special rules make the simplex sometimes shrink, and sometimes expand. Therefore, it is also called the flexible simplex algorithm.

The idea is easiest understood graphically in a case with 2 variables: Fig. 1 depicts an ideal response surface the yield of a batch reactor with respect to the batch length in minutes and the reactor temperature in °C. The model is ideal in the sense that the response values are free from experimental error. We can see that first the simplex expands because the surface around the starting simplex is quite planar. Once the chain of simplexes attains the ridge going approximately from right, some of the simplexes are contracted, i.e. they shrink considerably. You can easily see, how a reflection would worsen the response (this is depicted as an arrow in the upper left panel). Once the chain finds the direction of the ridge, the simplexes expand again and approach the optimum effectively. The Nelder-Mead simplex algorithm is not very effective in final positioning of the optimal point, because that would require many contractions.

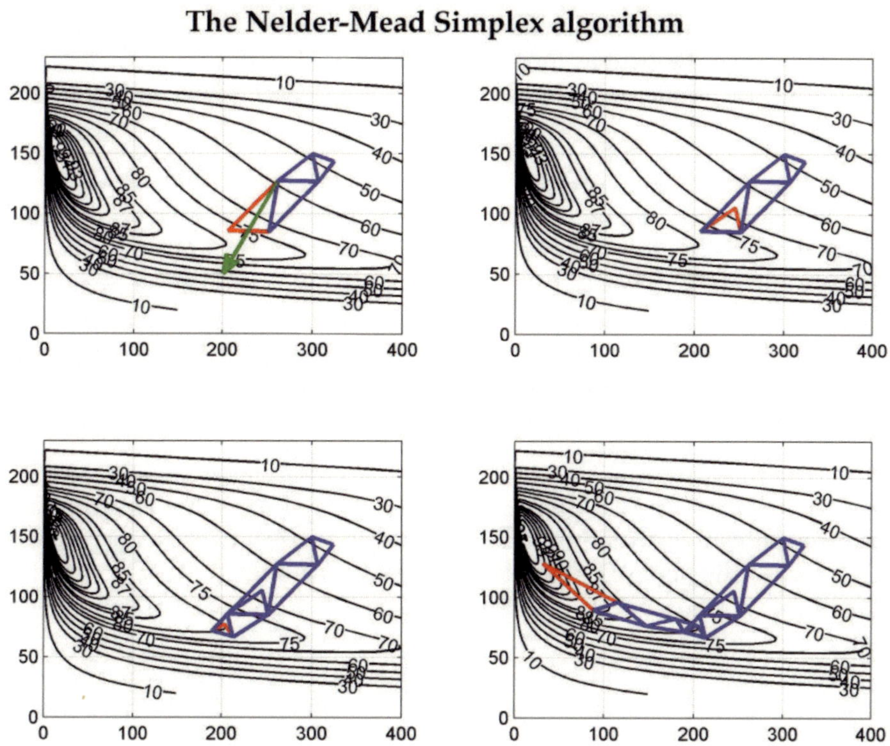

Fig. 1. Sequences of Nelder-Mead simplex experiments with respect to time and temperature based on an errorless reactor model. In all panels, the x-axis corresponds to reaction time in minutes and y-axis corresponds to reactor temperature in °C. Two edges of the last simplex are in red in all panels. Upper left panel: the first 4 simplexes and the reflection of the last simplex. Upper right panel: the first 4 simplexes and the first contraction of the last simplex. Lower right panel: the first 7 simplexes and the second contraction of the last simplex. Lower right panel: the first 12 simplexes and the expanded reflection of the last simplex.

The Nelder-Mead algorithm has been used successfully e.g. in optimizing chromatographic columns. However, its applicability is restricted by the fact that it doesn't work well if the results contain substantial experimental error. Therefore, in most cases another type of a strategy is a better choice, presented in the next section.

2.2 The Box-Wilson strategy (the gradient method)

In this section we try to give an overall picture of the Box-Wilson strategy, and the different types of designs used within the strategy will be explained in subsequent sections; the focus is on the strategy itself.

The basic idea behind the Box-Wilson strategy is to follow the path of the steepest ascent towards the optimal point. In determining the direction of the steepest ascent, mathematically speaking, the gradient vector, the method uses local polynomial modelling. It is a sequential method, where the sequence of main steps is: 1) make a design around the current best point, 2) make a polynomial model, 3) determine the gradient path, and 4) carry out experiments along the path as long as the results will improve. After step 4, return to step 1, and repeat the sequence of steps. Typically the steps 1-4 have to be repeated 2 to 3 times.

Normally the first design is a 2^N factorial design (see section 3.1) with an additional centre point, possibly replicated one or more times. The idea is that, at the beginning of the optimization, the surface within the design area is approximately linear, i.e. a hyperplane. A 2^N factorial design allows also modelling of interaction effects. Interactions are common in problems of chemical or biological origin. The additional centre point can be used to check for curvature. If the curvature is found to be statistically significant, the design should be upgraded into a second order design (see section 5), allowing building of a quadratic model. The replicate experiments are used to estimate the mean experimental error, and for testing model adequacy, i.e. the lack-of-fit in the model.

After the first round of steps 1-4 (see also Fig. 4), it is clear that a linear or linear plus interactions model cannot fit the results anymore, as the results first get better and then worse. Therefore, at this point, an appropriate design is a second order design, typically a central composite or a Box-Behnken design (explained in section 5), both allowing building of a quadratic polynomial model. The analysis of the quadratic model lets us estimate whether the optimum is located near the design area or further away. In the latter case, new experiments are again conducted along the gradient path, but in the first case, the new experiments will be located around the optimum predicted by the model.

The idea is best grasped by a graphical illustration given in Figs. 2 and 3 using the same reactor model as in section 2.1. Fig. 2 shows the theoretical errorless response surface, the region of the initial design (the black rectangle), and the theoretical gradient path. The contours inside the rectangle show that the response behaves approximately linearly inside the region of the first design (the contours are approximately parallel straight lines).

It is important to understand that the gradient path must be calculated using small enough steps. This is best seen graphically: Fig. 3 shows what happens using too large a step size: too large a step size creates the typical zigzag pattern. Obviously, this is inefficient, and such a path also misses the optimum.

Next we shall try to illustrate how the gradient method works in practice. In order to make the situation more realistic, we have added Gaussian noise ($\mu = 0, \sigma^2 = 1$) to the yield, given by the simulation model of the reactor, i.e. instead of carrying out real experiments, the results are obtained from the model. In addition, random experimental error is added to the modelled responses. The sequence of designs following the box-Wilson strategy and the corresponding gradient path experiments are depicted in Fig. 4. Notice that the gradient path based on the model of the first design is slightly curved due to the interaction between time and temperature.

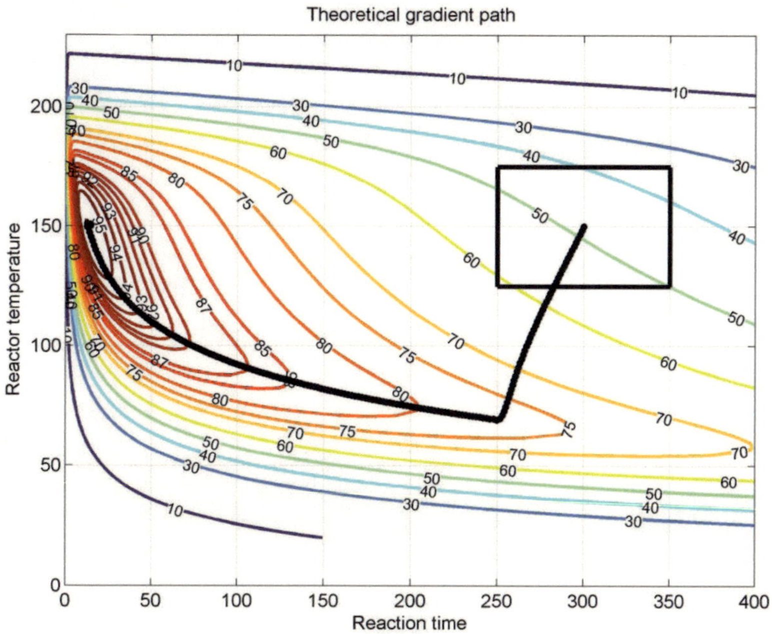

Fig. 2. The gradient path (black solid line with dots at points of calculation) of a reactor model yield surface. The solid line cannot be distinguished due to the small step size between the points of calculation.

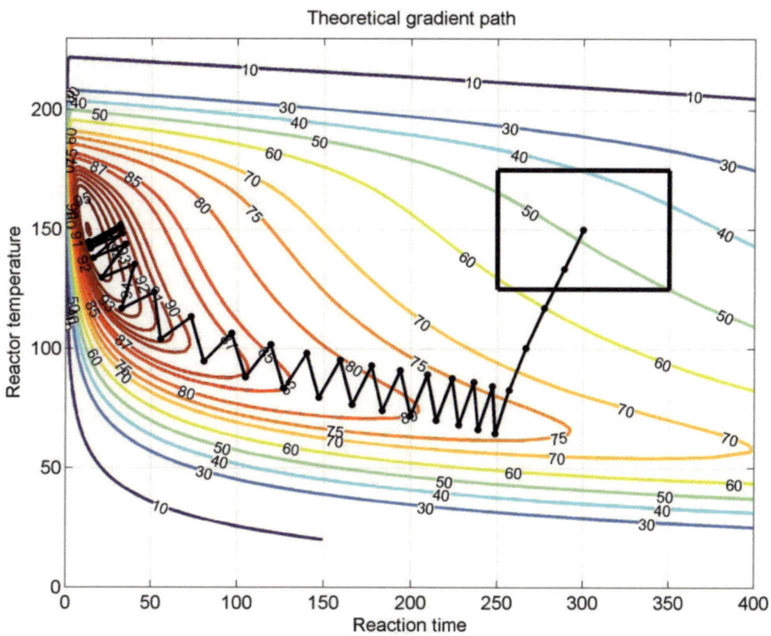

Fig. 3. The gradient path (black solid line with dots at the points of calculation) of a reactor model yield surface. The path is calculated using too large a step size causing the zigzag pattern.

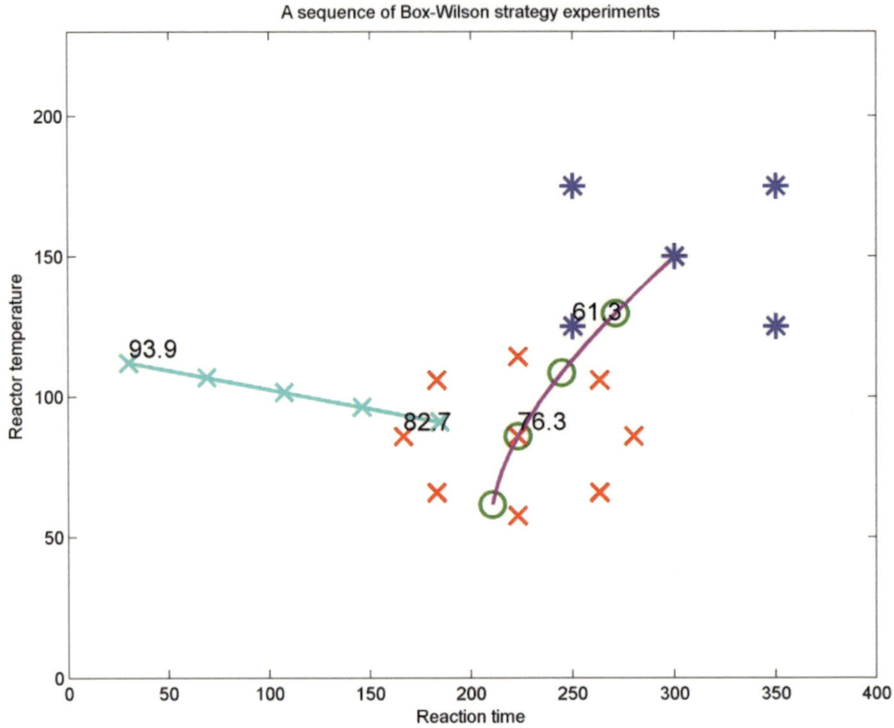

Fig. 4. The first design (blue asterisks) and the gradient path (magenta solid line) based on an empirical model estimated from the results of the experiments. Four points from the gradient path are chosen for the experiments (green circles). A second order design (red x's) and the gradient path based on its modelled results (turquoise solid line with x's at the experimental points). The best results of the four sets of experiments, showing excellent improvement, are 61.3, 76.3, 82.7 and 93.9.

Typically, the next step would be making of a new second order design around the best point. However, one should keep in mind that the sensitivity to changes in the design variables decreases. As a consequence, any systematic changes may be hidden under experimental errors. Therefore, the accurate location of the optimum is difficult to find, perhaps requiring repetitions of the whole design.

Simulation models like the one used in this example are very useful in practising Box-Wilson strategy. It can be obtained upon request from the author in the form of a Matlab, R or Excel VBA. For maximal learning, the user is advised to start the procedure at different locations.

3. Factorial designs

Factorial designs make the basis of all most common designs. The idea of factorial designs is simple: a factorial design is made up of all possible combinations of all chosen values, often called levels, of all design variables. Factorial designs can be used both for qualitative and quantitative variables. If variables $x_1, x_2, \dots, x_N$ have $m_1, m_2, \dots, m_N$ different levels, the number of experiments is $m_1 \cdot m_2 \cdot \dots \cdot m_N$. As a simple example, let us consider a case where the variables and their levels are: x_1 the type of a catalyst (A, B and C), x_2 the catalyst concentration (1 ppm and 2 ppm), and x_3 the reaction temperature (60 °C, 70 °C and 80 °C). The corresponding factorial design is given in Table 1.

x_1	x_2	x_3
A	1	60
B	1	60
C	1	60
A	2	60
B	2	60
C	2	60
A	1	70
B	1	70
C	1	70
A	2	70
B	2	70
C	2	70
A	1	80
B	1	80
C	1	80
A	2	80
B	2	80
C	2	80

Table 1. A simple factorial design of three variables.

It is good to understand why factorial designs are good designs. The main reasons are that they are *orthogonal* and *balanced*. Orthogonality means that the factor (variable) effects can be estimated independently. For example, in the previous example the effect of the catalyst can be estimated independently of the catalyst concentration effect. In a balanced design, each variable combination appears equally many times. In order to understand why orthogonality is important, let us study an example of a design that is not orthogonal. This design, given in Table 2 below, has two design variables, x_1 and x_2, and one response variable, y.

x_1	x_2	y
1.18	0.96	0.91
1.90	2.12	1.98
3.27	2.98	2.99
4.04	3.88	3.97
4.84	5.10	5.03
5.88	6.01	5.96
7.14	7.14	7.07
8.05	8.08	7.92
9.04	8.96	9.09
9.98	10.19	10.02
2.30	2.96	3.76
2.94	4.10	4.85
4.29	5.01	5.95
5.18	5.80	6.88
5.84	7.14	7.98
6.85	7.90	9.16
8.33	8.98	10.26
8.96	10.05	10.98
10.00	11.09	12.21
10.82	12.08	12.94
10.13	8.88	8.20
10.91	9.94	9.15
12.10	10.83	10.30
12.57	11.56	11.16
13.53	13.04	12.13
14.85	14.00	12.72
16.19	15.16	13.99
17.01	16.22	14.72
18.31	16.86	16.28
19.21	18.47	17.35

Table 2. A non-orthogonal design.

Now, if we plot the response against the design variables we get the following plot:

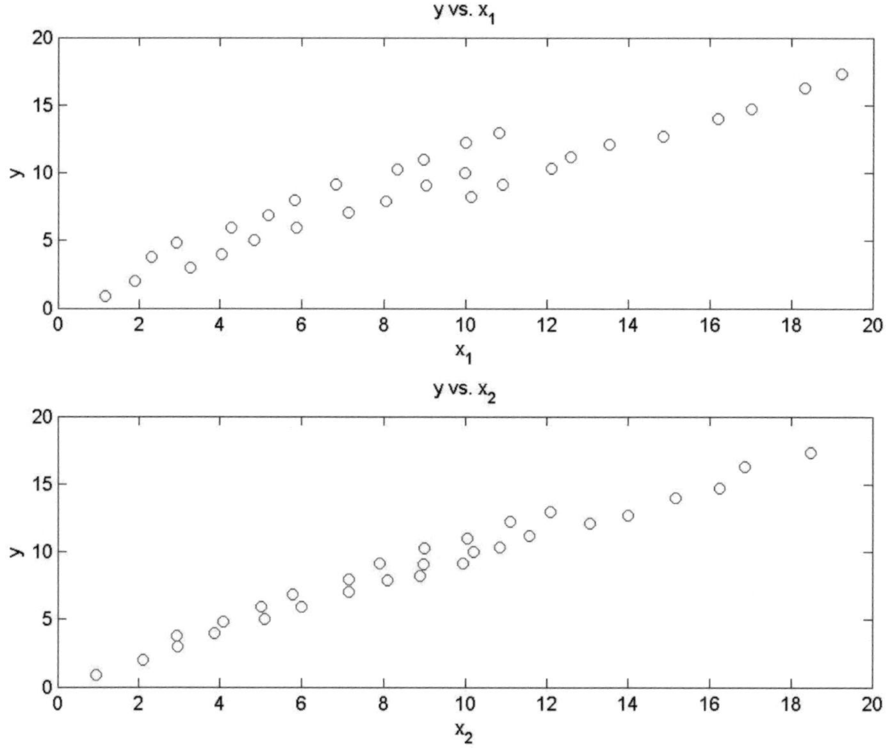

Fig. 5. Response (y) against the design variables (x_1 and x_2) in a non-orthogonal design; upper panel: y against, lower panel: y against.

Now, Fig. 5 clearly gives the illusion that the response depends approximately linearly both on x_1 and x_2 with positive slopes. However, the true slope between y and x_1 is negative. To see this, let us plot the design variable against each other and show the response values as text, as shown in Fig. 6.

Now, careful inspection of Fig. 6 reveals that actually yield decreases when x_1 increases. The reason for the wrong illusion that Fig. 5 gives is that x_1 and x_2 are strongly correlated with each other, i.e. the design variables are collinear. Although fitting a linear regression model using both design variables would give correct signs for the regression coefficients, collinearity will increase the confidence intervals of the regression coefficients. Problems of this kind can be avoided by using factorial designs.

After this example, it is obvious that orthogonality, or near orthogonality, is a desired property of a good experimental design. Other desired properties are

- The design contains as few experiments as possible for reliable results.
- The design gives reliable estimates for the empirical model fitted to the data.

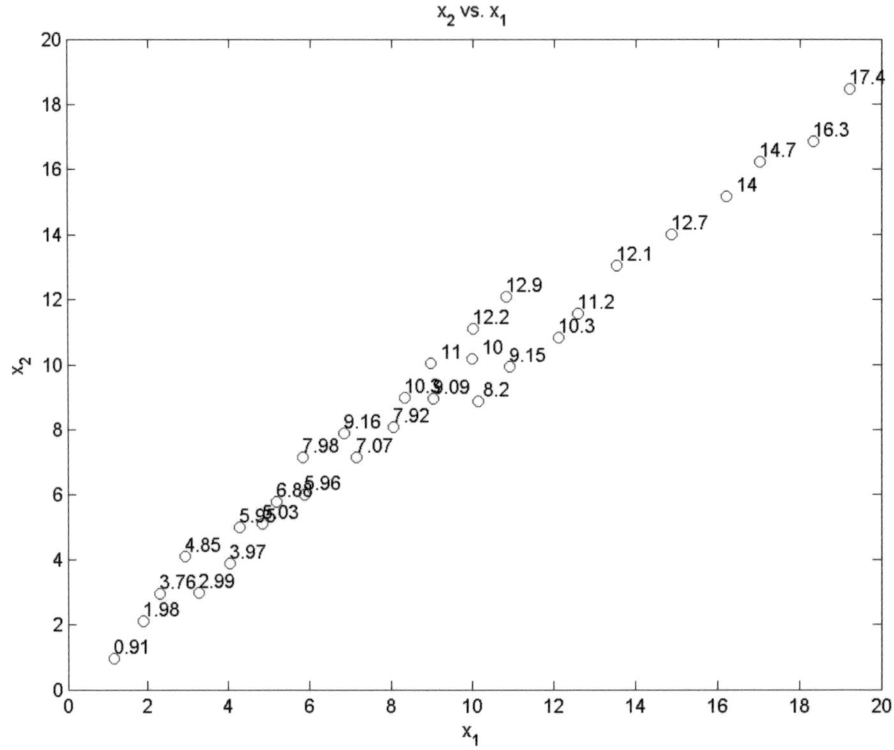

Fig. 6. Response (y) against the design variables (x_1 and x_2) in a non-orthogonal design; the response values (y values) are shown on upper right position with respect to each point.

- The design allows checking the reliability of the model fitted to the data, typically by statistical tests about the model parameters and the model adequacy (lack-of-fit), and by cross validation.

In general, factorial designs have these properties, except for the minimal number of the experiments. This topic will be treated in the subsequent sections.

3.1 Two level factorial designs

Factorial designs with only two values (levels) for all design variables are the most frequently used designs. This is mainly due to the following facts: 1) the number of experiments is less than with more levels, and 2) the results can be analysed using regression analysis for both qualitative and quantitative variables. The natural limitation is that only linear (main) effects and interactions effects can be detected. A drawback of full two-level factorial designs with a high number of variables is that the number of experiments is also high. Fortunately, this problem can be solved by using so-called fractional factorial designs, explained in section 3.3.

Two-level factorial designs, usually called 2^N designs, are typically tabulated using dimensionless coded variables having only values -1 or +1. For example, for a variable that represents a catalyst type, type A might correspond to -1 and type B might correspond to +1, or coarse raw material might be -1 and fine raw material might be +1. For quantitative variables, coding can be performed by the formula

$$X_i = \frac{x_i - \overline{x}_i}{\frac{1}{2}\Delta x_i} \tag{1}$$

where X_i stands for the coded value of the i'th variable, x_1 stands for the original value of the i'th variable, $\overline{x}_i$ stands for the centre point value of the original i'th variable, and Δx_i stands for the difference of the original two values of the i'th variable. The half value of the difference is called the step size. All statistical analyses of the results are usually carried out using the coded variables. Quite often, we need to convert also coded dimensionless values into the original physical values. For quantitative variables, we can simply use the inverse of Eq. 1, i.e.

$$x_i = \overline{x}_i + \tfrac{1}{2}\Delta x_i \cdot X_i \tag{2}$$

Tables of two-level factorial designs can be found in most textbooks of DOE. A good source is also NIST SEMATECH e-Handbook of Statistical Methods (NIST SEMATCH). Another way to create such tables is to use DOE-software e.g. (JMP, MODDE, MiniTab,…). It is also very easy to create tables of two-level factorial designs in any spreadsheet program. For example in Excel, you can simply enter the somewhat hideous formula

```
=2*MOD(FLOOR((ROW($B3)-ROW($B$3))/2^(COLUMN(C$2)-COLUMN($C$2));1);2)-1
```

into the cell C3, and then first copy the formula to the right as many time as there are variables in the design (*N*) and finally copy the whole first row down 2^N times. Of course, you can enter the formula anywhere in the spreadsheet, e.g. if you enter it into the cell D7 the references in the ROW functions must be changed into $C7 and C7, and the references in the column function must be changed into D$6 and D6, respectively.

If all variables are quantitative it is advisable to add a centre point into the design, i.e. an experiment where all variables are set to their mean values. Consequently, in coded units, all variables have value 0. The centre point experiment can be used to detect nonlinearities within the design area. If the mean experimental error is not known, usually the most effective way to find it out is to repeat the centre point experiment. All experiments, including the possible centre point replicates, should be carried out in random order. The importance of randomization is well explained in e.g. (Box, Hunter & Hunter).

3.1.1 Empirical models related to two-level factorial designs

2^N designs can be used only for linear models with optional interaction terms up order N. By experience, it is known that interaction of order higher than two are seldom significant. Therefore, it is common to consider those terms as random noise, giving extra degrees of freedom for error estimation. However, one should be careful about such interpretations, and models should always be carefully validated. It should also be noted that the residual errors always contain both experimental error and modelling error. For this reason, independent replicate experiments are of utmost importance, and only having a reliable estimate of the experimental error gives a possibility to check for lack-of-fit, i.e. the model adequacy.

The general form of a model for a response variable y with linear terms and interaction terms up to order N is

$$y = \sum_{i=1}^{N} b_i X_i + \sum_{i<j}^{N} b_{ij} X_i X_j + \sum_{i<j<k}^{N} b_{ijk} X_i X_j X_k + \cdots \tag{3}$$

The number of terms in the second sum is $\binom{N}{2}$, and in the third sum it is $\binom{N}{3}$, and so on.

The most common model types used are models with linear terms only, or models with linear terms and pairwise interaction terms.

If all terms of model (3) are used, and there are no replicate experiments in the corresponding 2^N design, there are as many unknown parameters in the model as there are experiments in the design, leaving no degrees of freedom for the residual error, i.e. all residuals are zero. In such cases, the design is called saturated with respect to the model, or just saturated, if it is obvious what the model is. In these cases traditional statistical tests cannot be employed. Instead, the significant terms can often be detected by inspecting the estimated model parameter values using normal probability plots.

Later we need to differentiate between the terms "design matrix" and "model matrix". A design matrix is a $N_{exp} \times N$ matrix whose columns are the values of design variables where N_{exp} is the number of experiments. A model matrix is a $N_{exp} \times p$ matrix that is the design matrix appended with columns corresponding to the model terms. For example, a model matrix for a linear plus interaction model for two variables has a column of ones (corresponding to the intercept), columns for values of X_1 and X_2 and a column for values of the product $X_1 X_2$.

It is good to understand the nature of the pairwise interaction terms. Let us consider a model for two variables, i.e. $y = b_0 + b_1 X_1 + b_2 X_2 + b_{12} X_1 X_2$, and let us rearrange the terms as $y = b_0 + b_1 X_1 + (b_2 + b_{12} X_1) X_2$. This form reveals that the interaction actually means that the slope of X_2 depends linearly on X_1. Taking X_1 as the common factor instead of X_2 shows that the slope of X_1 depends linearly on X_2. In other words, a pairwise interaction between two variables means that the other variable affects the effect of the other one. If two variables don't interact, their effects are said to be additive. Fig. 7 depicts additive and interacting variables.

In problems of chemical or biological nature, it is more a rule than an exception that interactions between variables exist. Therefore, main effect models serve only as rough approximations, and are used typically in cases with a very high number of variables. It is also quite often useful to try to model some transformation of the response variable,

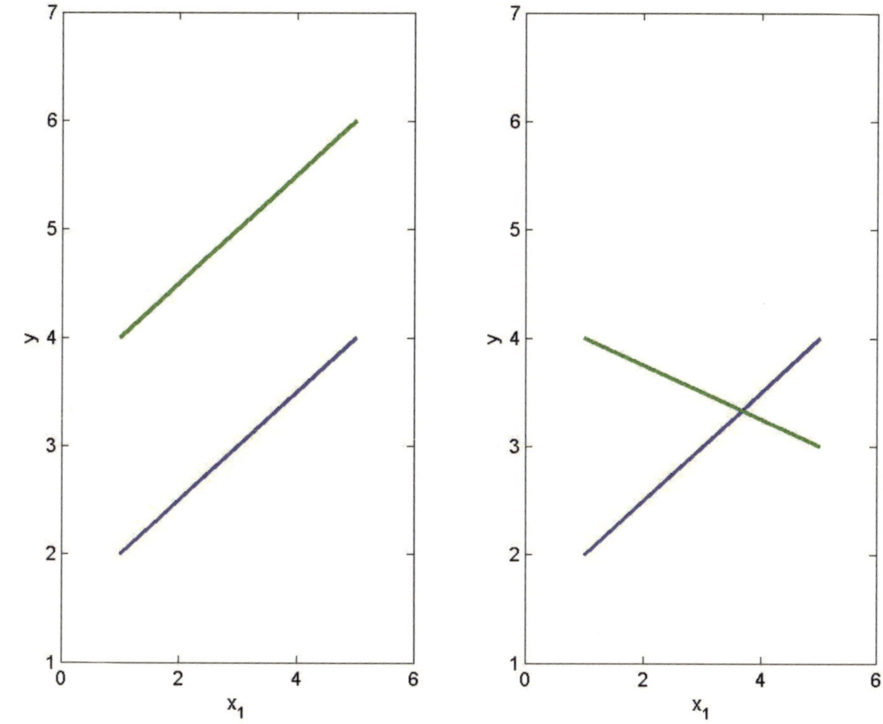

Fig. 7. Linear dependency of y on x_1 and on x_2 . Left panel: an additive case without interaction, right panel: a non-additive case. Blue line: x_2 has a constant value, green line: x_2 has another constant value.

typically a logarithm, or a Box-Cox transformation. Usually the aim is to find a transformation that makes the residuals as normal as possible.

3.1.2 Analysing the results of two level factorial designs

Two level factorial designs can be analysed either by analysis of variance (ANOVA) or by regression analysis. Using regression analysis is more straightforward, and we shall concentrate on it in the sequel. However, one should bear in mind that the interpretation of the estimated model parameter is different between quantitative and qualitative variables. Actually, due to the orthogonality of 2^N designs, regression analysis could be carried out quite easily even by hand using the well-known Yates algorithm, see e.g. (Box & Draper, 2007).

Ordinary least squares regression (OLS) is the most common way to analyse results of orthogonal designs, but sometimes more robust techniques, e.g. minimizing the median of absolute values of the residuals, can be employed. Using latent variable techniques, e.g. PLS, doesn't give any extra benefit with orthogonal designs. However, in some other kind of designs, typically optimal designs or mixture designs, latent variable techniques can be useful.

The mathematics and statistical theory behind regression analysis can be found in any basic textbook about regression analysis or statistical linear models see e.g. (Weisberg, 1985) or (Neter et. al., 1996). In this chapter we shall concentrate on applying and interpreting the results of OLS in DOE problems. However, it is always good to bear in mind the statistical assumptions behind the classical regression tests, i.e. approximately normally distributed and independent errors. The latter is more important, and dependencies between random errors can make the results of the tests completely useless. This fact is well illustrated in (Box et. al., 2005). Even moderate deviations from normality are usually not too harmful, unless they are caused by gross errors, i.e. by outliers. For this reason, normal probability plots, or other tools for detecting outliers, should always be included in validating the model.

Since OLS is such a standard technique, a plethora of software alternatives exists for carrying out the regression analyses of the results of a given design. One can use general mathematics software like Matlab, Octave, Mathematica, Maple etc., or general purpose statistical software like S-plus, R, Statistica, MiniTab, SPSS, etc, or even spreadsheet programs like Excel or Open Office Calc. However, it is advisable to use software that contains those model validation tools that are commonly used with designed experiments. Practically all general mathematical or statistical software packages contain such tools.

Quite often there are more than one response variables. In such cases, it is typical to estimate models for each response separately. If a multivariate response is a 'curve', e.g. a spectrum or a distribution, it may be simpler to use latent variable methods, typically PLS or PCR.

This example is taken from Box & Draper (Box & Draper, 2007).

3.2 Model validation

Model validation is an essential part of analysing the results of a design. It should be noted that most of the techniques presented in this section can be used with all kinds of designs, not only with 2^N designs.

In the worst case, the validation yields the conclusion that the design variables have no effect on the response(s), significantly different from random variation. In such a case, one has to consider the following alternatives: 1) to increase the step sizes in the design variables, 2) to replicate the experiments one or more times, or 3) to make a new design with new design variables. In the opposite case, i.e. the model and at least some of the design variables are found to be statistically significant, the continuation depends on the scope of the design, and on the results of the (regression) analysis. The techniques used for optimization tasks are presented in subsequent sections.

3.2.1 Classical statistical tests

Classical statistical tests can be applied mainly to validate regression models that are linear with respect to the model parameters. The most common empirical models used in DOE are linear models (main effect models), linear plus interactions models, and quadratic models. They all are linear with respect to the parameters. The most useful of these (in DOE context) are 1) t-tests for testing the significance of the individual terms of the model, 2) the lack-of-fit test for testing the model adequacy, and 3) outlier tests based on so-called externally studentized residuals, see e.g. (Neter et. al., 1996).

The t-test for testing the significance of the individual terms of the model is based on the test statistic that is calculated by dividing a regression coefficient by its standard deviation. This statistic can be shown to follow the t-distribution with $n-p-1$ degrees of freedom where n is the number experiments, and p is the number of model parameters. If the model doesn't contain an intercept, the number of degrees of freedom is $n-p$. Typically, a term in the model is considered significant if the p-value of the test statistic is below 0.05.

The standard errors of the coefficients are usually based on the residual error. If the design contains a reasonable number of replicates this estimate can also be based on the standard error of the replicates. The residual based standard error of the i'th regression coefficient s_{b_i} can be easily transformed into replicate error based ones by the formula $\sqrt{MS_E / MS_R} \cdot s_{b_i}$. In this case the degrees of freedom are $n_r - 1$ (the symbols are explained in the next paragraph).

The lack-of fit test can be applied only if the design contains replicate experiments which permit estimation of the so-called pure error, i.e. an error term that is free from modelling errors. Assuming that the replicate experiments are included in regression, the calculations are carried out according to the following equations. First calculate the pure error sum of squares SS_E

$$SS_E = \sum_{i=1}^{n_r} (y_i - \bar{y})^2, \tag{4}$$

where n_r is the number of replicates, y_i's are outcomes of the replicate experiments, and $\bar{y}$ is the mean value of the replicate experiments. The number of degrees of freedom of SS_E is $n_r - 1$. Then calculate the residual sum of squares SS_R:

$$SS_R = \sum_{i=1}^{n} (y_i - \hat{y})^2, \tag{5}$$

where n is the number of experiments and $\hat{y}$'s are the fitted values, i.e. the values calculated using the estimated model. The number of degrees of freedom of SS_R is $n-p-1$, or $n-p$ if the model doesn't contain an intercept. Then calculate the lack-of-fit sum of squares SS_{LOF}:

$$SS_{LOF} = SS_R - SS_E \tag{6}$$

The number of degrees of freedom of SS_{LOF} is $n-n_r-p-1$, or $n-n_r-p$ if the model doesn't contain an intercept. Then, calculate the lack-of-fit mean squares MS_{LOF} and the pure error mean squares MS_E by dividing the corresponding sums of squares by their degrees of freedom. Finally, calculate the lack-of-fit test statistic $F_{LOF} = MS_{LOF} / MS_E$. It can be shown that F_{LOF} follows an F-distribution with $n-n_r-p-1$ (or $n-n_r-p$) and n_r-1 degrees of freedom. If F_{LOF} is significantly greater than 1, it is said that the model suffers from lack-of-fit, and if it is significantly less than 1, it is said that the model suffers from over-fit.

An externally studentized residual is a deleted residual, i.e. residual calculated using leave-one-out cross-validation, divided the standard error of deleted residuals. It can be shown that the externally studentized residuals follow a t-distribution with $n-p-2$ degrees of freedom, or $n-p-1$ degrees of freedom if the model doesn't contain an intercept. If the p-value of an externally studentized residual is small enough, the result of the corresponding experiment is called an outlier. Typically, outliers should be removed, or the corresponding experiments should be repeated. If the result of a repeated experiment still gives an outlying value, it is likely that model suffers from lack-of-fit. Otherwise, the conclusion is that something went wrong in the original experiment.

3.2.2 Cross-validation

Cross-validation is familiar to all chemometricians. However, in using cross-validation for validating results of designed experiments some important issues should be considered. First, cross-validation requires extra degrees of freedom, and consequently all candidate models cannot be cross-validated. For example, in a 2^2 design, a model containing linear terms and the pairwise interaction cannot be cross-validated. Secondly, often the designs become severely unbalanced, when observations are left out. For example, in a 2^2 design with a centre point, the model containing linear terms and the pairwise interaction can be cross-validated, but when the corner point (+1, +1) is left out, the design is very weak for estimating the interaction term; in such cases the results of cross-validation can be too pessimistic. On the other hand, replicated experiments may give too optimistic results in cross-validation, as the design variable combinations corresponding to replicate experiments are never left out. This problem can be easily avoided by using the response averages instead of individual responses of the replicated experiments.

Usually only statistically significant terms are kept in the final model. However, it is also common to include mildly non-significant terms in the model, if keeping such terms improves cross-validated results.

3.2.3 Normal probability plots

Normal probability plots, also called normal qq-plots, can be used to study either the regression coefficients or the residuals (or deleted residuals). The former is typically used in saturated models where ordinary t-tests cannot be applied. Normal probability plots are constructed by first sorting the values from the smallest to largest. Then the proportions $p_i = (i-0.5)/n$ are calculated, where n is the number of the values, and i is the ordinal number of a sorted value, i.e. 1 for the smallest value and n for the largest value (subtracting 0.5 is called the continuity correction). Then the normal score, i.e. inverse of p_i using the standard normal distribution, is calculated. Finally, the values are plotted against the normal scores. If the distribution of the values is normal, the points lie approximately on a straight line. The interpretation in the former case is that the leftmost or the rightmost values that do not follow a linear pattern represent significant terms. In the latter case, the same kind of values represent outlying residuals.

3.2.4 Variable selection

If the design is orthogonal, or nearly orthogonal, removing or adding terms into the model doesn't affect the significance of the other terms. This is also the case if the estimates of the standard error of the coefficients are based on the standard error of the estimates (cf. 3.2.1). Therefore, variable selection based on significance is very simple; just take the variables significant enough, without worrying about e.g. the order of taking terms into a model. Because models based on designed experiments are often used for extrapolatory prediction, one should, whenever possible, test the models using cross-validation. However, one should bear in mind the limitations of cross-validation when it is applied to designed experiments (cf. 3.2.2). In addition, it is wise also to test models with almost significant variables using cross-validation, since sometimes such models have better predictive power.

If the design is not orthogonal, traditional variable (feature) selection techniques can be used, e.g. forward, backward, stepwise, or all checking possible models (total search). Naturally, the selection can be based on different criteria, e.g. Mallows C_p, *PRESS*, R^2, Q^2, Akaike's information etc., see e.g. (Weisberg, 1985). If models are used for extrapolatory prediction, a good choice for a criterion is to minimize *PRESS* or maximize Q^2. In many cases of DOE modelling, the number of possible model terms, typically linear, pair-wise interaction, and quadratic terms, is moderate. For example, a full quadratic model for 4 variables has 14 terms, plus the intercept. Thus the number of all possible sub-models is 2^{14} which is 16384. In such cases, with the speed of modern computers, it is easy to test all sub-models with respect to the chosen criterion. However, if the number of variables is greater, going through all possible regression models becomes impossible in practice. In such cases, one can use genetic algorithms, see e.g. (Koljonen & al., 2008).

Another approach is to use latent variable techniques, e.g. PLS or PCR, in which the selection of the dimension replaces the selection of variables. Although variable selection seems more natural, and is more commonly used in typical applications of DOE than latent variable methods, neither of the approaches have been proved generally better. Therefore, it is good to try out different approaches, combined with proper model validation techniques.

A third alternative is to use shrinkage methods, i.e. different forms of ridge regression. Recently, new algorithms based on L_1 norm have been developed, including such as LASSO (Tibshirani, 1996), LARS (Efron & al., 2004), or elastic nets (Zou & al., 2005). Elastic nets use combinations of L_1 and L_2 norm penalties. Penalizing the least squares solution by the L_1 norm of the regression coefficient tends to make the non-significant terms zero which effectively means selecting variables.

In a typical application of DOE, the responses are multivariate in a way that they represent individual features which, in turn, typically depend on different variable combinations of the design variables. In such cases, it is better to build separate models for each response, i.e. the significant variables have to be selected separately for each response. However, if the response is a spectrum, or an object of similar nature, variable selection should usually be carried out for all responses simultaneously, using e.g. PLS regression or some other multivariate regression technique. In such cases, there's an extra problem of combining the individual criteria of the goodness of fit into a single criterion. In many cases, a weighted average of e.g. the RMSEP values, i.e. the standard residual errors in cross-validation, of the individual responses is a good choice, e.g. using signal to noise ratios as weights.

3.2.5 An example of a 2^N design

As a simple example of a 2^N design we take a 2^2 design published in the Brazilian Journal of Chemical Engineering (Silva et. al., 2011). In this study the ethanol production by Pichia stipitis was evaluated in a stirred tank bioreactor using semi defined medium containing xylose (90.0 g/l) as the main carbon source. Experimental assays were performed according to a 2^2 full factorial design to evaluate the influence of aeration (0.25 to 0.75 vvm) and agitation (150 to 250 rpm) conditions on ethanol production. The design contains also a centre point (0.50 vvm and 200 rpm), and in a replication of the (+1, +1) experiment. It should be noted that this design is not fully orthogonal due the exceptional selection of the replication experiment (the design would have been orthogonal, if the centre point had been replicated).

The results of the design are given in Table 3 below (X_1 and X_2 refer to aeration and agitation in coded levels, respectively).

Assay	Aeration	Agitation	X_1	X_2	Production (g/l)
1	0.25	150	-1	-1	23.0
2	0.75	150	1	-1	17.7
3	0.25	250	-1	1	26.7
4	0.75	250	1	1	16.2
5	0.75	250	1	1	16.1
6	0.50	200	0	0	19.4

Table 3. A 2^2 design.

Fig. 8 shows the effect of aeration at the two levels of agitation. From the figure, it is clear that aeration has much greater influence on productivity (Production) than agitation. It also shows an interaction between the variables. Considering the very small difference in the response between the two replicate experiments, it is plausible to consider both aeration and the interaction between aeration and agitation significant effects.

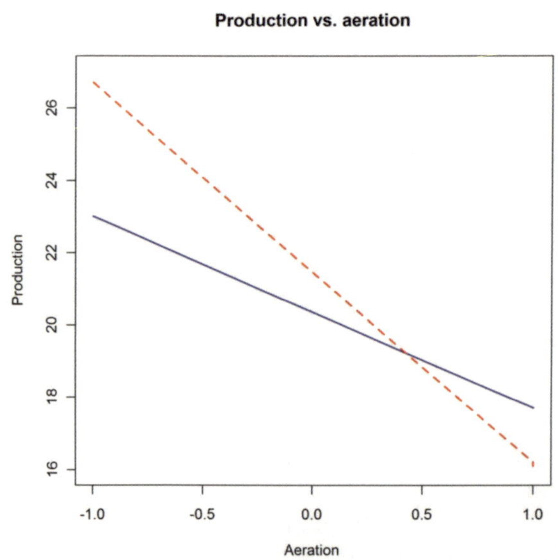

Fig. 8. Production vs. aeration. Blue solid line: Agitation = 150; red dashed line: Agitation = 250.

If we carry out classical statistical tests used in regression analysis, it should be remembered that the design has only one replicated experiment, and consequently very few degrees of freedom for the residual error. Testing lack-of-fit, or nonlinearity, is also unreliable because we can estimate the (pure) experimental error with only one degree of freedom. However, it is always possible to use common sense and investigations about the effects on relative basis. For example, the difference between the centre point result and the mean value of the other (corner point) results is only 0.45 which is relatively small compared to differences between the corner points. Therefore, it is highly unlikely that the behaviour would be nonlinear within the experimental region. Consequently, it is likely that the model doesn't suffer from lack-of-fit, and a linear plus interaction model should suffice.

Now, let us look at the results of the regression analyses of a linear plus interaction model (model 1), the same model without the agitation main effect (model 2), and the model with aeration only (model 3). The regression analyses are carried out using basic R and some additional DOE functions written by the author (these DOE functions, including the R-scripts of all examples of this chapter, are available from the author upon request).

The R listing of the summary of the regression models 1, 2 and 3 are given in Tables 4-6 below. Note that values of the regression coefficients of the same effects vary a little between the models. This is due to the fact that design is not fully orthogonal. In an orthogonal design, the estimates of the same regression coefficients will not change when terms are dropped out.

	Estimate	Std. Error	t value	p value
(Intercept)	20.6205	0.4042	51.015	0.000384
X1	-3.9244	0.4454	-8.811	0.012638
X2	0.5756	0.4454	1.292	0.325404
I(X1 * X2)	-1.2744	0.4454	-2.861	0.325404
Residual standard error: 0.9541 on 2 degrees of freedom Multiple R-squared: 0.9796, Adjusted R-squared: 0.9489 F-statistic: 31.95 on 3 and 2 DF, p-value: 0.03051				

Table 4. Regression summary of model 1 (I(X1 * X2) denotes interaction between X_1 and X_2).

	Estimate	Std. Error	t value	p value
(Intercept)	20.6882	0.4433	46.668	2.17e-05
X1	-3.8397	0.4873	-7.879	0.00426
I(X1 * X2)	-1.1897	0.4873	-2.441	0.09238
Residual standard error: 1.055 on 3 degrees of freedom Multiple R-squared: 0.9625, Adjusted R-squared: 0.9375 F-statistic: 38.48 on 2 and 3 DF, p-value: 0.007266				

Table 5. Regression summary of model 2 (I(X1 * X2) denotes interaction between X_1 and X_2).

	Estimate	Std. Error	t value	p value
(Intercept)	20.5241	0.6558	31.3	6.21e-06
X1	-4.0448	0.7184	-5.63	0.0049
Residual standard error: 1.579 on 4 degrees of freedom Multiple R-squared: 0.8879, Adjusted R-squared: 0.8599 F-statistic: 38.48 on 1 and 4 DF, p-value: 0.004896				

Table 6. Regression summary of model 3.

The residual standard error is approximately 1 g/l in models 1 and 2. This seems quite high compared to the variation in replicate experiments (16.2 and 16.1 g/l) corresponding to the pure experimental pure error standard deviation of ca. 0.071 g/l. The calculations of a lack-of-fit test for model 2 are the following: The residual sum of squares (SS_{RES}) is $3 \cdot 1.055^2 = 3.339$. The pure error sum of squares (SS_E) is $1 \cdot 0.071^2 = 0.005$. The lack-of-fit sum of squares (SS_{LOF}) is 3.339-0.005 = 3.334. The corresponding mean squares are $SS_{LOF}/(df_{RES} - df_E) = SS_{LOF}/(3-1) = 3.334/2 = 1.667$ and the lack-of-fit F-statistic is $MS_{LOF}/MS_E = 1.667/0.005 = 333.4$ having 2 and 1 degrees of freedom. The corresponding p-value is 0.039 which is significant at the 0.05 level of significance. Thus, a formal lack-of-fit test exhibits significant lack-of-fit, but one must keep in mind that estimating standard deviation from only two observations is very unreliable. The lack-of-fit p-values for models 1 and 3 are 0.033 and 0.028, respectively, i.e. the lack-of-fit is least significant in model 2.

The effect of aeration (X_1) is significant in all models, and according to model 1 it is obvious that agitation doesn't have a significant effect on productivity. The interaction term is not significant in any of the models; however, it is not uncommon to include terms whose p-values are between 0.05 and 0.10 in models used for designing new experiments. The results of the new experiments would then either support or contradict the existence of an interaction.

Carrying out the leave-one-out (loo) cross-validation, gives the following Q^2 values (Table 7).

Model	R^2	Q^2
1	98.0	-22.0
2	96.2	84.5
3	88.8	68.1

Table 7. Comparison of R^2 and Q^2 values between model 1-3.

Fig. 9 shows the fitted and CV-predicted production values and the corresponding residual normal probability plots of models 1-3. By cross-validation, the model 2, i.e. $y = b_0 + b_1X_1 + b_{12}X_1X_2$, is the best one. Finally, Fig. 10 shows the contour plot of the best model, model 2.

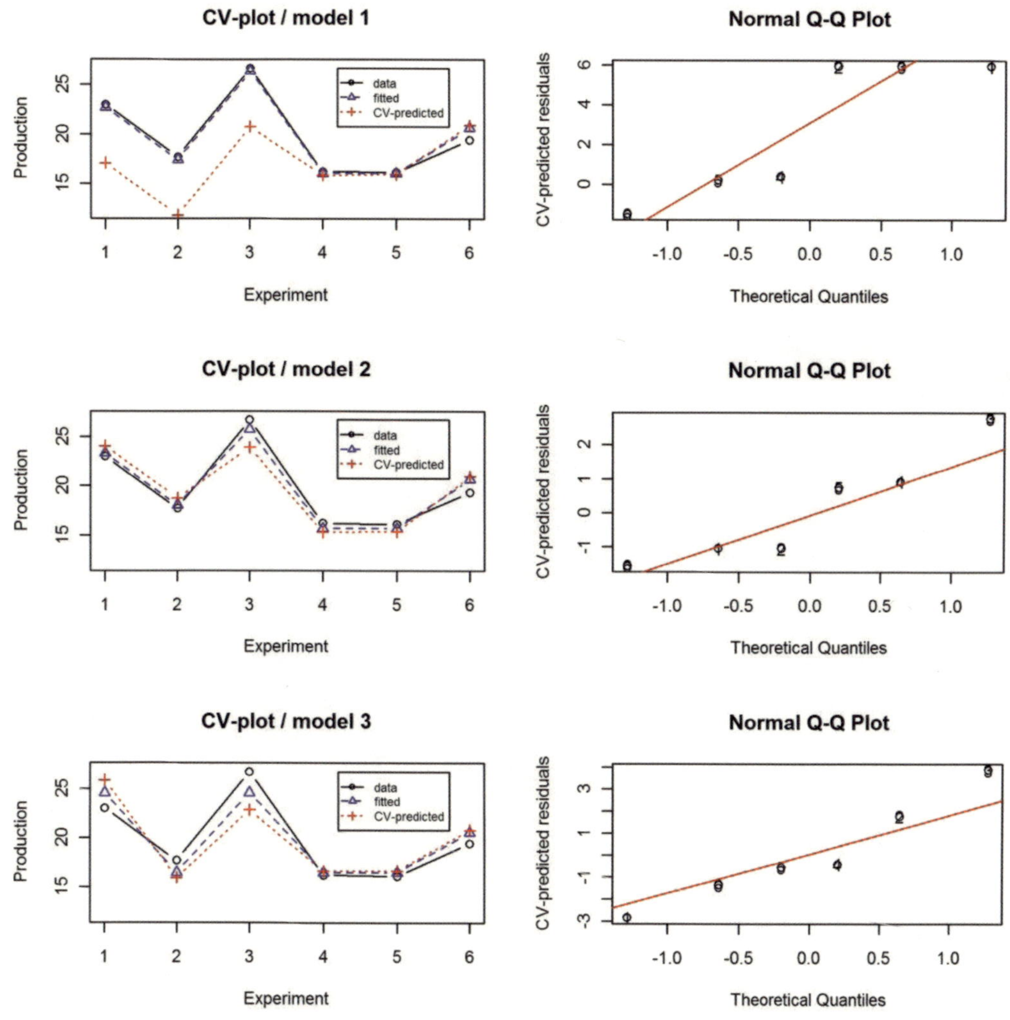

Fig. 9. Cross-validation of models 1-3. Left panel: Production vs. the number of experiment; black circles: data; blue triangles: fitted values; red pluses: cross-validated leave-one-out prediction. Right panel: Normal probability plots of the cross-validated leave-one-out residuals.

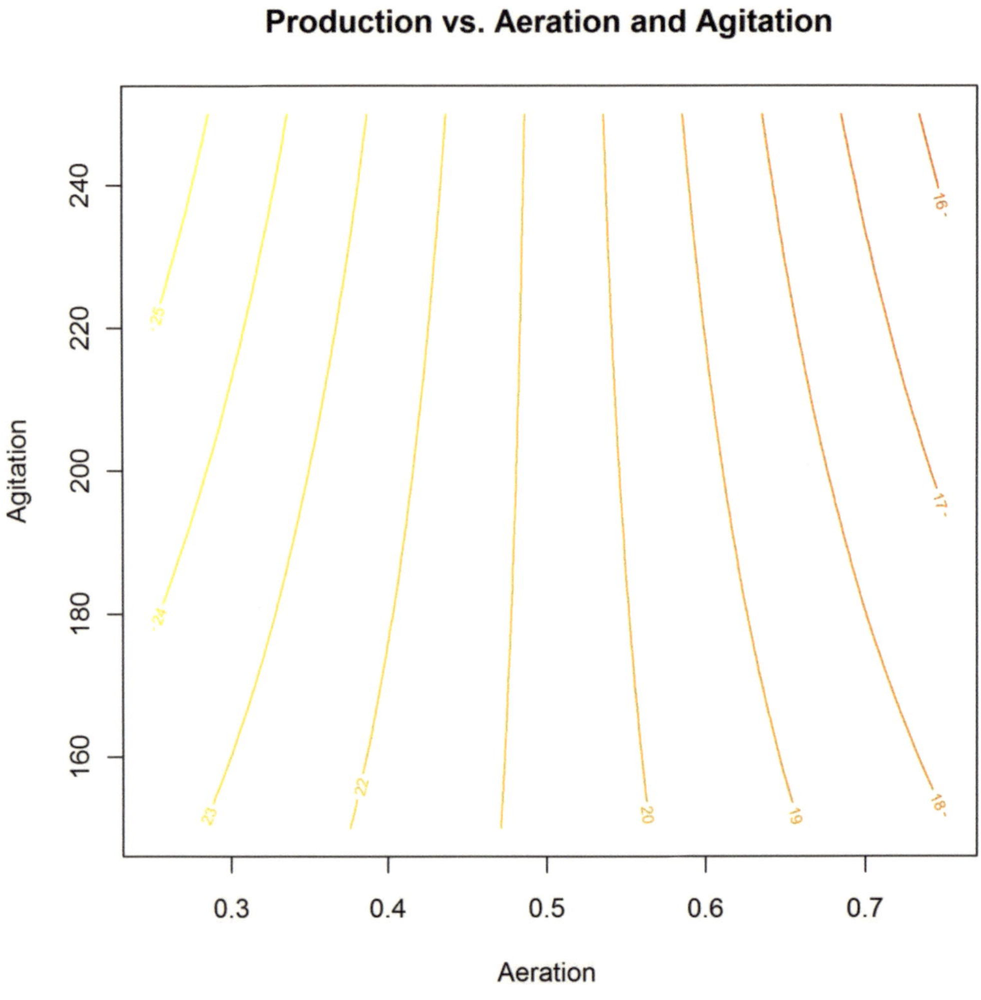

Fig. 10. Production vs. Aeration and Agitation.

3.3 Fractional 2^N designs (2^{N-p} designs)

The number of experiments in 2^N designs grows rapidly with the number of variables *N*. This problem can be avoided by choosing only part of the experiments of the full design. Naturally, using only a fraction of the full design, information is lost. The idea behind fractional 2^N designs is to select the experiments in a way that the information lost is related only to higher order interactions which seldom represent significant effects.

3.3.1 Generating 2^{N-p} designs

The selection of experiments in 2^{N-p} designs can be accomplished by using so-called generators (see e.g. Box & al., 2005). A generator is an equation between algebraic elements that represent variable effects, typically denoted by bold face numbers or upper case letters. For example **1** denotes the effect of variable 1. If there are more than 9 variables in the design, brackets are used to avoid confusion, i.e. we would use **(12)** instead of **12** to represent the effect of the variable 12. The bold face letter **I** represents the average response, i.e. the intercept of the model when coded variables are used. The generator elements (effects) follow the following algebraic rules of 'products' between the effects.

The effects are commutative, e.g. **12** = **21**
The effects are associative, e.g. **(12)3** = **1(23)**
I is a neutral element, e.g. **I2** = **2**
Even powers produce the neutral element, e.g. **22** = **I** or **2222** = **I** (naturally, for example **222** = **2**)

Now, a generator of a design is an equation between a product of effects and **I**, for example **123** = **I**. The interpretation of a product, also called a word, is that of a corresponding interaction between the effects. Thus, for example, **123** = **I** means that the third order interaction between variables 1-3 is confounded with the mean response in a design generated using this generator. Confounding (sometimes called aliasing) means that the confounding effects cannot be estimated unequivocally using this design. For example, in a design generated by **123** = **I** the model cannot contain both an intercept and a third order interaction. If the model is deliberately chosen to have both an intercept and the third order interaction term, there is no way to tell whether the estimate of the intercept really represents the intercept or the third order interaction.

Furthermore, any equation derived from the original generator, using the given algebraic rules, gives a confounding pattern. For example multiplying both sides of **123** = **I** by **1** gives **1123** = **1I**. Using the given rules this simplifies into **I23** = **1I** and then into **23** = **1**. Thus, in a design with this generator the pairwise interaction between variable 2 and 3 is confounded with variable 1. Multiplying the original generator by **2** and **3** it is easy to see that all pairwise interactions are confounded with main effects (**2** with **13** and **3** with **12**) in this design. Consequently, the only reasonable model whose parameters can be estimated unequivocally, is the main effect model $y = b_0 + b_1X_1 + b_2X_2 + b_3X_3$. Technically possible alternative models, but hardly useful in practice, would be e.g. $y = b_0 + b_1X_1 + b_2X_2 + b_{12}X_1X_2$ or $y = b_{123}X_1X_2X_3 + b_1X_1 + b_2X_2 + b_3X_3$.

A design can be generated using more than one generator. Each generator halves the number of experiments. For example, a design with two generators has only ¼ of the original number of experiments in the corresponding full 2^N design. If p is the number of generators, the corresponding fractional 2^N design is denoted by 2^{N-p}.

In practice, 2^{N-p} designs are constructed by first making a full 2^N design table and then adding columns that contain the interaction terms corresponding to the generator words. Then only those experiments (rows) are selected where all interaction terms are +1. Alternatively one can choose the experiments where all interaction terms are -1. As an example, let us construct a 2^{3-1} design with the generator **123** = **I**. The full design table with an additional column containing the three-way interaction term is given in Table 8.

x^1	x^2	x^3	$x_1x_2x_3$
-1	-1	-1	-1
-1	-1	+1	+1
-1	+1	-1	+1
-1	+1	+1	-1
+1	-1	-1	+1
+1	-1	+1	-1
+1	+1	-1	-1
+1	+1	+1	+1

Table 8. A table for constructing a 2^{3-1} design.

Now, the desired design table is obtained by deleting the rows 1, 4, 6 and 7. An alternative design is obtained by deleting the rows 2, 3, 5 and 8.

3.3.2 Confounding (aliasing) and resolution

An important concept related to 2^{N-p} designs is the resolution of a design, denoted by roman numerals. Technically, resolution is the minimum word length of all possible generators derived from the original set of generators. For design with a single generator, finding out the resolution is easy. For example, the resolution of the 2^{3-1} design with the generator **123** = **I** is III because the length of the word **123** is 3 (note that e.g. **(12)** would be counted as a single letter in a generator word). If there are more generators than one, the situation is more complicated. For example, if the generators in a 2^{5-2} design were **1234** = **I** and **1235** = **I,** then the equation **1234** = **1235** would be true which after multiplying both sides **1235** gives **45** = **I**. Thus the resolution of this design would be II. Naturally, this would be a really bad design with confounding main effects.

The interpretation of the resolution of a design is (designs of resolution below III are normally not used)

- If the resolution is III, only a main effect model can be used
- If the resolution is IV, a main effect model with half of all the pairwise interaction effects can be used
- If the resolution is V or higher, a main effect model with all pairwise interaction effects can be used

If the resolution is higher than V also at least some of the higher order interaction can be estimated. There are many sources of tables listing 2^{N-p} designs and their confounding patterns, e.g. Table 3.17 in (NIST SEMATCH). Usually these tables give so-called minimum aberration designs, i.e. designs that minimize the number of short words in all possible generators of a design with given N and p.

3.3.3 Example

This example is taken from (Box & Draper, 2007) (Example 5.2 p. 189), but the analysis is not completely identical to the one given in the book.

The task was to improve the yield (y) (in percentage) of a laboratory scale drug synthesis. The five design variables were the reaction time (t), the reactor temperature (T), the amount of reagent B (B), the amount of reagent C (C), and the amount of reagent D (D). The chosen design levels in a two level fractional factorial design are given in Table 9 below.

Coded	Original	Lower (-1)	Upper (+1)	Formula
X_1	t	6 h	10 h	$X_1 = \frac{t-8}{2}$
X_2	T	85°C	90°C	$X_2 = \frac{T-87.5}{2.5}$
X_3	B	30 ml	60 ml	$X_3 = \frac{B-45}{15}$
X_4	C	90 ml	115 ml	$X_4 = \frac{C-102.5}{12.5}$
X_5	D	40 g	50 g	$X_5 = \frac{D-45}{5}$

Table 9. The variable levels of example 3.3.3.

The design was chosen to be a fractional resolution V design (2^{5-1}) with the generator **I = 12345**. The design table in coded units, including the yields and the run order of the experiments is given in Table 10 (y stands for the yield).

order	X_1	X_2	X_3	X_4	X_5	y
16	-1	-1	-1	-1	1	51.8
2	1	-1	-1	-1	-1	56.3
10	-1	1	-1	-1	-1	56.8
1	1	1	-1	-1	1	48.3
14	-1	-1	1	-1	-1	62.3
8	1	-1	1	-1	1	49.8
9	-1	1	1	-1	1	49.0
7	1	1	1	-1	-1	46.0
4	-1	-1	-1	1	-1	72.6
15	1	-1	-1	1	1	49.5
13	-1	1	-1	1	1	56.8
3	1	1	-1	1	-1	63.1
12	-1	-1	1	1	1	64.6
6	1	-1	1	1	-1	67.8
5	-1	1	1	1	-1	70.3
11	1	1	1	1	1	49.8

Table 10. The design of example 3.3.3 in coded units.

Since the resolution of this design is V, we can estimate a model containing linear and pair-wise interaction effects. However the design is saturated with respect to this model, and thus the model cannot be validated by statistical tests, or by cross-validation. The regression summary is given Table 11.

	Estimate	Std. Error	t value	p value
(Intercept)	57.1750	NA	NA	NA
t	-3.3500	NA	NA	NA
T	-2.1625	NA	NA	NA
B	0.2750	NA	NA	NA
C	4.6375	NA	NA	NA
D	-4.7250	NA	NA	NA
I(t * T)	0.1375	NA	NA	NA
I(t * B)	-0.7500	NA	NA	NA
I(T * B)	-1.5125	NA	NA	NA
I(t * C)	-0.9125	NA	NA	NA
I(T * C)	0.3500	NA	NA	NA
I(B * C)	1.0375	NA	NA	NA
I(t * D)	0.2500	NA	NA	NA
I(T * D)	0.6875	NA	NA	NA
I(B * D)	0.5750	NA	NA	NA
I(C * D)	-1.9125	NA	NA	NA
Residual standard error: NaN on 0 degrees of freedom Multiple R-squared: 1, Adjusted R-squared: NaN F-statistic: NaN on 15 and 0 DF, p-value: NA				

Table 11. Regression summary of the linear plus pairwise interactions model. NA stands for "not available".

Because the design is saturated with respect to the linear plus pairwise interactions model there are no degrees of freedom for any regression statistics. Therefore, for selecting the significant terms we have to use either a normal probability plot of the estimated values of the regression coefficient or variable selection techniques. We chose to use forward selection based on the Q^2 value. This technique gave the maximum Q^2 value in a model with 4 linear terms and 7 pairwise interaction terms. However, after 6 terms the increase in the Q^2 value is minimal, and in order to avoid over-fitting we chose to use the model with 6 terms. The chosen terms were the main effects of t, T, C and D, and the interaction effects between C and D and between T and B. This model has a Q^2 value 83.8 % and the regression summary for this model is given in Table 12.

All terms in the model are now statistically significant at 5 % significance level, and the predictive power of the model is fairly good according the Q^2 value . Section 4.3 shows how this model has been used in search for improvement.

3.4 Plackett-Burman (screening) designs

If the number of variables is high, and the aim is to select the most important variables for further experimentation, usually only the main effects are of interest. In such cases the most cost effective choice is to use designs that have as many experiments as there are parameters in

	Estimate	Std. Error	t value	p value
(Intercept)	57.1750	0.6284	90.980	1.19e-14
t	-3.3500	0.6284	-5.331	0.000474
T	-2.1625	0.6284	-3.441	0.007378
C	4.6375	0.6284	7.379	4.19e-05
D	-4.7250	0.6284	-7.519	3.62e-05
I(T * B)	-1.5125	0.6284	-2.407	0.039457
I(C * D)	1.9125	0.6284	-3.043	0.013944
Residual standard error: NaN on 0 degrees of freedom Multiple R-squared: 1, Adjusted R-squared: NaN F-statistic: NaN on 15 and 0 DF, p-value: NA				

Table 12. Regression summary of the 6 terms model.

the corresponding main effect model, i.e. $N+1$ experiments. It can be proved that such designs that are also orthogonal exist in multiples of 4, i.e. for 3, 7, 11, … variables having 4, 8, 12, … experiments respectively. The ones in which the number of experiments is a power of 2 are actually 2^{N-p} designs. Thus for example a Plackett-Burman design for 3 variables that has $8 = 2^3$ experiments is a 2^{3-1} design. General construction of Plackett-Burman designs is beyond the scope of this chapter. The interested reader can refer to e.g. section 5.3.3.5 in (NIST SEMATECH). Plackett-Burman designs are also called 2-level Taguchi designs or Hadamard matrices.

3.5 Blocking

Sometimes uncontrolled factors, such as work shifts, raw material batches, differences in pieces of equipment, etc., may affect the results. In such cases the effects of such variables should be taken into account in the design. If the design variables are qualitative, such classical designs as randomized blocks design, Latin square design, or Graeco-Latin square design can be used, see e.g. (Montgomery, 1991). If the design variables are quantitative, a common technique is to have extra columns (variables) for the uncontrolled variables. For 2^N and CC-designs, tables of different blocking schemes exist, see e.g. section 5.3.3.3.3. in (NIST SEMATECH).

3.6 Sizing designs

An important issue in DOE is the total number of experiments, i.e. the size of a design. Sizing can be based on predictive power, or on the power of detecting differences of predefined size Δ. The latter is more commonly used, and many commercial DOE software packages have tools for determining the required number of estimates in such a way that the statistical power, i.e. $1-\beta$ (β is the probability of type II error), has a desired value at a given level of significance α. For pairwise comparisons, exact methods based on the non-central t-distribution exist. For example, in R the function called power.t.test can be used to find the number of experiments needed in pairwise comparisons. For multiple comparisons, one can use the so-called Wheeler's formula (Wheeler, 1974) for an estimate of the required number of experiments n: $n = (4r\sigma / \Delta)^2$ where r is the number of levels of a factor, σ is the experimental standard deviation, and Δ is size of the difference. The formula assumes that the level of significance α is 0.05, and the power $1-\beta$ is 0.90. Wheeler gives also formulas for several other common design/model combinations (Wheeler, 1974).

4. Improving results by steepest ascent

If the goal of the experimentation has been to optimize something, the next step after analysing the results of a 2^N or a fractional 2^N design is to try to make improvement using knowledge provided by the analysis. The most common technique is the method of steepest ascent, also called the gradient (path) method.

4.1 Calculating the gradient path

It is well known from calculus that the direction of the steepest ascent on a response surface is given by the gradient vector, i.e. the vector of partial derivatives with respect to the design variables at a given point. The basic idea has been presented in section 3.2, and now we shall present the technical details.

In principle, the procedure is simple. First we choose a starting point, say $\mathbf{X}_0$, which typically is the centre point of the design. Then we calculate the gradient vector, say ∇ at this point. Note that it is important to use coded variables in gradient calculations. Next, the gradient vector has to be scaled small enough in order to avoid zigzagging (see 2.2). This can be done by multiplying the corresponding unit vector, $\nabla^0 = \nabla / \|\nabla\|$, by a scaling factor, say c. Now, the gradient path points are obtained by calculating $\mathbf{X}_i = \mathbf{X}_{(i-1)} + c\nabla^0, i = 1,2,\ldots,n$ where n is the number of points. Once the points have been calculated, the experimental points are chosen from the path so that the distance between the points matches the desired step size, typically 1 in coded units. Naturally, the coded values have to be decoded into physical values having the original units before experimentation.

4.2 Alternative improvement techniques

Another principle in searching optimal new experiments is to use direct optimization techniques using the current model. In this approach, first the search region has to be defined. There are basically two different alternatives: 1) a hypercube whose centre is at the design centre with a given length for the sides of the hypercube, or 2) a hypersphere whose centre is at the design centre with a given radius. In the first alternative, typically the length of the side is first set to a value slightly over 2, say 3, giving mild extrapolation outside the experimental region. In the latter, typically the length of the radius is first set to a value slightly over 1, say 1.5, giving mild extrapolation outside the experimental region.

If the model is a linear plus pair-wise interactions model, the solution can easily be shown to be one of the vertices of the hypercube in the hypercube approach. If the model is a quadratic one, and the optimum (according to the model) is not inside the hypercube, the solution is a point on one of the edges of the hypercube and a point on the hypersphere in the hypersphere approach. In both approaches, the solution is found most easily using some iterative constrained optimization tool, e.g. Excel's Solver Tool. In the latter (hypersphere) approach, it is easy to show, using the Lagrange multiplier technique of constrained optimization, that the optimal point $\mathbf{X}_{opt}$ on the hypersphere of radius r is obtained by $\mathbf{X}_{opt} = (\mathbf{B} + 2\lambda\mathbf{I})^{-1}\mathbf{b}$, where λ is solved from the equation $(\mathbf{B} + 2\lambda\mathbf{I})^{-1}\mathbf{b}^2 = r^2$. The notation is explained in section 5.2. Unfortunately, λ must be solved numerically unless the model is linear. The benefit of using (numerical) iterative optimization in both approaches, or using the gradient path technique, is that they all work for all kind of models, not only for quadratic ones.

4.3 Example

Let us continue with the example of section 3.3.3 and see how the model can be used to design new experiments along the gradient path. The model of the previous example can be written (in coded variables)

$$y = b_0 + b_1X_1 + b_2X_2 + b_4X_4 + b_5X_5 + b_{23}X_2X_3 + b_{45}X_4X_5 \tag{7}$$

The coefficients (b's) refer to the values given in Table 12. The gradient, i.e. the direction of steepest ascent, is the vector of partial derivatives of the model with respect to the variables. Differentiating the expression given in Eq. 7 gives in matrix notation

$$\nabla = \begin{bmatrix} b_1 & b_2 + b_{23}X_3 & b_{23}X_2 & b_4 + b_{45}X_5 & b_5 + b_{45}X_4 \end{bmatrix}^T \tag{8}$$

Because this is a directional vector, it can be scaled to have any length. If we want it to have unit length, it must be divided by its norm, i.e. we use $\nabla / \|\nabla\|$. Now, let us start the calculation of the gradient path from the centre of the design, where all coded values are zeros. Substituting numerical values into Eq. 8 gives

$$\nabla \approx \begin{bmatrix} -3.35 -2.16 & 0.00 & 4.64 & -4.72 \end{bmatrix}^T \tag{9}$$

The norm of this vector is ca. 7.73. Dividing Eq. 9 by its norm gives

$$\frac{\nabla}{\|\nabla\|} \approx \begin{bmatrix} -0.43 -0.28 & 0.00 & 0.60 & -0.61 \end{bmatrix}^T \tag{10}$$

These are almost the same values as in the example 6.3.2 in (Box & Draper, 2007) though we have used a different model with significant interaction terms included. The reason for this is that the starting point is the centre point where the interaction terms vanish because the coefficients are multiplied by zeros.

The vector of Eq. 10 tells us that we should decrease the time by 0.43 coded units, the temperature by 0.28 coded units, and the amount of reagent D by 0.61 coded units and increase the amount of reagent C by 0.60 coded units. Of course, there isn't much sense to carry out this experiment because it is inside the experimental region. Therefore we shall continue from this point onwards in the direction of the gradient. Now, because of the interactions, we have to recalculate the normed gradient at the new point where $X_1 = -0.43$, $X_2 = -0.28$, $X_3 = 0.00$, $X_4 = 0.60$, and $X_5 = -0.61$.

When this is added to the previous values, we get $X_1 = -0.80$, $X_2 = -0.52$, $X_3 = 0.05$, $X_4 = 1.23$, and $X_5 = -1.25$. These values differ slightly more from the values in the original source; however the difference has hardly any significance. The difference becomes more substantial if we continue the procedure because the interactions start to bend the gradient path. Box and Draper calculated the new points at distances 2, 4 , 6 and 8 in coded units from the centre point. If we do the same we get the points given in Table 13.

Distance	X_1	X_2	X_3	X_4	X_5
2	-0.80	-0.52	0.05	1.23	-1.25
4	-1.38	-0.91	0.21	2.55	-2.57
6	-1.82	-1.25	0.40	3.90	-3.93
8	-2.18	-1.55	0.62	5.26	-5.23

Table 13. Four new experiments (in coded units) along the gradient path.

For comparison, the values in the book together with the reported yields of these experiments are given in Table 14.

Distance	X_1	X_2	X_3	X_4	X_5	yield
2	-0.86	-0.56	0.08	1.20	-1.22	72.6
4	-1.72	-1.12	0.16	2.40	-2.44	85.1
6	-2.58	-1.68	0.24	3.60	-3.66	82.4
8	-3.44	-2.24	0.32	4.80	-4.88	80.8

Table 14. Four new experiments (in coded units) along the gradient path given in (Box & Draper, 2007).

The differences in the design variables in the last two rows start to be significant, but unfortunately we cannot check whether they had been any better than the ones used in the actual experiments. The actual experiments really gave substantial improvement; see Example 6.3.2 in (Box & Draper, 2007).

Before going to quadratic designs and models, let us recall what was said about calculation step in section 3.2 and let us calculate the gradient using a step size 0.1 instead of 1.0, but tabulating only those points where the sum of the steps is two, i.e. the arc length along the path between two sequential points is approximately 2. These points are given in Table 15.

Distance	X_1	X_2	X_3	X_4	X_5
2	-0.74	-0.48	0.08	1.26	-1.27
4	-1.28	-0.87	0.24	2.58	-2.61
6	-1.70	-1.20	0.43	3.94	-3.96
8	-2.04	-1.50	0.64	5.30	-5.34

Table 15. Four new experiments (in coded units) along the gradient path using a small step size in the gradient path calculation.

If you compare these values with our first table, the differences are not big. The reason is that the model has not quadratic terms and the zigzag effect, explained in section 3.2, would take place only with really large step sizes. In any case, the best way to do these calculations is to use appropriate software, and then it doesn't matter if you calculate more accurately using a small step size.

Of course, before experimentation, one has to convert the coded units back to physical units. This could be easily done by solving for the variables in physical units from the equations given in the column Formula in Table 9. However, the easiest way is to use appropriate software. Table 16 gives the values in Table 15 in physical units.

Distance	t	T	B	C	D
2	6.5	86.3	46.1	118.2	38.6
4	5.4	85.3	48.6	134.8	32.0
6	4.6	84.5	51.5	151.7	25.2
8	3.9	83.7	54.6	168.8	18.3

Table 16. Experiments of Table 15 in physical units.

5. Second and higher order designs and response surface modelling

When the response doesn't depend linearly on the design variables, two-level designs are not adequate. Nonlinear behaviour can typically be detected by comparing the centre point results with the actual 2^N design point results. Alternatively, a nonlinear region is found by steepest ascent experiments. Sometimes nonlinearity can be assumed by prior knowledge about the system under study. There are several alternative designs for empirical nonlinear modelling. Before going to the different design alternatives let us review the most common nonlinear empirical model types. The emphasis is on so-called quadratic models, commonly used in the Box-Wilson strategy of empirical optimization. We shall first introduce typical models used with these designs, and after that, introduce the most common designs used for creating such models.

5.1 Typical nonlinear empirical models

The most common nonlinear empirical model is a second order polynomial of the design variables, often called a quadratic response surface model, or simply, a quadratic model. It is a linear plus pairwise interactions model added with quadratic terms, i.e. design variables raised to power 2. For example, a quadratic model for two variables is $y = b_0 + b_1X_1 + b_2X_2 + b_{12}X_1X_2 + b_{11}X_1^2 + b_{22}X_2^2$. In general, we use the notation that b_i is the coefficient of X_i, b_{ii} is the coefficient of X_i^2, and $b_{ij}, i<j$ is the coefficient of X_iX_j. Fig. 11 depicts typical quadratic surfaces of two variables X_1 and X_2. Now, let **B** be a matrix whose diagonal elements B_{ii} are defined by $B_{ii} = 2b_{ii}$, and the other elements $B_{ij}, i \neq j$ are defined by $B_{ij} = b_{ij}, i<j$ and $B_{ji} = b_{ij}, i<j$. By definition, the matrix **B** is symmetric. Also, let **b** be the vector of main effect coefficients, i.e. $\mathbf{b} = [b_1, b_2, \cdots, n_N]^T$. Using this notation, a quadratic model can expressed in matrix notation as $y = b_0 + \mathbf{x}^T\mathbf{b} + \frac{1}{2}\mathbf{x}^T\mathbf{B}\mathbf{x}$ where **x** is the vector $[X_1, X_2, \cdots, X_N]^T$.

If a quadratic model has more than 2 variables, any 2 variables can be chosen as free variables corresponding to the x and y axes of the plot, and the other variables are kept at constant levels. Varying the values of the other variables in a systematic way, a good overview of the dependencies can be obtained till up to 4 or 5 variables. With more variables, one must rely on computational techniques.

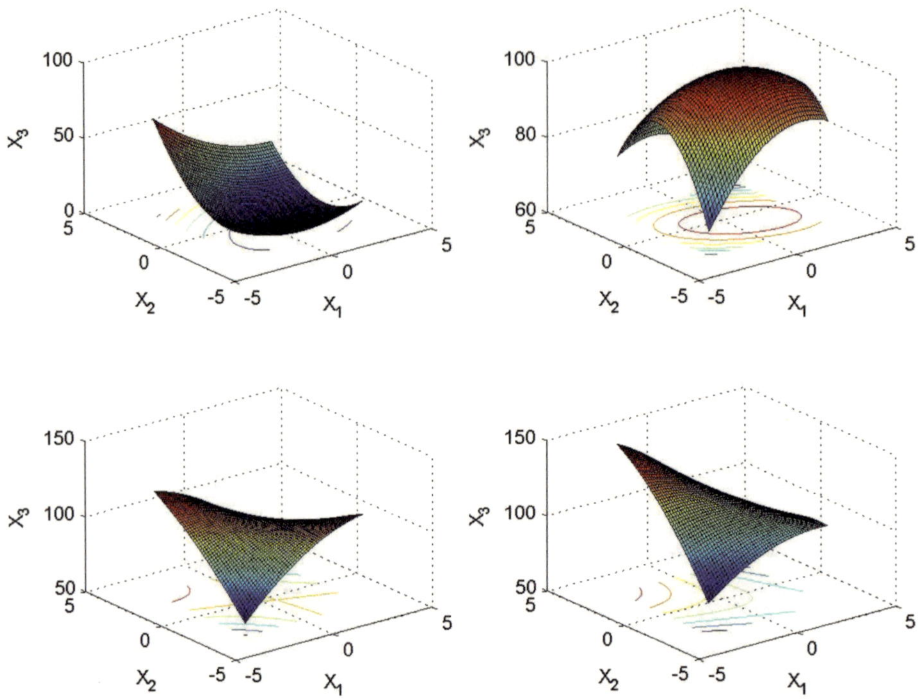

Fig. 11. Typical quadratic surfaces of 2 variables (X_1 and X_2): a surface having a minimum (upper left), a surface having a maximum (upper right), a surface having a saddle point (lower left), and a surface having a ridge (lower right).

Other model alternatives are higher order polynomials, rational functions of several variables, nonlinear PLS, neural networks, nonlinear SVM etc. With higher order polynomials, or with linearized rational functions, it advisable to use ridge regression, PLS, or some other constrained regression technique, see e.g. (Taavitsainen, 2010). These alternatives are useful typically in cases where the response is bounded in the experimental region; see e.g. (Taavitsainen et. al., 2010).

5.2 Estimation and validation of nonlinear empirical models

Basically the analyses and techniques presented in sections 3.1.2 and 3.2 are applicable to nonlinear models as well. Actually, polynomial models are linear in parameters, and thus the theory of linear regression applies. Normally, nonlinear regression refers to regression analysis of models that are nonlinear in parameters. This topic is not treated in this chapter, and the interested reader may see e.g. (Bard, 1973)

It should be noted that some of the designs presented in section 5.3 are not orthogonal, and therefore PLS or ridge regression are more appropriate methods than OLS for parameter estimation, especially in so-called mixture designs.

For quadratic models, a special form of analysis called canonical analysis is commonly used for gaining better understanding of the model. However, this topic is beyond the scope of this chapter, and the reader is advised to see e.g. (Box & Draper, 2007). Part of the canonical analysis is to calculate the so called stationary point of the model. A stationary point is a point where the gradient with respect to the design variables vanishes. Solving for the stationary point is straightforward. The stationary point is the solution of the linear system of equations $\mathbf{Bx} = -\mathbf{b}$, obtained by differentiation from the model in matrix form given in section 5.1. A stationary point can represent a minimum point, a maximum point, or a saddle point depending on the model coefficients.

5.3 Common higher order designs

Next we shall introduce the most common designs used for response surface modelling (RSM).

5.3.1 Factorial M^N designs

Full factorial designs with M levels can be used for estimating polynomials of order at most $M-1$. Naturally, these designs are feasible only with very few variables, say maximum 3, and typically for only few levels, say at most 4. For example, a 4^4 design would contain 256 which would be seldom feasible. However, the recent development in parallel microreactor systems having e.g. 64 simultaneously operating reactors at different conditions can make such designs reasonable.

5.3.2 Fractional factorial M^N designs, and mixed level factorial design.

Sometimes it is known that one or more variables act nonlinearly and the others linearly. For such cases a mixed level factorial design is a good choice. A simple way to construct e.g. a 3 or a 4 level mixed level factorial design is to combine a pair of variables in a 2^N design into a single new variable (Z) having 3 or 4 levels using the coding given in Table 17 (x_1, x_2, x_3 and x_4 represent the levels of the variable constructed from a pair of variables (X_i, X_j) in the original 2^N design).

X_i	X_j	Z (3 levels)	Z (4 levels)
-1	-1	x_1	x_1
-1	+1	x_2	x_2
+1	-1	x_2	x_3
+1	+1	x_3	x_4

Table 17. Construction of a 3, or 4 level variable from two variables of a 2^N design.

There are also fractional factorial designs which are commonly used in Taguchi methodology. The most common such designs are the so-called Taguchi L9 and L27 orthogonal arrays, see e.g. (NIST SEMATECH).

5.3.3 Box-Behnken designs

The structure and construction of Box-Behnken designs (Box & Behnken, 1960) is simple.

First, a 2^{N-1} design is constructed, say $\mathbf{X_0}$, then a $N \cdot 2^{N-1}$ by N matrix of zeros $\mathbf{X}$ is created. After this, $\mathbf{X}$ is divided into N blocks of 2^{N-1} rows and all columns, and in each block the columns, omitting the i'th column, is replaced by $\mathbf{X_0}$. Finally one or more rows of N zeros are appended to $\mathbf{X}$. This is easy to program e.g. in Matlab or R starting from a 2^{N-1} design. The following R commands will do the work for any number of variables (mton is a function that generates M^N designs, and nrep is the number of replicates at the centre point):

```
X0 <- as.matrix(mton(2,N-1))
M  <- 2^(N-1)
X  <- matrix(0,N*M,N)
for(i in 1:N) X[((i-1)*M+1):(M*i),(1:N)[-i]] <- X0
X  <- rbind(X,rep(nrep,N))
```

As an example, Table 18 shows a Box-Behnken design of 3 variables and 3 centre point replicate experiments.

X_1	X_2	X_3
0	-1	-1
0	-1	+1
0	+1	-1
0	+1	+1
-1	0	-1
-1	0	+1
+1	0	-1
+1	0	+1
-1	-1	0
-1	+1	0
+1	-1	0
+1	+1	0
0	0	0
0	0	0
0	0	0

Table 18. A Box-Behnken design with 3 variables.

5.3.4 Central composite designs

The so-called central composite (CC) designs are perhaps the most common ones used in RSM, perhaps due their simple structure (for other possible reasons, see section 5.3). As the name suggests they are composed of other designs, namely, of a factorial or fractional 2^N

part, of so-called axial points, and of centre points. Sometimes the latter two parts together are called a star design. As an example, Table 19 shows a CC design for two variables.

X_1	X_2	
-1	-1	Factorial 2^2 part
-1	+1	
+1	-1	
+1	+1	
$-\alpha$	0	Axial points
0	$-\alpha$	
$+\alpha$	0	
0	$+\alpha$	
0	0	Centre points
0	0	

Table 19. A CC design with 2 variables.

The value α depends on the kind of properties we want the design to have. Typical desired properties are orthogonality, rotatability, and symmetry. A rotatable design is such that the prediction variance of a point in the design space does depend only on its distance from. design centre, not on its direction. Let us denote the number of the centre points by N_{cp}. Then, for an orthogonal design α is given by the following Eq. 11.

$$\alpha = \left[\left(\sqrt{2^N + 2N + N_{cp}} - \sqrt{2^N} \right)^2 \cdot \frac{2^N}{4} \right]^{\frac{1}{4}} \tag{11}$$

The derivation of this rather formidable looking equation, and of the two following ones, are given e.g. in (Box & Draper, 2007). It should be noted that the model matrix obtained with this choice for α is not strictly orthogonal, because the intercept column (the column of ones) vector is not orthogonal to the column vector of the quadratic terms. However, all the other columns of the model matrix are orthogonal to each other. This can also be expressed by saying that the quadratic effects are partially confounded with the intercept.

For a rotatable design, the appropriate value for α is given by the equation Eq. 12.

$$\alpha = 2^{\frac{N}{4}} \tag{12}$$

For maximal symmetry, i.e. all points except for the centre point, lie on a hypersphere of radius α, the appropriate value for α is given by the equation Eq. 13

$$\alpha = \sqrt{N} \tag{13}$$

A common fourth choice is to set α to 1. Such CC design is called a face centred CC design (CCF). For α's greater than 1 the designs are called circumscribed, CCC. Some other alternatives, e.g. compromising between orthogonality and rotatability, exist too. Sometimes a CCC design is scaled so that α is scale to 1, and the coordinates of the factorial points are scaled to $1/\alpha$. Such designs are called inscribed (CCI), though they actually are CCC designs with a different coding of variables.

Figs. 12 and 13 depict a CCC and a CCF design of 3 variables.

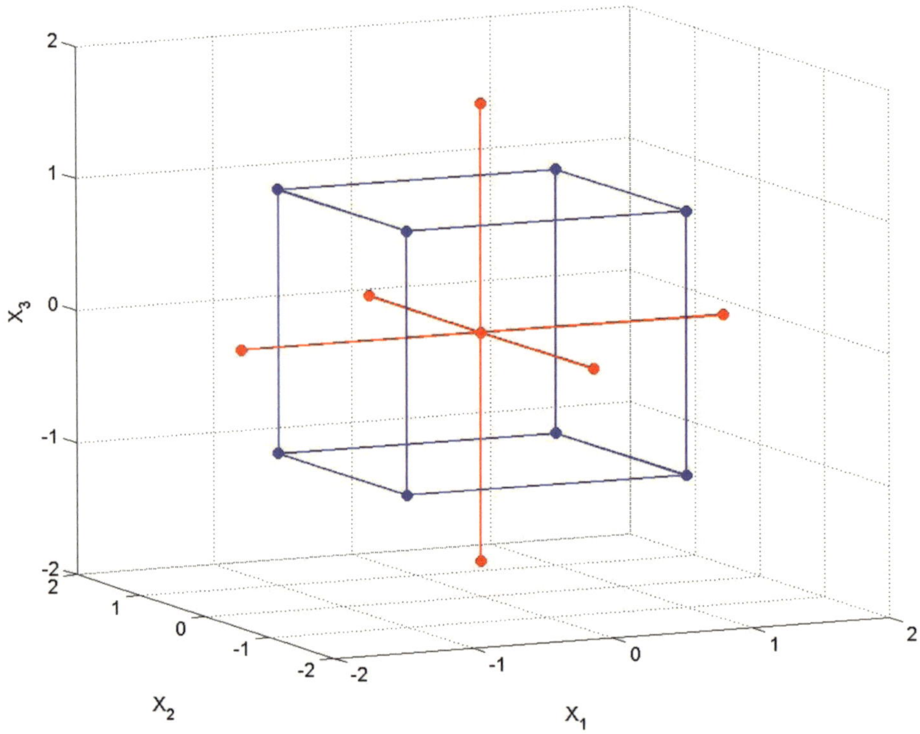

Fig. 12. A CCC design in coded units for 3 variables (X_1 , X_2 and X_3) with $\alpha = 2^{\frac{3}{4}}$.

5.3.5 Doehlert designs

Doehlert designs are constructed from so-called regular simplexes. For example, a regular triangle and a regular tetrahedron represent regular simplexes in 2D and 3D, respectively. A Doehlert design for two variables consists of the vertexes of 6 adjacent regular triangles. Thus it comprises the vertexes of a regular hexagon plus the centre point. Doehlert designs fill the experimental space in a regular way in the sense that distances between the experimental points are constant. Doehlert designs have $1+N+N^2$ experimental points, which is less than in CC designs. Thus they are typically used in cases where the experiments are either very expensive or time consuming. The interested reader may refer to e.g. (Doehlert, 1970). Construction of Doehlert designs for more than 2 variables is rather tedious, and use of appropriate software, or tables of Doehlert designs, are recommended, see e.g. (Bruns & al, 2006)

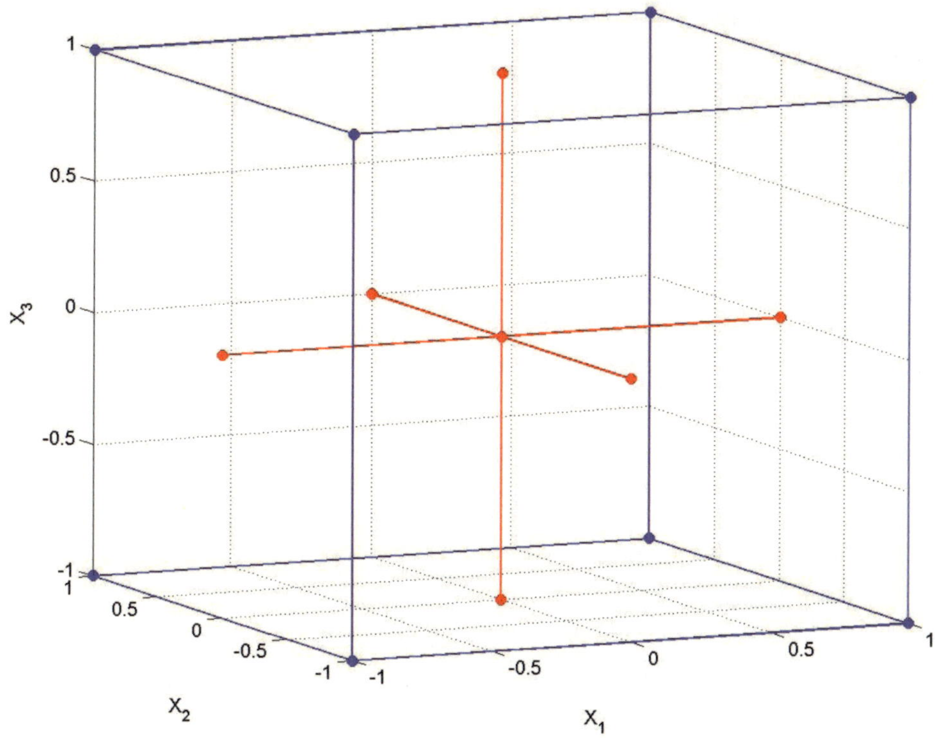

Fig. 13. A CCF design in coded units for 3 variables (X_1 , X_2 and X_3) with $\alpha = 1$.

5.3.6 Mixture designs

If the design variables are proportions of constituents in a mixture, then in each experiment the values of the design variables sum up to 1 (100 %). In such cases, ordinary designs cannot be applied, since the row sums of ordinary designs vary irrespective of the coding used. If there are no other constraints than the closure constraint, the most common designs are the so-called simplex lattice, and simplex centroid designs. If some constraints are imposed as well, a good choice is to use optimal designs (see the next section), though other alternatives exist as well; see e.g. (Cornell, 1981), or (Montgomery, 1991).

The closure constraint has to be taken into account also in modelling results of mixture experiments. The closure means that the columns of the model matrix are linearly dependent making the matrix singular. One way to overcome this problem is to make the model using only $N-1$ variables, because we need to know only the values of $N-1$ variables, and the value of the N'th variable is one minus the sum of the others. However, this may make the interpretation of the model coefficients quite difficult. Another alternative is to use the so-called Scheffe polynomials, i.e. polynomials without the intercept and the quadratic terms. It can be shown that Scheffe polynomials of N variables represent the same model as an ordinary polynomial of $N-1$ variables, naturally with different values for the polynomial coefficients. For example the quadratic polynomial of two variables $y = b_1X_1 + b_2X_2 + b_{12}X_1X_2$ can be simplified into $y = b_2 + (b_1 - b_2 + b_{12})X_1 - b_{12}X_1^2$ if $X_2 = 1 - X_1$. This shows that it is a quadratic function of X_1 only; for more details see e.g. (Cornell, 1981).

The model matrices of mixture designs are not orthogonal, and they are usually quite ill-conditioned. For this reason, it is commonly recommended to use PLS or ridge regression for estimating the model parameters.

5.3.7 Optimal designs

The idea behind so-called optimal designs is to select the experimental points so that they satisfy some optimality criterion about the model to be used. It is important to notice that the optimality of such designs is always dependent on the model. For this reason, optimal designs are often used in designing experiments for mechanistic modelling problems. In empirical modelling we don't know the model representing the 'true' behaviour, and even a good empirical model is just an approximation of the true behaviour. Of course, if it has been decided to use e.g. a quadratic approximation, using a design that is optimal for a quadratic model is perfectly logical. However, the design still should have extra experiments that allow assessing the lack-of-fit.

Typically optimal designs are planned for quadratic models. Probably the most common optimality criterion is the D-optimality criterion. A D-optimal design is a design that minimizes the determinant of the information matrix, i.e. $\left|\mathbf{X}^T\mathbf{X}\right|^{-1}$ where $\mathbf{X}$ is the model matrix. There are several other optimality criteria, typically related to minimizing the variance of predictions, or to minimizing the variances of the model parameter estimates. In many cases, a design that is optimal according to one criterion is also optimal or nearly optimal according to several other criteria as well.

A nice feature in optimal designs is that it is easy take into account constraints in the design space, e.g. a mixture constraint, or a constraint in which one variable always has to have a greater value than some other variable. Constraints can sometimes be handled by some 'tricks', e.g. instead of using x_1 and x_2 when $x_1 < x_2$, one could use in design x_1 and x_3 and set $x_2 = x_1 + x_3$, i.e. to use a variable that tells how much greater to the value of x_1 the value of x_2 is. In general, using optimal designs is the most straightforward approach for constrained problems.

In practice, constructing optimal designs requires suitable software. Optimal design routines are available in most commercial statistical software packages containing tools for DOE. There is also an R package for creating optimal designs, called AlgDesign (http://cran.r-project.org/web/packages/AlgDesign/index.html). See also (Fedorov, 1972) or (Atkinson et. al., 2007).

5.4 Choosing an appropriate second order design

As we have seen, there are many types of designs for nonlinear empirical (usually quadratic) models. How does a practitioner know which one to choose? A good strategy is to try first a simple design that has extra degrees of freedom for validation and for checking model adequacy. Of course, if the problem at hand is a mixture problem, one has to rely on mixture designs or optimal designs. If the experiments are very expensive, one may have to use saturated, or almost saturated designs, e.g. optimal designs or Doehlert designs. In other cases CC or Box-Behnken designs are better choices. For 3 variables, a Box-Behnken design contains fewer experiments than a CC design for 3 variables, but for more variables it is the other way round. For example, a 4 variable Box-Behnken design (without replicates) contains 33 experiments, as the corresponding CC design contains 25 experiments. Thus, except for mixture problems or constrained problems, a CC design is usually the best choice. In general, CCC designs should be preferred to CCF designs, but otherwise choosing the value for α is usually not a big issue from the practical point of view; the differences in performance are minor. CCF designs should be used only in cases where there is a real benefit of having fewer variable levels than the 5 variable levels of CCC designs (CCF designs use only 3 variable levels).

5.5 Example: Analysis of a Doehlert design for two variables

This example comes from (Dos Santos et. al., 2008). The aim was to optimize the recovery percentage of several elements with respect to the temperature and the volume of concentrated nitric acid from which we take only the recovery percentage of manganese (for details, see (Dos Santos et. al., 2008). The design is a Doehlert design with 3 replicates, and it is given in physical units in Table 20.

Temperature	Volume	Recovery %
135	5	89.0
165	5	90.2
120	3	90.4
150	3	94.3
150	3	91.6
150	3	91.2
180	3	91.0
135	1	82.6
165	1	88.0

Table 20. A Doehlert design with 2 variables.

Next, the variables are coded so that the maximum values are set to +1 and the minimum values are set to -1. Thus the coding formulas will be $X_1 = \frac{T-150}{30}$, and $X_2 = \frac{V-3}{2}$. The design in coded units is given in Table 21.

X_1	X_2	Recovery %
-0.5	+1	89.0
0.5	+1	90.2
-1	0	90.4
0	0	94.3
0	0	91.6
0	0	91.2
+1	0	91.0
-0.5	-1	82.6
+0.5	-1	88.0

Table 21. A Doehlert design with 2 variables in coded units.

Fig. 14 shows the design together with the recoveries visualizing the hexagonal structure of the design.

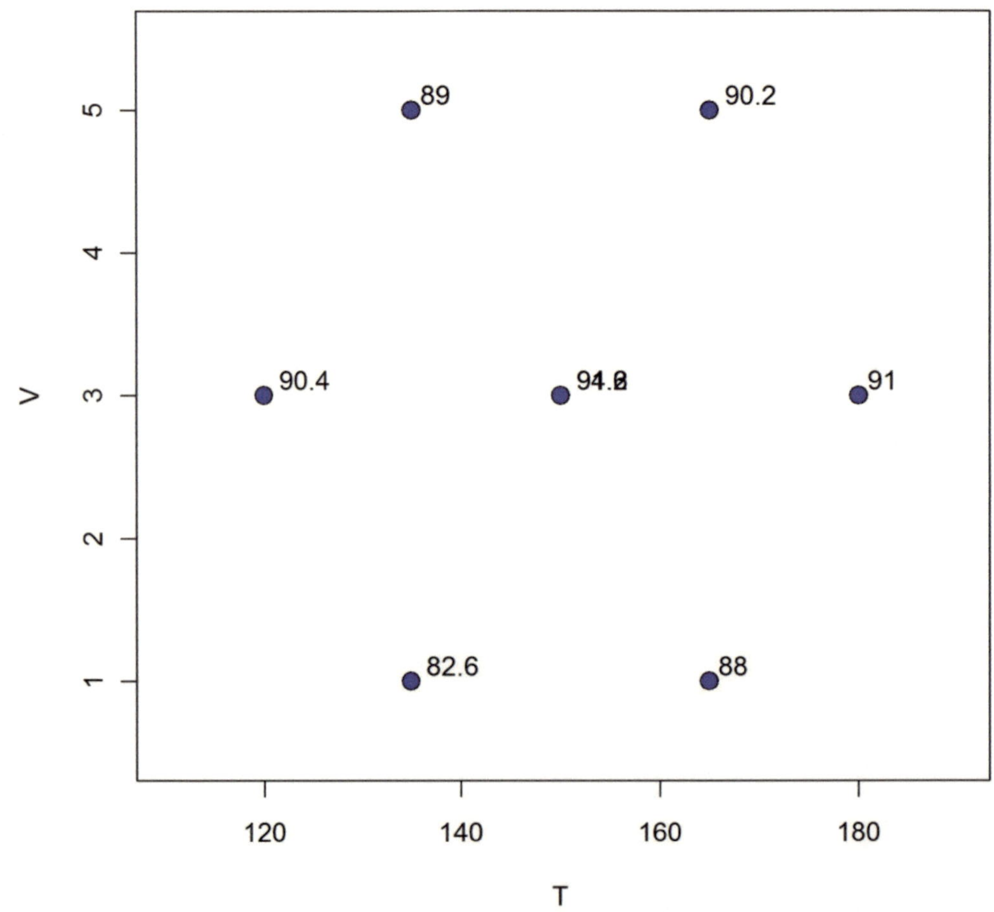

Fig. 14. The design of example 5.4.1 together with the measured recovery percentages.

Next a quadratic model is fitted to the data. The parameter estimates and the related statistics are given in Table 22.

	Estimate	Std. Error	t value	p value
(Intercept)	92.37	1.14	81.06	4.14e-06
X1	1.30	1.14	1.14	0.337
X2	2.15	0.99	2.18	0.118
I(X1^2)	-1.67	1.80	-0.93	0.423
I(X1 * X2)	-2.10	1.97	-1.06	0.365
I(X2^2)	-4.50	1.35	-3.33	0.045
Residual standard error: 1.974 on 3 degrees of freedom Multiple R-squared: 0.8594, Adjusted R-squared: 0.6251 F-statistic: 3.668 on 5 and 3 DF, p-value: 0.1569				

Table 22. Regression summary of the quadratic model.

According to Table 22 only the intercept and the quadratic effect of x_2 are significant. The p-value of the lack-of-fit test based on the 3 replicates is ca. 0.28. Thus the lack-of-fit is not significant. The apparent reason for the low significance is the rather poor repeatability of the experiments. The standard deviation of the recoveries of the replicate experiments is ca. 1.68 which is relatively high compared to the overall variation in the recoveries.

Next, let us see the results of cross-validation. Before cross-validation, the 3 replicates are replaced by the average of them. Fig. 15 shows the cross-validation results.

According to the cross-validation the predictions of the model are not very good. Due to the poor repeatability, i.e. large experimental error, it is hard to tell whether the reason for unreliable prediction is the large experimental error or something else, e.g. more complicated nonlinearity than quadratic one. According to the model, the optimum lies inside the experimental region and it corresponds to the stationary point. The optimal point in coded units is $X_1 = 0.25$ and $X_2 = 0.17$ which corresponds to T = 158 and V = 3.35 in physical units. This should be compared to Fig. 16 which shows the corresponding response surface.

6. Multi-response optimization

A common problem is to optimize the values of several responses simultaneously. This occurs quite frequently, because many products have to meet several different goodness criteria. The problem in such applications is that the individual optima can be contradictory, i.e. the optimal values for one of the responses may be far from the optimal values for some other response. Several different techniques, such as finding the so-called Pareto optimal result, exist. By far the simplest approach to this problem is to use so-called desirability functions, presented in the next section. The idea was first presented by (Derringer & Suich, 1980) in an application of product development in rubber industry.

Optimization using the desirability function technique can be divided into the following steps:

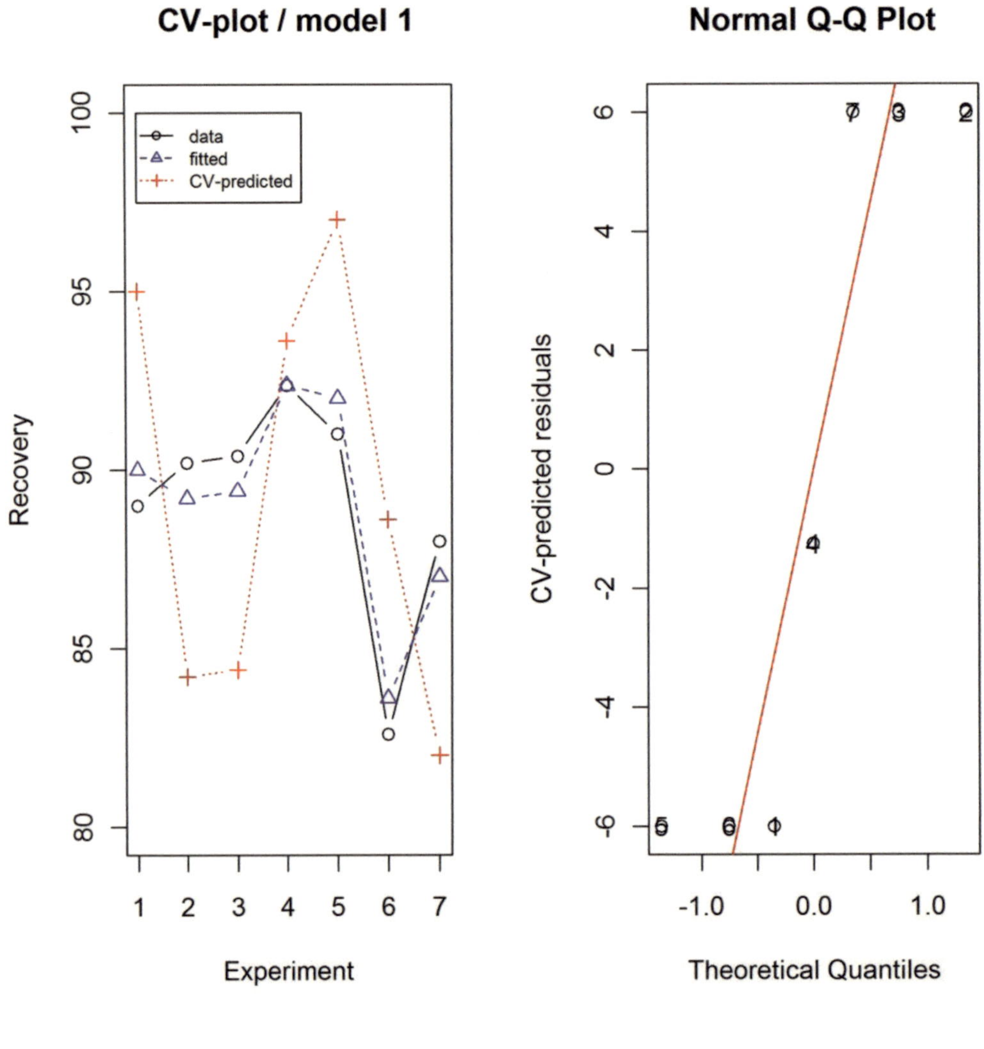

Fig. 15. Cross-validation of the quadratic model. Left panel: Recovery % vs. the number of experiment; black circles: data; blue triangles: fitted values; red pluses: cross-validated leave-one-out prediction. Right panel: Normal probability plots of the cross-validated leave-one-out residuals.

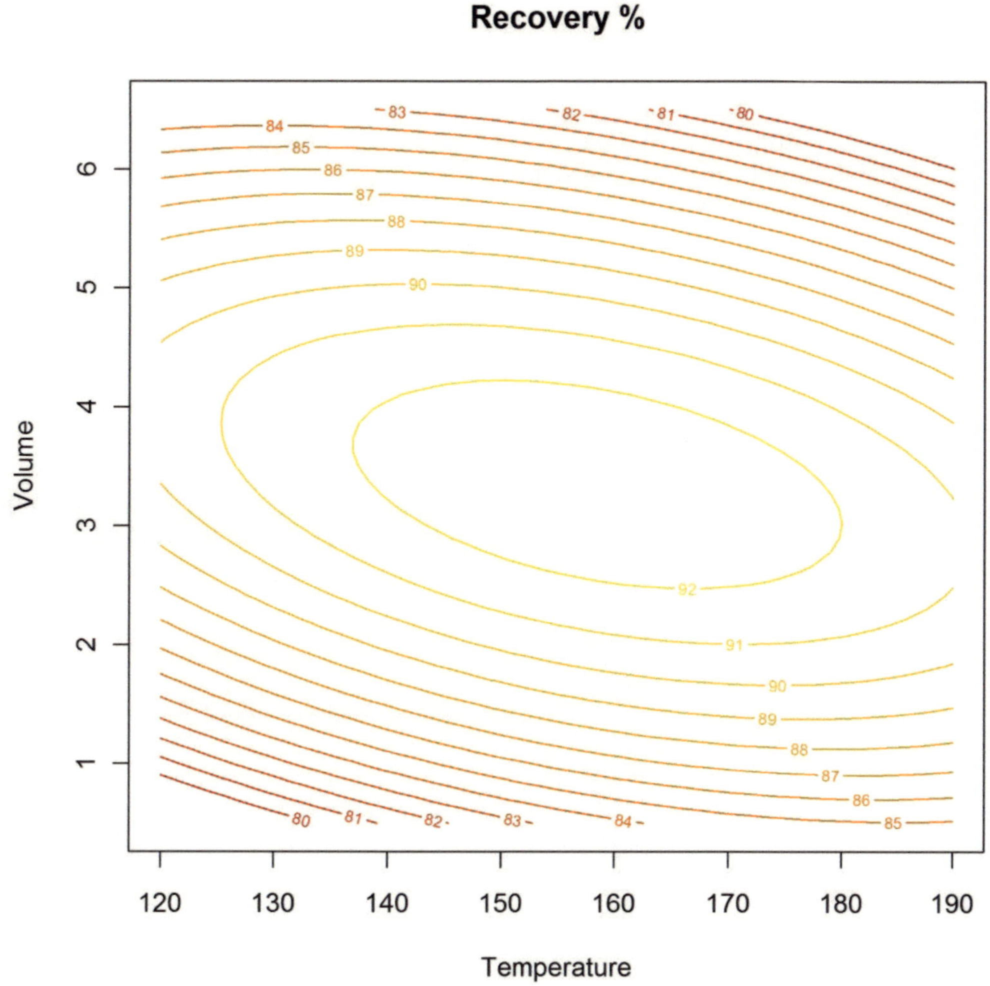

Fig. 16. Recovery % vs. Temperature and Volume.

First make a regression model, based on the designed experiments, individually for each response. Validate the models and proceed to step 2 after all models are satisfactory.

Make a desirability function $d_i = D_i(y_i)$ for each response separately (i goes from 1 to the number responses, say q). Remember that the responses have been modelled as functions of the design variables, i.e. $y_i = f_i(X_1, X_2, \cdots, X_N)$.

Building the desirabilities should be done together with a person who knows what the customers want from the product, and it is typically team work. How to build such functions in practice is explained later. Note that combining the two functions, desirabilities can be expressed as functions of design variables only.

Use an optimizer to maximize the combined desirability D which is the geometric mean of the individual desirabilities, i.e. $D = \left(D_1 \cdot D_2 \cdot \cdots \cdot D_q\right)^{\frac{1}{q}}$, with respect to the design variables.

Check by experimentation that the found optimum really gives a good product.

There are many ways to produce suitable desirability functions, one of which is explained in (Derringer & Suich, 1980). Any function that gives the 1 value for a perfect response and the value 0 for an unacceptable product and continuously values between 0 and 1 for responses whose goodness is in-between unacceptable and perfect can be used. One of the simplest alternatives is to use the following functions: $d_i = \left(1 + e^{-\frac{y_i - a}{b}}\right)^{-1}$ for one-sided desirabilities, and $d_i = e^{-\left|\frac{y_i - a}{b}\right|^c}$ for two-sided desirabilities. The parameters a, b and c are user-defined parameters chosen with the help of an expert on the product quality.

The idea is best illustrated by an example. Let us consider an example where the product would be the better the higher its elasticity is. Let us also assume that elasticity from 0.60 upwards would mean a practically perfect product and elasticity below 0.30 would mean a totally unacceptable product. Then the one-sided desirability function looks like (with a = 0.46 and b = 0.028) the one given in Fig. 17.

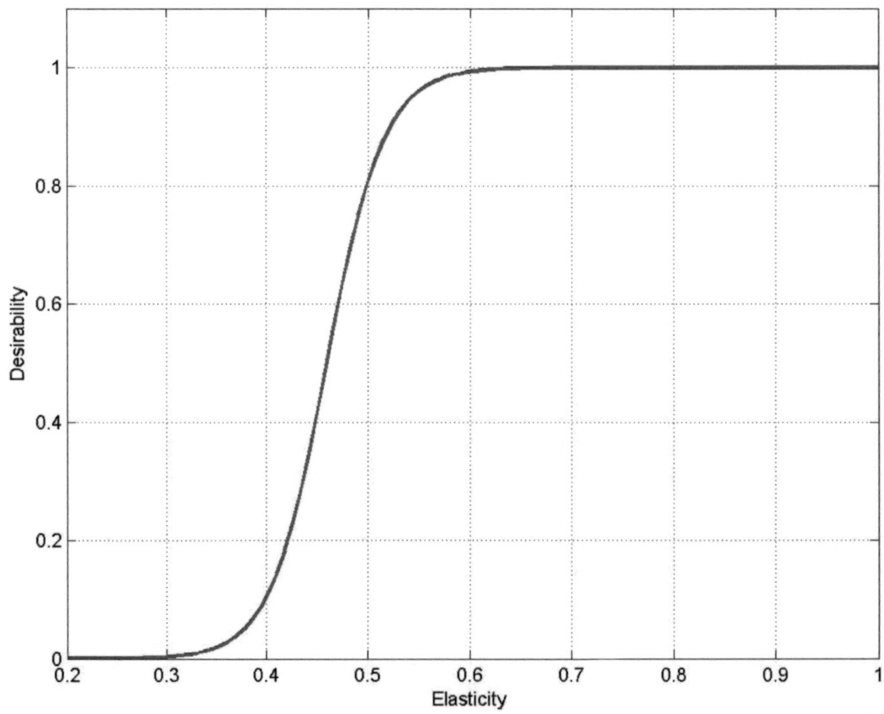

Fig. 17. A one-sided desirability function for elasticity that should be 0.60 or more and that would be totally unacceptable below 0.30.

If for some reason, the elasticity should not be higher than 0.60, and the elasticity over 0.90 or elasticity below 0.30 meant an unacceptable product, we would need a two-sided desirability function, e.g. like the one given in Fig. 18 (with a = 0.60, b = 0.028 and c = 2.5).

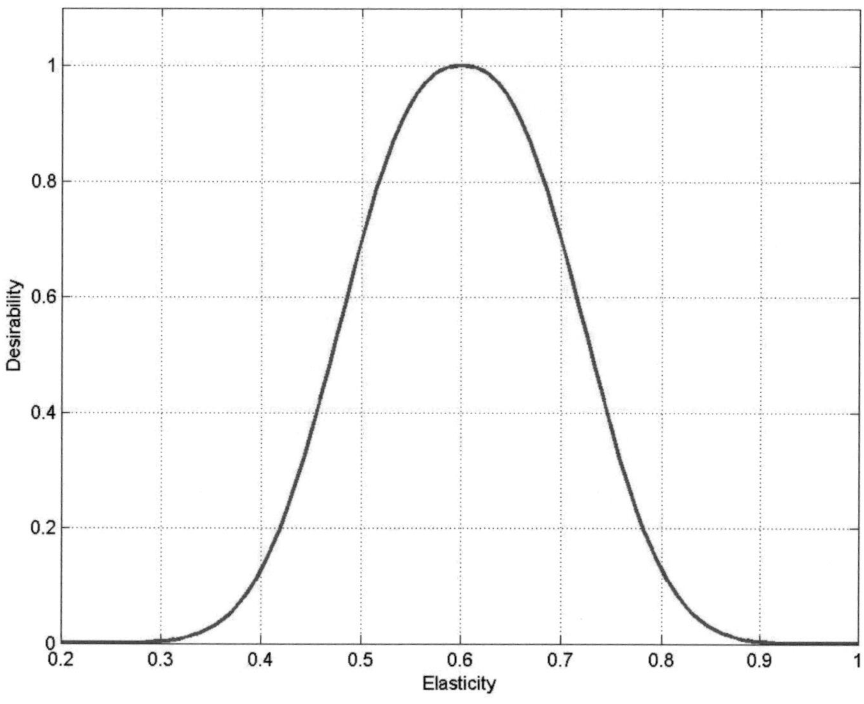

Fig. 18. A two-sided desirability function for elasticity that should be 0.60 and that would be totally unacceptable below 0.30 or above 0.90.

For practical examples see e.g. (Taavitsainen et. al., 2010) or (Laine et. al., 2011).

7. Conclusion

Design of experiments is as much of an art as of science. Becoming an expert in the field requires both theoretical studies and experience in practical applications. Although many problems can be solved in principle by hand calculations, in practice use of suitable software is needed. If the person involved is not familiar with command line style programs whose use is essentially that of programming, he or she is recommended to use some commercial software that typically also guide the user in the design and in the analysis of the results. The use of simulation models, where artificial experimental error is added into the results of the simulation, is highly recommended.

8. Acknowledgment

This work has been supported by the Helsinki Metropolia University of Applied Sciences.

9. References

Atkinson, A., Donev, A. & Tobias, R. (2007). *Optimum experimental designs, with SAS*. Oxford University Press, ISBN 978-0-19-929660-6, New York, USA

Bard, Y. (1973) *Nonlinear Parameter Estimation*, Academic Press, ISBN 978-0120782505, New York, USA

Bayne, K. & Rubin, I. (1986). *Practical Experimental Designs and Optimization Methods for Chemists*, VCH, ISBN 0-89573-136-3, Deerfield Beach, Florida, USA

Berthouex, P. & Brown, L. (2002). *Statistics for Environmental Engineers*, Lewis Publishers (CRC), ISBN 1-56670-592-4, Boca Raton, Florida, USA

Box, G. & Behnken, D. (1960). A simplex method for function minimization. *Technometrics*, Vol.2, No.4, pp. 455–475

Box, G. & Draper. (2007). *Response Surfaces, Mixtures, and Ridge Analyses*, Wiley (2nd ed.), ISBN 978-0-470-05357-7, Hoboken, New Jersey, USA

Box, G., Hunter, Hunter. (2005). *Statistics for Experimenters: Design, Innovation, and Discovery* (2nd ed.),Wiley, ISBN 978-0-471-71813-0, New York, USA

Bruns, R., Scarmiano, I. & de Barros Neto, B. (2006). *Statistical Design - Chemometrics* , Elsevier, ISBN 978-0-444-52181-1, Amsterdam, The Netherlands

Carlson, R. & Carlson, R. (2005). *Design and optimization in organic synthesis*, Elsevier

Cornell, J. (1981). *Experiments with Mixtures: Designs, Models, and the Analysis of Mixture Data* (3rd ed.), Wiley, ISBN 0-471-07916-2, New York, USA

Derringer, G. & Suich, R. (1980), Simultaneous Optimization of Several Response Variables. *Journal of Quality Technology*, Vol.12, pp. 214-219

Doehlert, D. (1970) Uniform shell designs. *Applied Statistics*, Vol.19, pp.231-239.

Dos Santos, W., Gramacho, D., Teixeira, A., Costa,A. & Korn, M. (2008), Use of Doehlert Design for Optimizing the Digestion of Beans for Multi-Element Determination by

Inductively Coupled Plasma Optical Emission Spectrometry, *J. Braz. Chem. Soc.*, Vol. 19, No. 1, pp.1-10.

Efron, B., Hastie, T., Johnstone. I. & Tibshirani, R. (2004), Least angle regression, *Ann. Statist.* Vol. 32, No. 2, pp. 407-499.

Fedorov, V. (1972). *Theory of Optimal Experiments*. Academic Press, ISBN 978-0824778811, New York, USA

Haaland, P. (1989). *Experimental Design in Biotechnology*, Marcel Dekker, ISBN 978-0824778811, New York, USA

Hanrahan, G. (2009). *Environmental Chemometrics*, CRC Press, ISBN 978-1420067965, Florida, USA

Himmelblau, D. (1970). *Process Analysis by Statistical Methods*, Wiley, ISBN 978-0471399858 , New York, USA

JMP, release 6, Design of Experiments, http://www.jmp.com/support/downloads/pdf/jmp_design_of_experiments.pdf, 9.9.2011

Kolarik, W. (1995). *Creating Quality*, McGraw-Hill, ISBN 0-07-113935-4

Koljonen, J., Nordling, T. & Alander, J. (2008), A review of genetic algorithms in near infrared spectroscopy and chemometrics: past and future, *Journal Of Near Infrared Spectroscopy*, Vol. 16, No. 3, pp. 189-197

Laine, P., Toppinen, E., Kivelä, R., Taavitsainen, V-M., Knuutila, O., Sontag-Strohm, T.,Jouppila, K. & Loponen, J.(2011), Emulsion preparation with modified oat bran: Optimization of the emulsification process for microencapsulation purposes, *Journal of Food Engineering*, Vol.104, pp.538-547

Montgomery, D. (1991). *Design and Analysis of Experiments* (3rd ed.), Wiley, ISBN 0-471-52994-X, Singapore

Nelder, J. & Mead, R. (1965). A simplex method for function minimization. *Computer Journal* Vol.7, No.4, pp. 308–313, ISSN 0010-4620

Neter, J., Kutner, M., Nachtsheim, C. & Wasserman, W. (1996), *Applied Linear Statistical Models* (4th ed.), WCB/McGraw-Hill, ISBN 0-256-11736-5, Boston, Massachusetts, USA

NIST/SEMATECH e-Handbook of Statistical Methods, http://www.itl.nist.gov/div898/handbook/, 9.8.2011

Taavitsainen, V-M. Ridge and PLS based rational function regression (2010), *Journal of Chemometrics*. Vol.24, No. 11-12, pp.665-673

Taavitsainen, V-M., Lehtovaara, A. & Lähteenmäki, M. (2010), Response surfaces, desirabilities and rational functions in optimizing sugar production, *Journal of Chemometrics*. Vol. 24, No. 7-8, pp. 505-513

Tibshirani, R. (1996), Regression shrinkage and selection via the lasso, *J. Royal. Statist. Soc. B.* Vol. 58, pp. 267–288

Vardeman, S. (1994). *Statistics for Engineering Problem Solving*, PWS , ISBN 978-0780311183, Boston, Massachusetts, USA

Weisberg, S. (1985). *Applied Linear Regression*, Wiley, ISBN 0-471-87957-6, New York, USA

Wheeler, R. (1974) Portable power , *Technometrics*, Vol. 16, No. 2, pp. 193-201
Zou, H. & Hastie, T. (2005), Regularization and Variable Selection via the Elastic Net, *Journal of the Royal Statistical Society, Series B*, Vol. 76, pp. 301-320.

Part 2

Biochemistry

6

Metabolic Biomarker Identification with Few Samples

Pietro Franceschi, Urska Vrhovsek, Fulvio Mattivi and Ron Wehrens
IASMA Research and Innovation Centre
Via E. Mach, 1 38010 S. Michele all'Adige (TN)
Italy

1. Introduction

Biomarker selection represents a key step in bioinformatic data processing pipelines; examples range from DNA microarrays (Tusher et al., 2001; Yousef et al., 2009) to proteomics (Araki et al., 2010; Oh et al., 2011) to metabolomics (Chadeau-Hyam et al., 2010). Meaningful biological interpretation is greatly aided by identification of a "short-list" of features – biomarkers – characterizing the main differences between several states in a biological system. In a two-class setting the biomarkers are those variables (metabolites, proteins, genes ...) that allow discrimination between the classes. A class or group tag can be used to distinguish many situations: it can be used to discriminate between treated and non-treated samples, to mark different varieties of the same organism, etcetera. In the following, we will – for clarity – restrict the discussion to metabolomics, and the variables will constitute concentration levels of metabolites, but similar arguments hold *mutatis mutandis* for other -omics sciences, such as proteomics and transcriptomics, where the variables correspond to protein levels or expression levels, respectively.

There are several reasons why the selection of biomarker short-lists can be beneficial:

- Predictive purposes: using only a small number of biomarkers in predictive class modeling in general leads to better, i.e., more robust and more accurate predictions.
- Interpretative purposes: it makes sense to first concentrate on those metabolites that show clear differences in levels in the different classes, since our knowledge of metabolic networks in many cases is only scratching the surface.
- Discovery purposes: the complete characterization of unknown compounds identified in untargeted experiments is time- and resource-consuming. The primary focus should thus be placed on a carefully selected group of "unknowns" to be characterized at structural and functional level.

Two fundamentally different statistical approaches to biomarker selection are possible. With the first, experimental data can be used to construct multivariate statistical models of increasing complexity and predictive power – well-known examples are Partial Least Square Discriminant Analysis (PLS-DA) (Barker & Rayens, 2003; Kemsley, 1996; Szymanska et al., 2011) or Principal Component Linear Discriminant Analysis (PC-LDA) (Smit et al., 2007; Werf et al., 2006). Inspection of the model coefficients then should point to those variables that are important for class discrimination. As an alternative, univariate statistical tests can be applied to individual variables, treating each one independent of the others and indicating which of them show significant differences between groups (see, e.g., Guo et al. (2007); Reiner et al. (2003); Zuber & Strimmer (2009)). Multivariate techniques are potentially more powerful in pin-pointing weak differences because they take into account correlation among the variables, but the models can be too much adapted to the experimental data, leading to poor generalization capacity. Univariate approaches, in contrast, both could miss important "weak" details and could overestimate the importance of certain variables, because correlation between variables is not taken into account.

As for many sciences with the "omics" suffix, in metabolomics the number of experimental variables usually greatly exceeds the number of objects, especially with the development of new mass-spectrometry-based technologies. In MS-based metabolomics, high resolution mass spectrometers are often coupled with high performance chromatographic techniques, like Ultra Performance Liquid Chromatography (UPLC). In these experiments, the variables, i.e., the metabolites, are represented by mass/retention-time combinations, and it is typical to have numbers of features varying from several hundreds to several thousands, depending on the experimental and analytical conditions. This increase in experimental possibilities, however, does not correspond to a proportional increase in the number of available samples, which can be limited by the availability of biological samples, by laboratory practice, in particular when complex protocols are required, and also by ethical issues, when, for example, experiments on animals have to be planned.

All these constraints produce *small sample sets*, presenting serious challenges for the statistical analysis, mainly because there is simply not enough information to model the natural biological variability. The situation is critical for multivariate approaches where the parameters of the statistical model need to be optimized (e.g., the number of components in a PLS-DA model). For this purpose, the classical approach is to use sub-sampling in combination with estimates of predictive power, like crossvalidation (Stone, 1974). In extreme conditions, i.e., really small sample sizes, this sub-sampling can give rise to inconsistent sub-models and tuning in the classical way becomes virtually impossible. In Hanczar et al. (2010), as an example, conclusions are focussed on ROC-based statistics (see below), but they are equally relevant for classical error estimates like the root-mean-square error of prediction, RMSEP) multivariate techniques can be still applied to the full data set, but it is not possible to assess the reliability of the biomarker selection pipeline, even if it is still reasonable to think that the biomarkers are strongly contributing to the statistical model. In these situations, univariate methods seem the best solution, also considering the presence of several strategies able to determine cut-off values in *t*-test based techniques (e.g., thresholding of p values subjected to some form of multiple testing correction (Benjamini & Hochberg, 1995; Noble, 2009; Reiner et al., 2003)). Regardless of the statistical strategy, for the "biomarkers" extracted in these conditions there is no obvious validation possible in the statistical sense; however, the results of the experiments are extremely important in the hypothesis generation phase to plan more informative investigations.

Interestingly, there is no literature on the effect of sample size on biomarker identification in the "omics" sciences, and the objective of this contribution is to fill this gap. We focus on a two-class problem, and in particular on small data sets. In our approach, real class differences have been introduced by spiking apple extracts with selected compounds, analyzing them using UPLC-TOF mass spectrometry, and comparing the feature lists to those of unspiked apple extracts. Using these data we are able to run a comparison between two multivariate methods (PLS-DA and PC-LDA) and the univariate t-test, leading to at least a rough estimate of how consistent biomarker discovery can be when small sample sizes are considered. In particular, we compare the effect of sample size reduction on multivariate and univariate models on the basis of Receiver Operating Characteristics (ROC) (Brown & Davis, 2005).

2. Material and methods

2.1 Biomarker Identification

There are many strategies for identifying differentially expressed variables in two-class situations – a recent overview can be found in Saeys et al. (2007). A general approach is to construct a model with good predictive properties, and to see which variables are important in such a model. Given the low sample-to-variable ratio, however, one can not expect to be able to fit very complicated models, and in many cases a linear model is the best one can do (Hastie et al., 2001). The oldest, and most well-known technique is Linear Discriminant Analysis (LDA, McLachlan (2004)). One formulation of this technique, dating back to R.A. Fisher, is to find a linear combination of variables $\boldsymbol{a}$ that maximizes the ratio of the between-groups sums of squares, $\boldsymbol{B}$, and the within-groups sums of squares $\boldsymbol{W}$:

$$\boldsymbol{a}^T \boldsymbol{B} \boldsymbol{a} / \boldsymbol{a}^T \boldsymbol{W} \boldsymbol{a} \quad (1)$$

That is, $\boldsymbol{a}$ is the direction that maximizes the separation between the classes, both by having compact classes (a small within-groups variance) and by having the class centers far apart (a large between-groups variance). Large values in $\boldsymbol{a}$ indicate which variables are important in the discrimination. Another formulation is to calculate the Mahalanobis distance of a new sample $\boldsymbol{x}$ to the class centers $\boldsymbol{\mu}_i$:

$$d(\boldsymbol{x}, i) = (\boldsymbol{x} - \boldsymbol{\mu}_i)^T \boldsymbol{\Sigma}^{-1} (\boldsymbol{x} - \boldsymbol{\mu}_i) \quad (2)$$

The new sample is then assigned to the class of the closest center. This approach is equivalent to Fisher's criterion for two classes (but not for more than two classes). In this equation, $\boldsymbol{\Sigma}$ is the (estimated) pooled covariance matrix of the classes. If the Mahalanobis distance to each class center is calculated using the individual class covariance matrices, the result is Quadratic Discriminant Analysis (QDA), which as the name suggests, no longer leads to linear class boundaries. A final formulation is to use regression using indicator variables for the class. In a two-class situation one can use, e.g., the values of -1 and 1 for the two classes; positive predictions will be assigned to class one, and negative predictions to class -1. In many other cases, 0 and 1 are used, and the class threshold is put at 0.5. When there are more than two classes, one can use a separate column in the dependent variable for every class – if a sample belongs to that class the column should contain 1, else 0. Again, the size of the regression coefficients indicates which of the variables contribute most to the discrimination.

For most applications in the "omics" fields, even the most simple multivariate techniques such as Linear Discriminant Analysis (LDA) cannot be applied directly. From Equation 2 it is clear that an inverse of the the covariance matrix $\boldsymbol{\Sigma}$ needs to be calculated, which is impossible in cases where the number of variables exceeds the number of samples. In practice, the number of samples is nowhere near the number of variables. For QDA, the situation is even worse: to allow a stable matrix inversion, every single class should have at least as many samples as variables (and preferably quite a bit more). A common approach is to compress the information in the data into a low number of latent variables (LVs), either using PCA (leading to PC-LDA, e.g. Smit et al. (2007); Werf et al. (2006)) or PLS (which gives PLS-DA; see Barker & Rayens (2003); Kemsley (1996)), and to perform the discriminant analysis on the resulting score matrices. These are not only of low dimension, but also orthogonal so that the matrix inversion, the calculation of $\mathbf{\Sigma}^{-1}$, can be performed very fast and reliably. Both for PC-LDA and PLS-DA, the problem is more often usually cast in a regression context, where again the response variable $\boldsymbol{Y}$ can take values of either 0 or 1. The model thus becomes:

$$\boldsymbol{Y} = \boldsymbol{XB} + \mathcal{E} \approx \boldsymbol{TP}^T\boldsymbol{B} + \mathcal{E} \tag{3}$$

where $\mathcal{E}$ is the matrix of residuals. Matrix $\boldsymbol{X}$ is decomposed into a score matrix $\boldsymbol{T}$ and a loading matrix $\boldsymbol{P}$, both consisting of a very low number of latent variables, typically less than ten or twenty. The coefficients for the scores, $\boldsymbol{A} = \boldsymbol{P}^T\boldsymbol{B}$, can therefore be easily be calculated in the normal way of least-squares regression:

$$\boldsymbol{A} = (\boldsymbol{T}^T\boldsymbol{T})^{-1}\boldsymbol{T}^T\boldsymbol{Y} \tag{4}$$

which by premultiplication with $\boldsymbol{P}$ lead to estimates for the overall regression coefficients $\boldsymbol{B}$:

$$\boldsymbol{B} = \boldsymbol{PA} \tag{5}$$

These equations are the same for both PLS-DA and PC-LDA. The difference lies in the decomposition of $\boldsymbol{X}$. In PC-LDA, $\boldsymbol{T}$ and $\boldsymbol{P}$ correspond to the scores and loadings, respectively, from PCA. That is, the class of the samples is completely ignored, and the only criterion is to capture as much variance as possible from $\boldsymbol{X}$. In PLS-DA, on the other hand, the scores and loadings are taken from a PLS model and the decomposition of $\boldsymbol{X}$ *does* take into account class information: the first PLS components by definition explain more, often much more, variance of $\boldsymbol{Y}$ than the first PCA components.

Both methods, PC-LDA as well as PLS-DA, are usually very sensitive to the choice of the number of LVs. Taking too few LVs will lead to bad predictions since important information is missed. Taking too many, the model will be too flexible and will show a phenomenon known as *overtraining*: it is more or less learning all the examples in the training set by heart but is not able to generalize and to make good predictions for new, unseen samples. As discussed, the assessment of the optimal number of LVs is neigh impossible with small sample sets. In the case under consideration, the extent of this effect is investigated by constructing several models with increasing numbers of LVs. Using real and simulated data sets (see below), models with 1–4, 6, and 8 LVs, respectively, are compared.

A simplification of statistical modeling can be obtained by ignoring all possible correlations between variables and assuming a diagonal covariance matrix, which leads to diagonal discriminant analysis (DDA). It can be shown that using the latter for feature selection corresponds to examining regular t-statistics (Zuber & Strimmer, 2009), and this is the approach we will take in this paper. For each variable, the difference between the class means $\bar{x}_{1i}$ and $\bar{x}_{2i}$ is transformed into a z-score by dividing by the appropriate standard deviation estimate s_i:

$$z_i = |\bar{x}_{1i} - \bar{x}_{2i}|/s_i \tag{6}$$

Using the appropriate number of degrees of freedom, these z-scores can be transformed into p values, which have the usual interpretation of the probability under the null hypothesis of encountering an observation with a value that is at least as extreme. In biomarker identification, p values can be used to sort the variables in order of importance and it is also possible to decide a cut-off value to identify variables which show "significant" differences from the null hypothesis.

Generally speaking, the absolute size of coefficients is taken as a measure for the likelihood of being a true marker: the variable with the largest coefficient, in a PLS-DA model for example, is the first biomarker candidate, the second largest the second candidate, and so on. Note that this approach assumes that all variables have been standardized, i.e., scaled to mean zero and unit variance. This is often done in metabolomics to prevent dominance of highly abundant metabolites. Statistics from a t-test can be treated in the same way.

2.2 Quality assessment

To evaluate the performance of biomarker selections one typically relies on quantities like the fraction of true positives, i.e., that fraction of the real biomarkers that is actually identified by the selection method, and the false positives – those variables that have been selected but do not correspond to real differences. Similarly, true and false negatives can be defined. These statistics can be summarized graphically in an ROC plot (Brown & Davis, 2005), where the fraction of true positives (y-axis) is plotted against the fraction of false positives (x-axis). These two characteristics are also known as the sensitivity and the (complement of) specificity. An ideal biomarker identification method would lead to a position in the top left corner: all true biomarkers would be found (the fraction of true positives would be one, or close to one) with no or only very few false positives. Gradually relaxing the selection criterion, allowing more and more variables to be considered as biomarkers, generally leads to an increase in the true positive fraction (upwards in the plot), but also to an increase in the false positive fraction (in the plot to the right). The best biomarker selection method is obviously the one that finds all biomarkers very quickly, leading to a very steep ROC curve at the beginning.

A quantitative measure of the efficiency of a method can be obtained by calculating the area under the ROC curve (AUC). A value of one (or close to one) indicates that the method does a very good job in identifying biomarkers – all true biomarkers are found almost immediately. A value of one half indicates a completely random selection (this corresponds to the diagonal in the ROC plot). Values significantly lower than one half should not occur. In many cases, the most important area in the ROC plot is the left side, which indicates the efficiency of the model in selecting the most important biomarkers. Consequently, it is common to calculate a partial area under the curve (pAUC), for instance up to twenty percent of false positives (pAUC.2). In a method with higher pAUC, the true biomarkers will be present in the first positions of the candidate biomarkers list, hence this is the quantity that will be considered in the current paper.

2.3 Apple data set

Twenty apples, variety Golden Delicious, were purchased at the local store. Extracts of every single fruit were prepared according to Vrhovsek et al. (Vrhovsek et al., 2004). The core of the fruit was removed with a corer and each apple was cut into equal slices. Three slices (cortex and skin) from the opposite side of each fruit were used for the preparation of aqueous acetone extracts. The samples were homogenized in a blender Osterizer model 847-86 at speed one in a mixture of acetone/water (70/30 w/w). Before the injection, acetone was evaporated by rotary evaporation, the samples were brought back to the original volume with ethanol and were filtered with a 0.22 μm filter (Millipore, Bedford, USA). UPLC-MS spectra were

HPLC	ACQUITY UPLC (Waters)
Column	BEH C18 1.7 μm, 2.1*50 mm
Column temperature	40°C
Injection volume	5μl
Eluent flux	0.8 mlmin^{-1}
Solvent A	0.1% formic acid in H_2O
Solvent B	0.1% formic acid in MeOH
Gradient	linear gradient
	from 0 to 100% of solvent B in 10 minutes
	100% of B for 2 minutes
	100% A within 0.1 minutes
	Equilibration for 2.9 minutes.
Mass Spectrometer	SYNAPT Q-TOF (Waters)
Mass range	50-3000 Da.
Capillary	3 kV
Sampling cone	25 V
Extraction cone	3 V
Source temperatures	150°C
Desolvation temperatures	500°C
Cone gas flow	50 Lh^{-1}
Desolvation gas flow	1000 Lh^{-1}

Table 1. Chromatographic and spectrometric conditions of the spiked-apple data set.

acquired on a ACQUITY - SYNAPT Q-TOF (Waters, Milford, USA) in positive and negative ion mode with the chromatographic conditions summarized in Table 1. No technical replicates were performed. Raw data were transformed to the open NetCDF format by the DataBridge built-in utility of the MassLynx software.

Class differences were introduced by spiking ten of the twenty extracts with a number of selected compounds, leaving the other ten as "untreated" controls. The majority of the spiked compounds are known to be commonly present in apples, while two of them (*trans*-resveratrol and cyanidin-3-galactoside) are not naturally present in the chosen matrix. The concentrations of the specific compounds in the pooled extract are presented in Table 2; markers were added in different concentrations to test the identification pipeline in conditions which mimic those found in a typical metabolomic experiment, where variation is usually present at different concentration levels. As an example of what the data look like, the first control sample, measured in positive mode, is shown in Figure 1. The horizontal axis shows the chromatographic dimension, and the vertical axis the mass-to-charge ratio. Circles indicate features that have been identified in this plane. In the remainder only the extracted triplets for the features, consisting of retention time, mass-to-charge ratio and intensity, will be used.

Feature extraction is performed with XCMS (Smith et al., 2006) and all statistical analyses are carried out in R (R Development Core Team, 2011). The CentWave peak-picking algorithm (Tautenhahn et al., 2008) is applied, using the following parameter settings: ppm = 20, peakwidth = c(3,15), snthresh = 2, prefilter = c(3,5). The average numbers of detected features per chromatogram are 1179 and 610 for positive and negative ion mode, respectively.

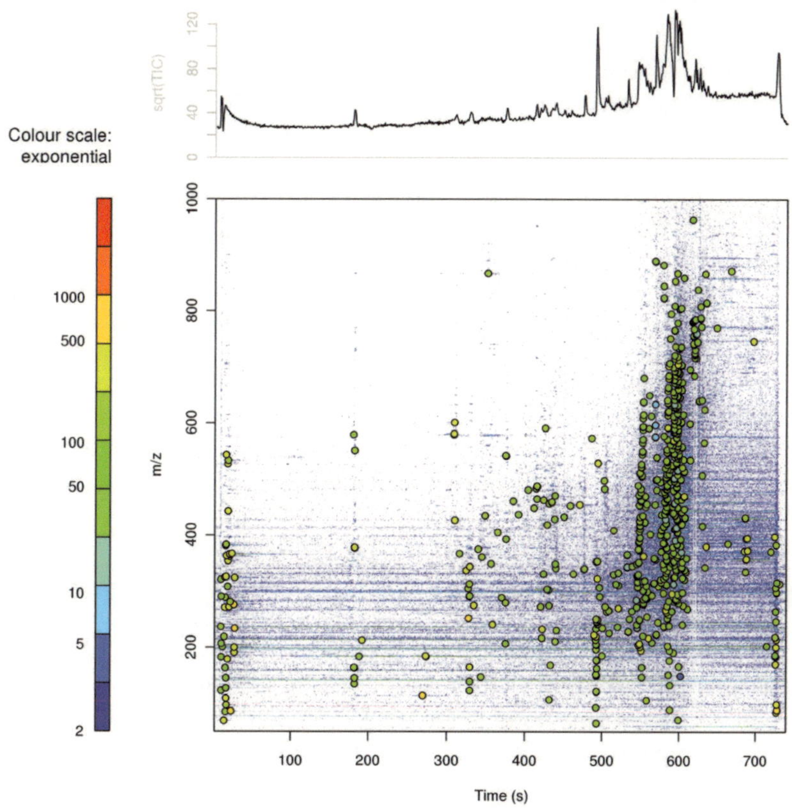

Fig. 1. Visualization of the data of the first control sample, measured in positive mode. The top of the figure shows the square root of the Total Ion Current (TIC); background color indicates the intensity of the signal in the plane formed by retention time and *m/z*axes. Circles indicate features found by the peak picking; the fill colour of these circles indicates the intensity of the features.

Compound	mgl^{-1} pool	Δ Conc. (mgl^{-1})
quercetin-3-galactoside (querc-3-gal)	5.69	1.48
quercetin	0.006	0.008
quercetin-3-glucoside (querc-3-glc)	1.05	0.3
quercetin-3-rhamnoside (querc-3rham)	3.64	3.55
phloridzin	2.92	2.3
cyanidin-3-galactoside (cy-3-gal)	n.d.	0.57
trans-resveratrol	n.d.	0.4

Table 2. Spiked compound summary. The difference in concentration is relative to the one measured in the pooled extract. Cyanidin-3-galactoside and *trans*-resveratrol are not normally found in Golden Delicious.

After grouping across samples, features are screened for isotopes, clusters and common adducts with in-house developed software.

Due to fragmentation occurring in the ionization source, it is common for a single neutral molecule to give rise to several ionic species. A single spiked compound can then generate several "biomarkers" in the MS peak table. Adducts, isotopes and common clusters are automatically screened, but fragments must be included in the biomarker list, as in real metabolomic experiments no a priori knowledge can be used to distinguish molecular from fragment ions. For the apple data set, the characteristic couples mass/retention time for all spiked metabolites were identified by manual inspection of the UPLC-MS profiles of standards. For negative ions the following numbers of features have been associated with the spike-in compounds: querc-3-gal/querc-3-glc (1 feature), phloridzin (2 features), *trans*-resveratrol (1), querc-3-rham (1). In the case of positive ion mode the numbers are cy-3-gal (1), *trans*-resveratrol (1), querc-3-rham (1), quercetin (1) and phloridzin (4). These feature are now taken to be the "true" biomarkers and they are used to construct ROC curves. The data set, as well as a more extended version including different concentrations of spiked-in compounds is publicly available in the R package BioMark (see `http://cran.r-project.org/web/packages/BioMark`, Wehrens & Franceschi (2011)) and has been used to evaluate a novel stability-based biomarker selection method (Wehrens et al., 2011).

In this application, the effects of decreasing sample size are investigated by subsampling the original set of twenty samples: sample sizes of 16, 12, 8 and 6 apples, respectively, are considered. In all cases, both classes (spiked and control) have equal sizes, which is the most easy case for detecting significant differences. Results are summarized by analysis of ROC curves – to prevent effects from accidentally easy or difficult subsets, the final ROC curves are obtained by averaging the results of 100 repeated re-samplings.

2.4 Simulated data sets

To assess the behaviour of biomarker selection for larger data sets, we resort to simulation. Simulated data sets have been constructed as multivariate normal distributions, using the means and covariance matrices of the experimental data: both classes (untreated and spiked) have been simulated separately. Simulations are performed for both positive and negative modes; in every simulation, one hundred data sets are created. The outcomes reported here are the averages of the results for the one hundred simulations. Data sets consisting of 10, 25, 50 and 200 biological samples per class have been synthesized.

3. Results and discussion

As a first step, the data are visualized using Principal Component Analysis (PCA). Since the intensities of the features can vary enormously, standardized data are used. The score plots for the positive and negative data sets are shown in Figure 2 for the positive ion mode, and in Figure 3 for the negative mode. In both cases, control and spiked data sets are not completely separated and the same is also true for the other PCs (not shown). This fact indicates that the "inherent" variability of the data set is not perturbed to a significant extent by spiking, as could be expected considering the small number of affected variables.

Even with this data structure, biomarker selection strategies can still perform efficiently. Figure 2 and Figure 3 also display the score plots of a PCA analysis performed considering only the top 10 variables selected by univariate t-testing. In these conditions, the separation

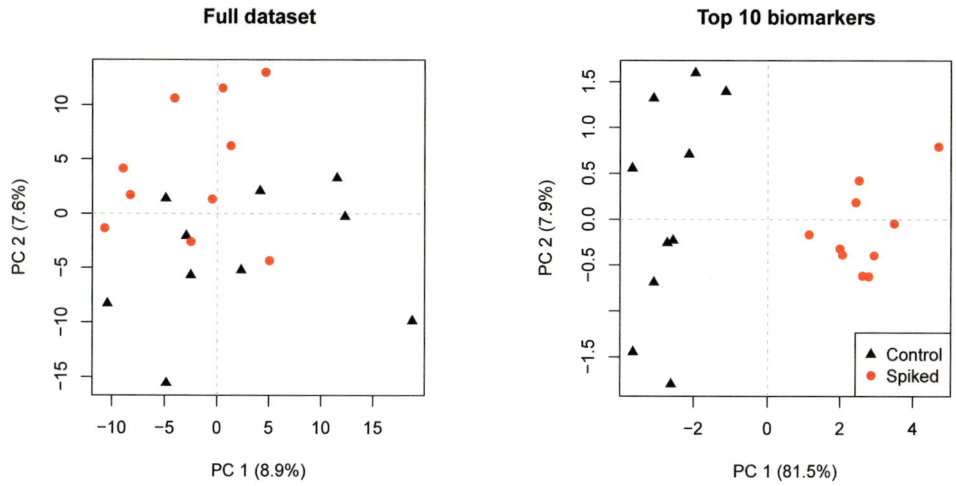

Fig. 2. PCA score plot (PC1 vs PC2) for the positive ion mode data set after standardization. In the left plot the principal components have been calculated on the full data set. In the right panel PCA analysis has been performed considering only the top 10 variables selected by a *t*-test.

between control and spiked samples is evident, thus indicating that this subset of the variables separates the two classes. Whether these ten variables contain the true biomarkers remains to be seen: especially in small data sets there may be chance correlations causing false positives, and seeing differences between the two groups in the score plots after *t*-testing in fact is trivial. The score plot is merely showing that the variables, selected on the bases of their discriminating power, are separating the two classes. As already discussed, small data sets will in general not capture all relevant biological variability, which implies that the predictive power of statistical models based on small data sets usually is very low. To illustrate this effect, the predictive power, i.e., the fraction of correct predictions for PC-LDA and PLS-DA models is presented in Figure 4. Four subsets of different sizes are considered as training sets, and the estimate of predictive power is based on predictions for the apples not in the training set. Again, the results are the average over 100 different subsamplings. Even if the control and spiked subsets are different, it can be seen that the predictive power of the multivariate methods is comparable to random selection, meaning that for every subset different variables will be important in the models and no consistency can be achieved. However, it is important to point out that this fact does not mean that some of the true biomarkers are not consistently selected upon subsetting, but rather that the more important variables are changing from a subset to another: even with models that are unpredictive it is possible to extract relatively good lists of putative biomarkers. Obviously, with very different characteristics for the two classes there *will* be predictive power, but for realistic data sets like the one used in this paper, where differences are small, it is unwise to focus solely on prediction.

To evaluate the efficiency of the different methods as far as biomarker selection is concerned, ROC curves for the *t*-test and two-component PLS-DA and PC-LDA models are presented in Figure 5, for 3, 4, 6 and 8 biological samples per class, respectively. The ROC curves indicate that all three variables selection methods perform significantly better than random selection.

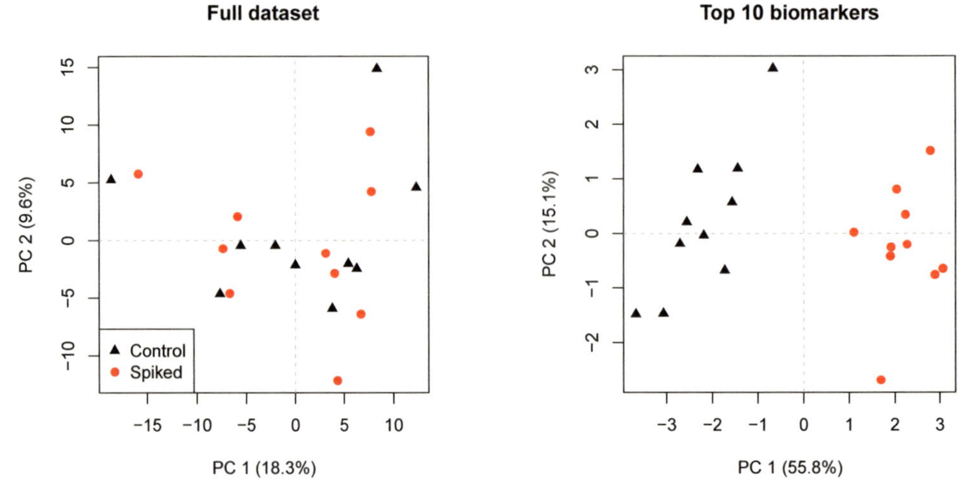

Fig. 3. PCA score plot (PC1 vs PC2) for the negative ion mode data set after standardization. In the left plot the Principal Components have been calculated on the full data set. In the right panel PCA analysis has been performed considering only the top 10 variables selected by a *t*-test.

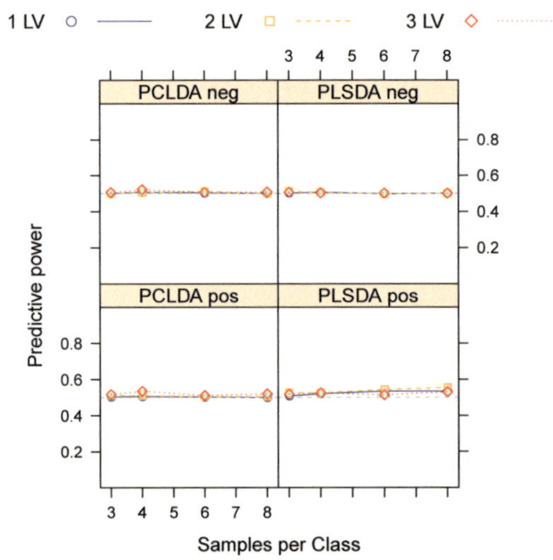

Fig. 4. Predictive power of multivariate PLS-DA and PC-LDA on a subset of the initial data set for positive and negative ion mode. Different lines are relative to models constructed with an increasing number of LVs. The horizontal dashed line indicates random selection.

Of the three, PC-LDA is always the least efficient, while PLS-DA and the *t*-test have a very similar performance. In absolute terms, the efficiency of the three methods increases with the number of biological samples. ROC curves for all possible conditions were constructed and the results are summarized in terms of early AUC (pAUC.2) in Figure 6, for positive and negative ion mode, respectively. From these figures it is possible to extract some clear trends:

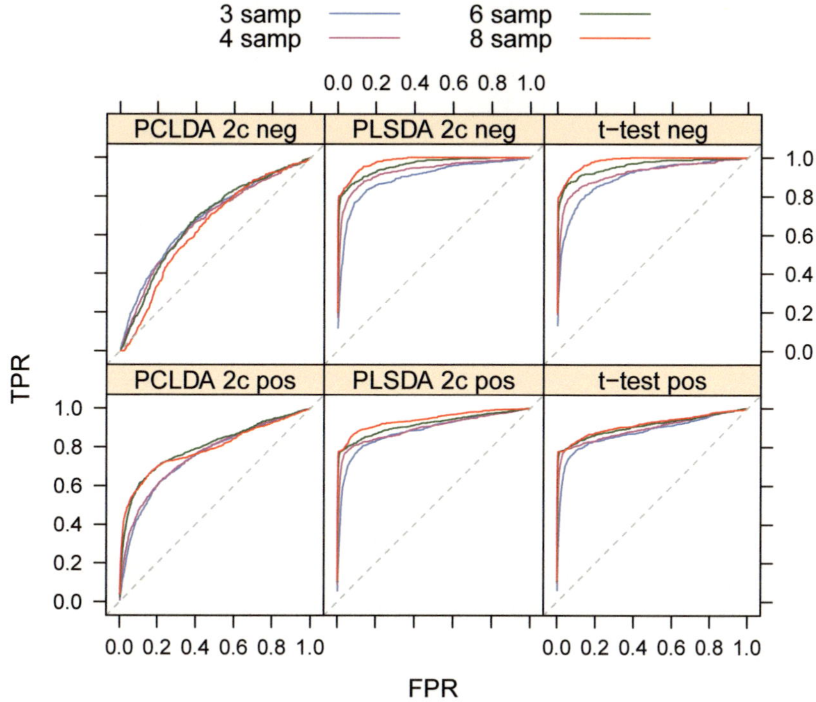

Fig. 5. ROC curves for the *t*-test and two component PLS-DA and PC-LDA as a function of the number of samples per class.

1. The performance of the methods improves by increasing the number samples per class.
2. The performance of PLS-DA is not particularly sensitive to the number of components.
3. PC-LDA does not show top class performance in any of the conditions considered.
4. The performance of PC-LDA is very much dependent on the number of components.
5. Multivariate approaches do not show a definitive advantage over univariate *t*-testing.

As expected, the performances of all the methods in terms of biomarker identification decrease with a reduction of the data set size. However, it is important to point out that even in the worst possible case (3 samples per class) early AUC for PLS-DA and the *t*-test are significantly greater than that obtained for completely random selection. This indicates that both methods can be used effectively in the biomarker selection phase, even with a low number of samples. In other words, features related to spiked compounds are consistently present in the top positions of the ordered list of experimental variables, which implies that also models constructed with very few samples can be relied upon to recognize these features.

The performance of PC-LDA is very much dependent on the number of components taken into account. This behavior can be explained by considering that in PC-LDA the variable reduction step is performed without any knowledge of class labels, only selecting the directions of greater variance. If these directions show little discriminating power, their supervised linear combination leads to poor modeling. However, performance improves with

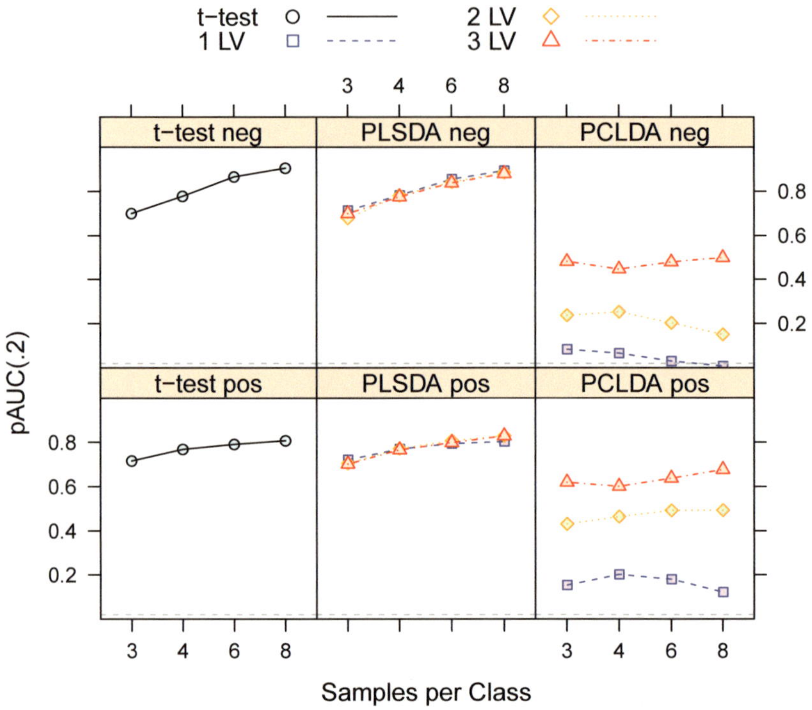

Fig. 6. pAUC.2 for PLS-DA, PC-LDA and *t*-test as a function of the number of samples per class and the number of LVs. The gray dashed line indicates the pAUC.2 of random selection.

the number of components, as an increase of the number of LVs leads to a better "coverage" of the data space. These limitations do not affect PLS-DA, as the variable reduction step is already performed in a supervised framework, where discriminating power is the main request. This means that the first PLS components are by definition more relevant than the first PCA components in biomarker identification. The other side of the coin is the danger of overfitting, very real in the application of PLS-DA (Westerhuis et al., 2008) – we will come back to this point later.

In this small-sample set, the *t*-test does as well as the best multivariate methods. This shows that modeling the correlation structure is not necessarily an advantage if the number of samples is low, or, alternatively, that the true correlation structure has not been captured well enough from the few samples that are available to allow meaningful inference. A definite advantage of the *t*-test is that it has no tunable parameters and can be applied without further optimization. It should be noted that we do not need to apply multiple-testing corrections in this context since we only use the order of the absolute size of the *t*-statistics to construct the ROC curves, and not a specific cut-off level α. In other applications, however, this aspect should be taken into account.

To extend the comparison between different models beyond the limits imposed by the apple experiment, ROC curves and early AUC were calculated for the simulated data using larger sample sizes (10, 25, 50, 200), both for positive and negative ion modes. The dependence of

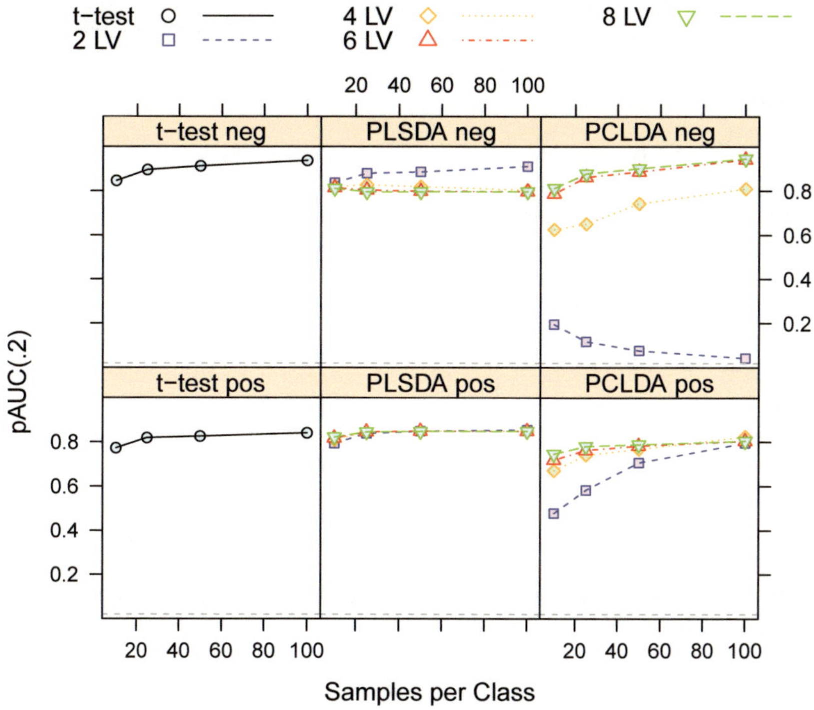

Fig. 7. pAUC.2 for PLS-DA, PC-LDA and *t*-test as a function of the number of samples per class and the number of LVs. Simulated data set. Gray dashed line indicates the pAUC.2 of random selection

pAUC.2 on the number of replicates and of components is presented in Figure 7, comparing the multivariate methods to the *t*-test and to "random" selection.

This analysis shows that PC-LDA only becomes effective if a large number of LVs is considered: the true biomarkers should have appreciable weight in the latent variables and it is by no means certain that this is the case for the first couple of LVs. Is it worth noting that for negative ion mode, the model with 2 LVs is comparable to random selection. In the case of PLS-DA, this dependence on the number of LVs is less evident and shows an opposite trend: the best performance is obtained with the smallest number of LVs. This is in agreement with the explanation given earlier: the relevant variables are captured in the very first PLS components, and the effect of overtraining leads to deterioration if more components are added. If anything, it is surprising that the overtraining effect is relatively small for these data.

The results on the simulated data sets are in agreement with the conclusions from the apple data. Differences between the methods decrease with increasing sample sizes, but even with the largest number of objects (200 in each group) the *t*-test still performs as well as PLS-DA. Multivariate testing is slightly more effective for the positive ion mode, while the *t*-test shows a slight advantage for the negative ion mode. This behaviour is probably due to the different characteristics of both ionization modes, leading to different levels of correlation between biomarkers. Indeed, in positive ion mode, the ionization shows a more pronounced fragmentation (phloridzine, for example, gives rise to four different biomarkers).

4. Conclusions

In this paper we have investigated the effects of sample set size on the performance of some popular strategies for biomarker identification (PLS-DA, PC-LDA and the *t*-test). The experiments are performed on a spiked metabolomic data set measured in apple extracts by UPLC-QTOF. The efficiency of the different statistical approaches is compared in terms of ROC curves, and in order to assess general trends, simulated data have been used to extend the data set. The experimental results clearly show that Linear Discriminant Analysis carried out on the Principal Components (PC-LDA) is the least efficient strategy for biomarker identification among the ones we considered. PLS-DA and the *t*-test show comparable performances in all the considered conditions. These results, and the observation that PLS-DA based selection is relatively consistent for different numbers of components, indicate that multivariate and univariate approaches are equally efficient for the apple data set. It is perhaps surprising that relatively good results in terms of biomarker selection are obtained, even for models that have very poor predictive performance. One should realise, however, that this is not a paradox at all: it merely is the result from the low sample-to-variable ratio, leading to chance correlations of intensities of metabolite signals with class. The true biomarkers are often present among the most significant variables in, e.g., a PLS-DA model, but many other false positives are, too, destroying the predictive power. One recently published approach actually utilizes this variability by focusing only on those variables that are *consistently* present in the most important variables upon disturbance of the data by jackkifing or bootstrapping (Wehrens et al., 2011).

The main point of this contribution, however, is the relation between data set size and reliability of biomarker identification. As expected, all the methods become less efficient as the number of biological replicates decreases, but even in these conditions the use of PLS-DA and the *t*-test offer effective biomarker identification strategies. This observation is fundamentally important in all studies where it is impossible to acquire more samples, and suggests that small sample sizes can still allow reliable selection of biomarkers.

5. References

Araki, Y., Yoshikawa, K., Okamoto, S., Sumitomo, M., Maruwaka, M. & Wakabayashi, T. (2010). Identification of novel biomarker candidates by proteomic analysis of cerebrospinal fluid from patients with moyamoya disease using SELDI-TOF-MS, *BMC Neurology* 10: 112.

Barker, M. & Rayens, W. (2003). Partial least squares for discrimination, *J. Chemom.* 17: 166–173.

Benjamini, Y. & Hochberg, Y. (1995). Controlling the false discovery rate: a practical and powerful approach to multiple testing, *J. Royal. Stat. Soc. B* 57: 289–300.

Brown, C. D. & Davis, H. T. (2005). Receiver operating characteristics curves and related decision measures: A tutorial, *Chemom. Intell. Lab. Syst.* 80: 24–38.

Chadeau-Hyam, M., Ebbels, T., Brown, I., Chan, Q., Stamler, J., Huang, C., Daviglus, M., Ueshima, H., Zhao, L., Holmes, E., Nicholson, J., Elliott, P. & Iorio, M. D. (2010). Metabolic profiling and the metabolome-wide association study: significance level for biomarker identification, *J. Proteome Res.* 9(9): 4620–4627.

Guo, Y., Hastie, T. & Tibshirani, R. (2007). Regularized discriminant analysis and its application in microarrays, *Biostatistics* 8: 86–100.

Hanczar, B., Hua, J., Sima, C., Weinstein, J., Bittner, M. & Dougherty, E. (2010). Small-sample precision of ROC-related estimates, *Bioinformatics* 28: 822–830.

Hastie, T., Tibshirani, R. & Friedman, J. (2001). *The Elements of Statistical Learning*, Springer Series in Statistics, Springer, New York.

Kemsley, E. K. (1996). Discriminant analysis of high-dimensional data: a comparison of principal components analysis and partial least squares data reduction methods, *Chemom. Intell. Lab. Syst.* 33: 47–61.

McLachlan, G. (2004). *Discriminant Analysis and Statistical Pattern Recognition*, Wiley-Interscience.

Noble, W. S. (2009). How does multiple testing correction work?, *Nat. Biotechnol.* 27: 1135–1137.

Oh, J., Craft, J., Townsend, R., Deasy, J., Bradley, J. & Naqa, I. E. (2011). A bioinformatics approach for biomarker identification in radiation-induced lung inflammation from limited proteomics data, *J. Proteome Res.* 10(3): 1406–1415.

R Development Core Team (2011). *R: A Language and Environment for Statistical Computing*, R Foundation for Statistical Computing, Vienna, Austria. ISBN 3-900051-07-0. URL: *http://www.R-project.org*

Reiner, A., Yekutieli, D. & Benjamini, Y. (2003). Identifying differentially expressed genes using false discovery rate controlling procedures, *Bioinformatics* 19(3): 368–375.

Saeys, Y., Inza, I. & Larranaga, P. (2007). A review of feature selection techniques in bioinformatics, *Bioinformatics* 23: 2507–2517.

Smit, S., Breemen, M. J. v., Hoefsloot, H. C. J., Aerts, J. M. F. G., Koster, C. G. d. & Smilde, A. K. (2007). Assessing the statistical validity of proteomics based biomarkers, *Anal. Chim. Acta* 592: 210–217.

Smith, C. A., Want, E. J., Tong, G. C., Abagyan, R. & Siuzdak, G. (2006). XCMS: Processing Mass Spectrometry Data for Metabolite Profiling Using Nonlinear Peak Alignment, Matching, and Identification, *Anal. Chem.* 78: 779–787.

Stone, M. (1974). Cross-validatory choice and assessment of statistical predictions, *J. R. Statist. Soc. B* 36: 111–147. Including discussion.

Szymanska, E., Saccenti, E., Smilde, A. & Westerhuis, J. (2011). Double-check: validation of diagnostic statistics for PLS-DA models in metabolomics studies, *Metabolomics* .

Tautenhahn, R., Bottcher, C. & Neumann, S. (2008). Highly sensitive feature detection for high resolution LC/MS, *BMC Bioinformatics* 9: 504.

Tusher, V., Tibshirani, R. & Chu, G. (2001). Significance analysis of microarrays applied to the ionizing radiation response, *PNAS* 98: 5116–5121.

Vrhovsek, U., Rigo, A., Tonon, D. & Mattivi, F. (2004). Quantitation of polyphenols in different apple varieties, *J. Agr. Food. Chem.* 52(21): 6532–6538.

Wehrens, R. & Franceschi, P. (2011). *BioMark: finding biomarkers in two-class discrimination problems*. R package version 0.3.0.

Wehrens, R., Franceschi, P., Vrhovsek, U. & Mattivi, F. (2011). Stability-based biomarker selection, *Anal. Chim. Acta* 705: 15–23.

Werf, M. J. v. d., Pieterse, B., Luijk, N. v., Schuren, F., Vat, B. v. d. W.-v. d., Overkamp, K. & Jellema, R. H. (2006). Multivariate analysis of microarray data by principal component discriminant analysis: prioritizing relevant transcripts linked to the degradation of different carbohydrates in Pseudomonas putida S12, *Microbiology* 152: 257–272.

Westerhuis, J., Hoefsloot, H., Smit, S., D.J., V., Smilde, A. K., van Velzen, E., van Duijnhoven, J. & van Dorsten, F. A. (2008). Assessment of PLSDA cross validation, *Metabolomics* 4: 81–89.

Yousef, M., Ketany, M., Manevitz, L., Showe, L. & Showe, M. (2009). Classification and biomarker identification using gene network modules and support vector machines, *BMC Bioinformatics* 10: 337.

Zuber, V. & Strimmer, K. (2009). Gene ranking and biomarker discovery under correlation, *Bioinformatics* 25: 2700–2707.

7

Kinetic Analyses of Enzyme Reaction Curves with New Integrated Rate Equations and Applications

Xiaolan Yang, Gaobo Long, Hua Zhao and Fei Liao*
College of Laboratory Medicine, Chongqing Medical University, Chongqing, China

1. Introduction

A reaction system of Michaelis-Menten enzyme on single substrate can be characterized by the initial substrate concentration before enzyme action (S_0), the maximal reaction rate (V_m) and Michaelis-Menten constant (K_m), besides some other required parameters. The estimations of S_0, V_m and K_m can be used to measure enzyme substrates, enzyme activities, epitope or hapten (enzyme-immunoassay), irreversible inhibitors and so on. During enzyme reaction, the changes of substrate or product concentrations can be monitored; continuous monitor of such changes provides a reaction curve while discontinuous monitor of such changes provides signals just for the starting point and the terminating point of enzyme reaction. It is an end-point method when only signals for the starting point and the terminating point are analyzed. It is a kinetic method when a range of data from a reaction curve are analyzed, and can be classifieid into the initial rate method and kinetic analysis of reaction curve. The initial rate method only analyzes data for initial rate reaction whose instantaneous rates are constants; kinetic analysis of reaction curve analyzes data whose instantaneous rates show obvious deviations from the initial rate (Bergmeyer, 1983; Guilbault, 1976; Marangoni, 2003). To estimate those parameters of an enzyme reaction system, kinetic analysis of reaction curve is favoured because the analysis of one reaction curve can concomitantly provide V_m, S_0 and K_m. Hence, methods for kinetic analysis of reaction curve to estimate parameters of enzyme reaction systems are widely studied.

An enzyme reaction curve is a function of dependent variables, which are proportional to concentrations of a substrate or product, with respect to reaction time as the predictor variable. In general, there are two types of enzyme reaction curves. The first type involves the action of just one enzyme, and employs either a selective substrate to detect the activity of one enzyme of interest or a specific enzyme to act on a unique substrate of interest. The second type involves the actions of at least two enzymes, and requires at least one auxiliary enzyme as a tool to continuously monitor a reaction curve. The second type is an enzyme-coupled reaction system. For kinetic analysis of reaction curve, there are many reports on

* Corresponding Author

one enzyme reaction system, but are just a few reports on enzyme-coupled reaction system (Atkins & Nimmo, 1973; Liao, et al., 2005; Duggleby, 1983, 1985, 1994; Walsh, 2010).

In theory, enzyme reactions may tolerate reversibility, the activation/inhibition by substrates/products, and even thermo-inactivation of enzyme. From a mathematic view, it is still feasible to estimate parameters of an enzyme reaction system by kinetic analysis of reaction curve if the roles of all those factors mentioned above are included in a kinetic model (Baywenton, 1986; Duggleby, 1983, 1994; Moruno-Davila, et al., 2001; Varon, et al., 1998). However, enzyme kinetics is usually so complex due to the effects of those mentioned factors that there are always some technical challenges for kinetic analysis of reaction curve. Hence, most methods for kinetic analysis of reaction curve are reported for enzymes whose actions suffer alterations by those mentioned factors as few as possible.

In practice, kinetic analysis of reaction curve usually employs nonlinear-least-square-fitting (NLSF) of the differential or integrated rate equation(s) to either the reaction curve *per se* or data set(s) transformed from the reaction curve (Cornish-Bowden, 1995; Duggleby, 1983, 1994; Orsi & Tipton, 1979). The use of NLSF rather than matrix inversion is due to the existence of multiple minima of the sum of residual squares with respect to some nonlinear parameters (Liao, et al., 2003a, 2007a). When a differential rate equation is used, numerical differentiation of data from the reaction curve has to be employed to derive instantsneous reaction rates. In this case, there must be intervals as short as possible to monitor reaction curves (Burden & Faires, 2001; Dagys, 1990; Hasinoff, 1985; Koerber & Fink, 1987). However, the instantaneous reaction rates from reaction curves inherenetly exhibit narrow distribution ranges and large errors; the strategy by numerical differentiation of data in a reaction curve is unfavourable for estimating V_m and S_0 because of their low reliaiblity and unsatisfactory working ranges. On the other hand, when an integrated rate equation of an enzyme reaction is used for kinetic analysis of reaction curve, there is no prerequisites of short intervals to record reaction curves so that automated analyses in parallel can be realized for enhanced performance with a large number of samples. As a result, integrated rate equations of enzymes are widely studied for kinetic analysis of reaction curve to estimate parameters of enzyme reaction systems (Duggleby, 1994;Liao, et al, 2003a, 2005a; Orsi & Tipton, 1979).

Due possibly to the limitation on computation resources, integrated rate equations of enzymes in such methods are usually rearranged into special forms to facilitate NLSF after data transformation (Atkins & Nimmo, 1973; Orsi & Tipton, 1979). In appearance, the uses of different forms of the same integrated rate equation for NLSF to data sets transformed from the same reaction curve can give the same parameters. However, kinetic analysis of reaction curve with rearranged forms of an integrated rate equation always gives parameters with uncertainty too large to have practical roles (Newman, et al, 1974). Therefore, proper forms of an integrated rate equation should be selected carefully for estimating parameters by kinetic analysis of reaction curve.

In the past ten years, our group studied chemometrics for kinetic analysis of reaction curve to estimate parameters of enzyme reaction systems; the following results were found. (a) In terms of reliability and performance for estimating parameters, the use of the integrated rate equations with the predictor variable of reaction time is superior to the use of the integrated rate equations with predictor variables other than reaction time (Liao, et al., 2005a); (b) the integration of kinetic analysis of reaction curve with other methods to quantify initial rates and substrates has more absorbing advantages (Liu, et al., 2009; Yang, et al., 2010); such integration strategies can be applied to enzyme-coupled reaction systems and enzymes sufferring inhibition by substrates/products. Herein, we discuss chemometrics for both kinetic analysis of reaction curve and its integration with other methods, and demonstrate their applications to quantify enzyme initial rates and substrates with some typical enzymes.

2. Kinetic analysis of enzyme reaction curve: chemometrics and application

To estimate parameters by kinetic analysis of reaction curve, the desired parameters are included in a set of parameters for the best fitting. Regardless of the number of enzymes involved in a reaction curve, there are the following two approaches for kinetic analysis of reaction curve based on different ways to realize NLSF and their data transformation.

In the first approach, with a differential or integrated rate equation, a series of dependent variables are derived from data in a reaction curve with each set of preset parameters. Such dependent variables should follow a predetermined response to predictor variables that are either reaction time or data transformed from those in the reaction curve. The goodness of the predetermined response is the criterion for the best fitting. In this approach, NLSF is realized with a model for data transformed from a reaction curve (Burguillo, 1983; Cornish-Bowden, 1995; Liao, 2005; Liao, et al., 2003a, 2003b, 2005a, 2005b; Orsi & Tipton, 1979).

In the second approach, reaction curves are calculated with sets of preset parameters by iterative numerical integration from a preset staring point. Such calculated reaction curves are fit to a reaction curve of interest; the least sum of residual squares indicates the best fitting (Duggleby, 1983, 1994; Moruno-Davila, et al., 2001; Varon, et al., 1998; Yang, et al., 2010). In this approach, calculated reaction curves still utilize reaction time as the predictor variable and become discrete at the same intervals as the reaction curve of interest. Clearly, there is no transformation of data from a reaction curve in this approach.

With any enzyme, iterative numerical integration of the differential rate equation(s) from a starting point with sets of preset parameters can be universally applicable regardless of the complexity of the kinetics. Thus, the second approach exhibits better universality and there are few technical challenges to kinetic analysis of reaction curve *via* NLSF. In fact, however, the second approach is occasionally utilized while the first approach is widely practiced.

In the following subsections, the differential rate equation of simple Michaelis-Menten kinetics on single substrate is integrated; the prerequisites for kinetic analysis of reaction curve with integrated rate equations, kinetic analysis of enzyme-coupled reaction curve, the integrations of kinetic analysis of reaction curve with other methods, and the applications of such integration strategies to some typical enzymes are discussed.

2.1 Integrated rate equation for one enzyme on single substrate

Assigning instantaneous substrate concentration to S, instantaneous reaction time to t, steady-state kinetics of Michaelis-Menten enzyme on single substrate follows Equ.(1).

$$-dS/dt = (V_{\mathrm{m}} \times S)/(K_{\mathrm{m}} + S) \tag{1}$$

Assigning the substrate concentration at the first point for analysis to S_1, Equ.(1) is integrated into Equ.(2) when the enzyme is stable, the substrate and product do not alter the intrinsic activity of the enzyme and the reaction is irreversible (Atkins & Nimmo, 1973; Marangoni, 2003; Orsi & Tipton, 1979; Zou & Zhu, 1997). In Equ.(2), t_{lag} accounts for the lag time of steady-state reaction. After transformation of data in a reaction curve according to Equ.(3), there should be a linear response of the left part in Equ.(2) to reaction time, as in Equ.(4). The goodness of this linear response is judged based on regression analysis. However, to estimate parameters by kinetic analysis of reaction curve *via* NLSF, there are the following general prerequisites for Equ.(2) or any of its equivalency.

$$(S_1 - S)/K_m + \ln(S_1/S) = (V_m/K_m) \times (t - t_{lag}) \tag{2}$$

$$y = (S_1 - S)/K_m + \ln(S_1/S) \tag{3}$$

$$y = a + b \times t \tag{4}$$

The first prerequisite is that enzyme reaction should apparently follow kinetics on single substrate. For enzyme reactions with multiple substrates whose concentrations are all changing during reactions, kinetic analysis of reaction curve always give parameters of too low reliability to have practical roles no matter what methods are used for NLSF (data unpublished). From our experiences to estimate parameters by kinetic analyses of reaction curves, any substrate at levels below 10% of its K_m can be considered negligible; the use of one substrate at levels below 10% of those of other substrates can make enzyme reactions follow single substrate kinetics (Liao, et al, 2001; Liao, et al, 2003a, 2003b; Li et al., 2011; Zhao et al., 2006). For any enzyme on multiple substrates, therefore, there are two approaches to make it apparently follow kinetics on single substrate. The first is the use of one substrate of interest at levels below 10% of those of the other substrates; this approach has universal applicability to common enzymes such as hydrolases in aqueous buffers and oxidases in air-saturated buffers. The second is the utilization of special reaction systems to regenerate the substrate of the enzyme of interest by actions of some auxiliary enzymes, and indeed this approach usually yields enzyme-coupled reaction curves of complicated kinetics.

The second prerequisite is that enzyme reaction should be irreversible. In theory, the estimation of parameters by kinetic analysis of reaction curve is still feasible when reaction reversibility is considered, but the estimated parameters possess too low reliability to have practical roles (data unpublished). Generally, a preparation of a substance with contaminants less than 1% in mass content can be taken as a pure substance. Namely, a reagent leftover in a reaction accounting for less than 1% of that before reaction can be negligible. For convenience, therefore, an enzyme reaction is considered irreversible when the leftover level of a substrate of interest in equilibrium is much less than 1%of its initial one. To promote the consumption of the substrate of interest, the concentrations of other substrates should be preset at levels much over 10 times the initial level of the substrate of interest. In this case, the enzyme reaction is apparently irreversible and follows kinetics on single substrate. Or else, the use of scavenging reactions to remove products can drive the reaction forward. The concurrent uses of both approaches are usually better.

The third prerequisite is that there should be steady-state data for analysis (Atkins & Nimmo, 1973; Dixon & Webb, 1979; Liao, et al, 2005a; Marangoni, 2003; Orsi & Tipton, 1979).

For this prerequisite, the first and the last points of data in a reaction curve for analysis should be carefully selected. The first point should exclude data within the lag time of steady-state reaction. The last point should ensure data for analyses to have substrate concentrations high enough for steady-state reaction. Namely, substrate concentrations should be much higher than the concentration of the active site of the enzyme (Dixon & Webb, 1979). The use of special weighting functions for NLSF can mitigate the contributions of residual squares at low substrate levels that potentially obviate steady-state reaction.

The forth prerequisite is that the enzyme should be stable to validate Equ.(2), or else the inactivation kinetics of the enzyme should be included in the kinetic model. Enzyme stability should be checked before kinetic analysis of reaction curve. When the inactivation kinetics of an enzyme is included in a kinetic model for kinetic analysis of reaction curve, the integrated rate equation is usually quite complex or even inaccessible if the inactivation kinetics is too complex. For kinetic analysis of reaction curve of complicated kinetics, numerical integration to produce calculated reaction curves for NLSF to a reaction curve of interest, instead of NLSF with Equ.(4), can be used to estimate parameters (Duggleby, 1983, 1994; Moruno-Davila, et al., 2001; Varon, et al., 1998; Yang, et al., 2010).

The fifth prerequisite is that there should be negligible inhibition/activation of activity of an enzyme by products/substrates, or else such inhibition/activation on the activity of the enzyme by its substrate/product should be included in an integrated rate equation for kinetic analysis of reaction curve (Zhao, L.N., et al., 2006). For validating Equ.(2), any substrate that alters enzyme activity should be preset at levels low enough to cause negligible alterations; any product that alters enzyme activity can be scavenged by proper reactions. When such alterations are complex, numerical integration of differential rate equations for NLSF to a reaction curve of interest can be used (Duggleby, 1983, 1994; Moruno-Davila, et al., 2001; Varon, et al., 1998).

Obviously, the first three prerequisites are mandatory for the inherent reliability of parameters estimated by kinetic analysis of reaction curve; the later two prerequisites are required for the validity of Equ.(2) or its equivalency for kinetic analysis of reaction curve.

2.2 Realization of NLSF and limitation on parameter estimation

To estimate parameters by kinetic analysis of reaction curve based on NLSF, the main concerns are the satisfaction to the prerequisites for the quality of data under analysis, the procedure to realize NLSF, and the reliability of parameters estimated thereby.

For the estimation of parameters by kinetic analysis of reaction curve, there are two general prerequisites for the quality of data under analysis: (a) there should be a minimum number of the effective data whose changes in signals are over three times the random error; (b) there should be a minimum consumption percentage of the substrate in such effective data for analysis. In general, at least two parameters like V_m and S_0 should be estimated; the minimum number of the effective data should be no less than 7 (Atkins & Nimmo, 1973; Baywenton, 1986; Miller, J. C. & Miller, J. N., 1993). The minimum consumption percentage of the substrate can be about 40% if only V_m and S_0 are estimated while other parameters are fixed as constants. In general, the estimation of more parameters requires higher consumption percentages of the substrate in the effective data for analysis.

With a valid Equ.(2), data in a reaction curve can be transformed according to Equ.(3) to realize NLSF with Equ.(4). The use of Equ.(4) for NLSF needs no special treatment of the unknown t_{lag}. For any method to continuously monitor reaction curve, there may be an unknown but constant background in signals (Newman, et al., 1974; Liao, et al., 2003a, 2005a; Yang, et al., 2010). Thus, the background in the signal for S_1 in Equ.(2) is better to be treated as a nonlinear parameter to realize NLSF; this procedure gives the term of NLSF but causes the burden of computation; as a result, a rearranged form of Equ.(2) is suggested for kinetic analysis of reaction curve (Atkins & Nimmo, 1973; Liao, et al., 2005a).

In theory, Equ.(2) can be rearranged into Equ.(5) as a linear function of V_m and K_m. In Equ.(5), the instantaneous reaction time at the moment for S_1 is preset as zero so that there is no treatment of t_{lag}. When the signal for S_1 is not treated as a nonlinear parameter, kinetic analysis of reaction curve by fitting with Equ.(5) can be finished within 1 s with a pocket calculator. However, parameters estimated with Equ.(5) always have so large errors that Equ.(5) is scarcely practiced in biomedical analyses. Hence, the proper form of an integrated rate equation after validating should be selected carefully.

$$(S_1 - S)/(t - t_{lag}) = V_m - K_m \times (\ln(S_1/S)/(t - t_{lag})) \qquad (5)$$

In principle, to reliably estimate parameters based on NLSF, the distribution ranges of both the dependent variables and the predictor variables in any kinetic model should be as wide as possible while their random errors should be as small as possible (Baywenton, 1986; del Rio, et al., 2001; Draper & Smith, 1998; Miller, J. C. & Miller, J. N., 1993). By serial studies with common enzymes, we found the use of Equ.(4) or similar forms of integrated rate equations with the predictor variables of reaction time for kinetic analysis of reaction curve could give reliable V_m and S_0, when K_m was fixed at a constant after optimization (Liao, 2005; Liao, et al, 2001, 2003a, 2003b, 2005a, 2005b, 2006, 2007b; Zhao, Y.S., et al., 2006, 2009). Reaction time as the predictor variable has the widest distribution and the smallest random errors, in comparison to the predictor variable in Equ.(5). The left part in Equ.(4) also possess a wider distribution range. Such differences in predictor variables and dependent variables should account for different reliability of parameters estimated with Equ.(2) and Equ.(5), and thus an integrated rate equation with the predictor variable of reaction time may be the proper form for kinetic analysis of reaction curve. However, when NLSF with Equ.(4) is realized with S_1 as a nonlinear parameter, there is nearly 10 s for computation with Celeron 300A CPU on a personal computer. Currently, computation resource is no longer a problem and thus Equ.(4) or its equivalent equations should always be adopted.

The selection of a weighting factor for kinetic analysis of reaction curve is also a concern. Based on error propagation and the principle for weighted NLSF with y defined in Equ.(3), squares of instantaneous rates can be the weighting factors (W_f) with Equ.(4) for NLSF to get the weighted sum of residual squares (Q), as described in Equ.(6), Equ.(7) and Equ.(8) (Baywenton, 1986; Draper & Smith, 1998; Gutierrez & Danielson, 2006; Miller, J. C. & Miller, J. N., 1993). The use of a weighting function like Equ.(7) can mitigate the effects of errors in substrate or product concentrations near the completion of reaction. The resistance of an estimated parameter (the variation within 3% in our studies) to reasonable changes in data ranges for analysis can be a criterion to judge the reliability of parameter estimated.

$$\partial y/\partial S = -(K_m + S)/(K_m \times S) \qquad (6)$$

$$W_f = \partial S/\partial y = -K_m \times S/(K_m + S) \quad (7)$$

$$Q = \sum W_f^2 \times (y_{predicted} - y_{calculated})^2 \quad (8)$$

It is also concerned which parameter is suitable for estimation by kinetic analysis of reaction curve. In theory, all parameters of an enzyme reaction system can be simultaneously estimated by kinetic analysis of reaction curve. However, there is unknown covariance among some parameters to devalue their reliability; there is the limited accuracy of original data for analyses and the estimation of some parameters with narrow working ranges will have negligible practical roles. V_m is independent of all other parameters and so is S_0, and the assay of V_m and S_0 are already routinely practiced in biomedical analyses. Therefore, V_m and S_0 may be the parameters suitable for estimation by kinetic analysis of reaction curve. Additionally, K_m is used for screening enzyme mutants and enzyme inhibitors; but K_m estimated by kinetic analysis of reaction curve usually exhibits lower reliability and is preferred to be fixed for estimating V_m and S_0. If K_m is estimated as well, S_1 should be at least 1.5-fold K_m and there should be more than 85% consumption of the substrate in the data selected for analysis (Atkins & Nimmo, 1973; Liao, et al., 2005a; Newman, et al., 1974; Orsi & Tipton, 1979). To estimate K_m, the initial datum (S_1) and its corresponding ending datum from a reaction curve for analysis should be tried sequentially till the requirements for data range are met concurrently. In this case, the estimation of S_1 has no practical roles. In general, the resistance of V_m and S_0 to reasonable changes in ranges of data for analyses can be a criterion to select the optimized set of parameters that are fixed as constants.

In comparison to the low reliability to estimate K_m independently for screening enzyme inhibitors and enzyme mutants, the ratio of V_m to K_m as an index of enzyme activity can be estimated robustly by kinetic analysis of reaction curve. Reversible inhibitors of Michaelis-Menten enzyme include competitive, noncompetitive, uncompetitive and mixed ones (Bergmeyer, 1983; Dixon & Webb, 1979; Marangoni, 2003). The ratios of V_m to K_m will respond to concentrations of common inhibitors except uncompetitive ones that are very rare in nature. Thus, the ratio of V_m to K_m can be used for screening common inhibitors. More importantly, the ratio of V_m to K_m is an index of the intrinsic activity of an enzyme and the estimation of the ratios of V_m to K_m can also be a promising strategy to screen enzyme mutants of powerful catalytic capacity (Fresht, 1985; Liao, et al., 2001; Northrop, 1983).

For robust estimation of the ratio of V_m to K_m of an enzyme, S_0 can be preset at a value below 10% of K_m to simplify Equ.(2) into Equ.(9). Steady-state data from a reaction curve can be analyzed after data transformation according to the left part in Equ.(9). For validating Equ.(9), it is proposed that S_0 should be below 1% of K_m (Mey1er-Almes & Auer, 2000). The use of extremely low S_0 requires special methods to monitor enzyme reaction curves and steady-state reaction can not always be achieved with enzymes of low intrinsic catalytic activities. On the other hand, the use of S_0 below 10% of K_m is reasonable to estimate the ratio of V_m to K_m (Liao, et al., 2001). To estimate the ratio of V_m to K_m, the use of Equ.(9) to analyze data is robust and resistant to variations of S_0 if Equ.(9) is valid; this property makes the estimation of the ratio of V_m to K_m for screening reversible inhibitors superior to the estimation of the half-inhibition concentrations (Cheng & Prusoff, 1973).

$$\ln(S_1/S) = a + (V_m/K_m) \times t \quad (9)$$

Kinetic analysis of reaction curve requires more considerations when activities of enzymes are altered by their substrates/products. In this case, more parameters can be included in kinetic models similar to Equ.(2) for kinetic analysis of reaction curve, but there must be complicated process to optimize reaction conditions and preset parameters. Based on the principle for kinetic analysis of reaction curve described above, we developed some new integration strategies to successfully quantify enzyme initial rates and substrate with absorbing performance even when the activities of enzymes of interest are altered significantly by substrates/products (Li, et al., 2011; Liao, 2007a; Zhao, L.N., et al., 2006).

2.3 Kinetic analysis of enzyme-coupled reaction curve

When neither substrate nor product is suitable for continuous monitor of reaction curve, a tool enzyme can be used to regenerate a substrate or consume a product of the enzyme of interest; the action of the tool enzyme should consume/generate a substrate/product as an indicator suitable for continuous monitor of reaction curve. Namely, the reaction of the tool enzyme is coupled to the reaction of an enzyme of interest for continuous monitor of reaction curve (Bergmeyer, 1983; Guilbault, 1976; Dixon & Webb, 1979). When such enzyme-coupled assays are used to measure initial rates of an enzyme, there are always unsatisfactory linear range because the activities of the tool enzyme is always limited and the concentration of the substrate of the tool enzyme is also limited (Bergmeyer, 1983; Dixon & Webb, 1979). It is expected that kinetic analysis of enzyme-coupled reaction curve may effectively enhance the upper limit of linear response. However, kinetics of enzyme-coupled reaction systems is described with a set of differential rate equations, which cause difficulty in accessing an integrated rate equation with the predictor variable of reaction time.

In this case, iterative numerical integration to obtain calculated reaction curves for NLSF to a reaction curve of interest can be used (Duggleby, 1983, 1994; Moruno-Davila, et al., 2001; Varon, et al., 1998; Yang, et al., 2010). Lactic dehydrogenase (LDH) is widely used as a tool enzyme for enzyme-coupled assay. The assay of activity of alanineaminotransferase (ALT) in sera has important biomedical roles and usually employs LDH-coupled assay. For LDH-coupled ALT assay, iterative numerical integration of the set of differential rate equations with each set of preset parameters from a preset starting point can produce a calculated reaction curve; such a calculated reaction curve can be made discrete at the same intervals as the reaction curve of interest and then be used for NLSF to the reaction curve of interest.

The process of iterative numerical integration for LDH-coupled ALT assay is given below (Yang, et al., 2010). In an LDH-coupled ALT reaction system, assigning instantaneous concentration of NADH to $C_{n,i}$, instantaneous concentration of pyruvate to $C_{p,i}$, instantaneous absorbance at 340 n for NADH to A_i, the molar absorptivity of NADH to ε, the initial rate of ALT under steady-state reaction to V_{1k}, the maximal activity of LDH to V_m, the integration step to Δt, there are Equ.(10), Equ.(11) and Equ.(12) to describe the iterative integration of the set of differential rate equations. Calculated reaction curves according to Equ. (12) using different sets of preset parameters become discrete and are fit to the reaction curve of interest, and background absorbance at 340 nm is treated as a parameter as well.

$$C_{n,i} = (A_i - A_b)/\varepsilon \tag{10}$$

$$C_{p,i+1} = C_{p,i} + V_{1k} \times \Delta t - V_m \times \Delta t / (1 + K_a/C_{n,i} + K_b/C_{p,i} + K_{ab}/(C_{n,i} \times C_{p,i})) \tag{11}$$

$$A_{i+1} = A_i - \varepsilon \times V_m \times \Delta t / (1 + K_a/C_{n,i} + K_b/C_{p,i} + K_{ab}/C_{n,i} \times C_{p,i}) \tag{12}$$

By simulation with such a new approach for kinetic analysis of enzyme-coupled reaction curve recorded at 1-s intervals, the upper limit of linear response for measuring ALT initial rates is increased to about five times that by the classical initial rate method. This new approach is resistant to reasonable variations in data range for analysis. By experimentation using the sampling intervals of 10 s, the upper limit is about three times that by the classical initial rate method. Therefore, this new approach for kinetic analysis of enzyme-coupled reaction curve is advantageous, and can potentially be a universal approach for kinetic analysis of reaction curve of any system of much complicated kinetics.

The computation time for numerical integration is inversely proportional to the integration step, Δt; the use of shorter Δt is always better but Δt of 0.20 s at low cost on computation is sufficient for a desirable upper limit of linear response. This new approach with Celeron 300A CPU on a personal computer needs about 10 min for just 30 data in a LDH-coupled reaction curve, but it consumes just about 5 s with Lenovo Notebook S10e. The advancement of personal computers surely can promote the practice of this approach.

2.4 Integration of kinetic analysis of reaction curve with other methods

Any analytical method should have favourable analysis efficiency, wide linear range, low cost and strong robustness. Kinetic analysis of reaction curve for V_m and S_0 assay can have much better upper limit of linear response, but inevitably tolerates low analysis efficiency when wide linear range is required. Based on kinetic analysis of reaction curve, however, our group developed two integration strategies for enzyme initial rate and substrate assay, respectively, with both favourable analysis efficiency and ideal linear ranges.

2.4.1 New integration strategy for enzyme initial rate assay

The classical initial rate method to measure enzyme initial rates requires S_0 much higher than K_m to have desirable linear ranges (Bergmeyer, 1983; Dixon & Webb, 1979; Guilbault, 1976; Marangoni, 2003). Due to substrate inhibition, limited solubility and other causes, practical substrate levels are always relatively low and thus the linear ranges by the classical initial rate method are always unsatisfactory (Li, et al., 2011; Morishita, et al., 2000; Stromme & Theodorsen, 1976). As described above, kinetic analysis of reaction curve can measure enzyme V_m, and many approaches based on kinetic analysis of reaction curve are already proposed (Cheng, et al., 2008; Claro, 2000; Cornish-Bowden 1975, 1995; Dagys, et al., 1986, 1990; Duggleby, 1983, 1985, 1994; Hasinoff, 1985; Koerber, & Fink, 1987; Liao, et al., 2001; Lu & Fei, 2003; Marangoni, 2003; Walsh, et al. 2010). Such approaches all require substrate consumption percentage over 40% with K_m preset as a constant. As a result, there should be intolerably long reaction duration to monitor reaction curves for samples of low enzyme activities, or else the lower limits of linear response are unfavourable.

The integration of kinetic analysis of reaction curve using proper integrated rate equations with the classical initial rate method gives an integration strategy to measure enzyme initial rates with expanded linear ranges and practical analysis efficiency. This integration strategy is effective at substrate concentrations from one-eighth of K_m to three-fold of K_m (Li, et al., 2011; Liao, et al., 2009; Liu, et al., 2009; Yang, et al., 2011). The integration strategy for enzyme initial rate assay uses a special method to convert V_m into initial rates so that the indexes of enzyme activities by both methods become the same; it is applicable to enzymes suffering strong inhibition by substrates/products (Li, et al., 2011). Walsh et al. proposed an integration strategy to measure enzyme initial rate but they employed Equ.(9) that requires substrate levels below 10% of K_m (Walsh, et al. 2010). Our integration strategy is valid at any substrate level to satisfy Equ.(2) and hence can be a universal approach to common enzymes of different K_m. The principles and applications of the integration strategy to one enzyme reaction systems and enzyme-coupled reaction systems are discussed below.

As for one enzyme reaction systems, kinetic analysis of reaction curve can be realized with an integrated rate equation after data transformation; the integration strategy for enzyme initial rate assay requires enzyme kinetics on single substrate and an integrated rate equation with the predictor variable of reaction time (Liao, et al., 2003a, 2005a, Zhao, L.N., et al., 2006). Moreover, the integration strategy should solve the following challenges: (a) there should be an overlapped range of enzyme activities measurable by both methods with consistent results; (b) there should be consistent slopes of linear response for enzyme activities to enzyme quantities by both methods (Figure 1). After these two challenges are solved, the linear segment of response by the classical initial rate method is an extended line of the linear segment of response by kinetic analysis of reaction curve (Liu, et al., 2009).

To solve the first challenge, a practical S_0 and reasonable duration to monitor reaction curve for favourable analysis efficiency are required as optimized experimental conditions. By mathematic derivation and simulation analyses to solve the first challenge, it is demonstrated that a ratio of S_0 to K_m from 0.5 to 2.5, the duration of 5.0 min to monitor reaction curves at intervals no longer than 10 s can solve the first challenge for most enzymes, any ratio of S_0 to K_m smaller than 0.5 or larger than 2.5 requires longer duration to monitor reaction curves. The use of S_0 about one-eighth of K_m requires no shorter than 8.0 min to monitor reaction curves at 10-s intervals to solve the first challenge (Li, et al., 2011; Liu, et al., 2009). When S_0 is too much larger than three times K_m, reaction time to record reaction curves for analysis to solve the first challenge should be much longer than 5 min. Clearly, the first challenge can be solved with practical S_0 for favourable analysis efficiency.

To solve the second challenge, K_m and other parameters should be optimized and fixed as constants to estimate V_m by kinetic analysis of reaction curve, and a preset substrate concentration (PSC) should be optimized to covert V_m into initial rates according to the differential rate equation. In theory, a reliable V_m should be independent of ranges of data when they are reasonably restricted, and CVs for estimating parameters by enzymatic analysis are usually about 5%. Hence, the estimation of V_m with variations below 3% for the changes of substrate consumption percentages from 60% to 90% can be a criterion to select the optimized set of preset parameters. For converting V_m into initial rates, the optimized PSC is usually about 93% of S_0 and can be refined for different enzymes (Li, et al., 2011; Liao, et al., 2009; Liu, et al., 2009; Yang, et al., 2011). Optimized K_m and PSC to solve the second challenge are parameters for data processing while optimized S_0 and reaction duration to solve the first challenge are experimental conditions. The concomitant solution of the two challenges provides feasibility and potential reliability to the integration strategy.

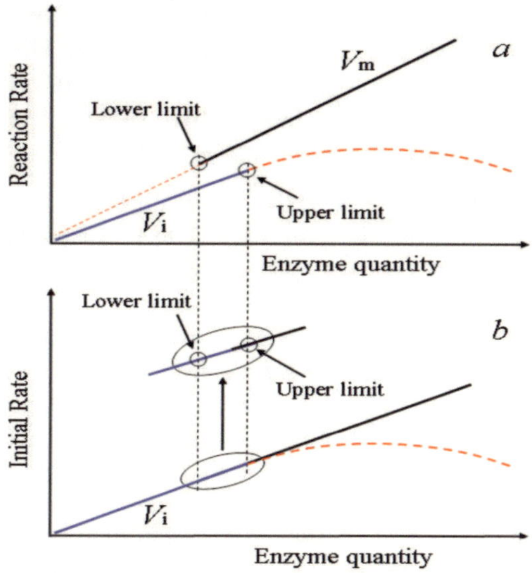

Fig. 1. The integration strategy for enzyme initial rate assay (Modified from Liu et al. (2009)).

After the integration strategy for enzyme initial rate assay is validated, a switch point should be determined for changing from the classical initial rate method to kinetic analysis of reaction curve. The estimation of V_m by kinetic analysis of reaction curve usually prefers substrate consumption percentages reasonably high. Therefore, the substrate consumption percentage that gives an enzyme activity from 90% to 100% of the upper limit of linear response by the classical initial rate method can be used as the switch point.

It should be noted that the lower limit of linear response is difficult to be defined for enzyme initial rate assay by an integration strategy. For most methods, their lower limits of linear response are usually defined as three times the standard errors of estimate (Miller, J. C. & Miller, J. N., 1993). Usually, enzyme initial rate assay utilizes just one method for data processing and the difference between the lower limit and the upper limit of linear response is seldom over 30-fold. By the integration strategy, the measurable ranges of enzyme quantities cover two magnitudes and the detection limit is reduced to that by the classical initial rate method. By manual operation, different dilution ratios of a stock solution of the enzyme have to be used and any dilution error will increase the standard error of estimate for regression analysis. The measurement of higher enzyme activities will inevitably have larger standard deviation. Thus, regression analysis of the response of all measurable enzyme initial rates by the integration strategy to quantities of the enzyme will give higher standard error of estimate and thus an unfavourable lower limit of linear response. By this new integration strategy, we arbitrarily use twice the lower limit of linear response by the classical initial rate method as the lower limit if the overall standard error of estimate is more than twice that by the classical initial rate method alone; or else, the lower limit of linear response is still three times the overall standard error of estimate.

Taken together, for measuring initial rates of enzyme acting on single substrate by the integration strategy based on NLSF and data transformation, there are the following basic steps different from those by the classical initial rate method. The first is to work out the integrated rate equation with the predictor variable of reaction time. The second is to optimize individually their parameters fixed as constants for kinetic analysis of reaction curve. The third is to optimize a ratio of S_0 to K_m and duration to monitor reaction curves; usually a ratio of S_0 to K_m from 0.5 to 2.5, the duration of 5.0 min and intervals of 10 s are effective. The forth is to refine PSC around 93% of S_0 to convert V_m into initial rates.

As for enzyme-coupled reaction system, initial rate itself is estimated by kinetic analysis of reaction curve based on numerical integration and NLSF of calculated reaction curves to a reaction curve of interest. Consequently, neither the conversion of indexes nor the optimization of parameters for such conversion is required and the integration strategy can be realized easily. By kinetic analysis of enzyme-coupled reaction curve, there still should be a minimum number of the effective data and a minimum substrate consumption percentage in the effective data for analysis; these prerequisites lead to unsatisfactory lower limits of linear response for favourable analysis efficiency (the use of reaction duration within 5.0 min). The classical initial rate method is effective to enzyme-coupled reaction systems when activities of the enzyme of interest are not too high. Therefore, this new approach for kinetic analysis of enzyme-coupled reaction curve can be integrated with the classical initial rate method to quantify enzyme initial rates potentially for wider linear ranges.

With enzyme-coupled reaction systems, only the first challenge should be solved to practice the integration strategy. Namely, reaction duration and sampling intervals to record reaction curve should be optimized so that there is an overlapped region of enzyme initial rates measurable by both methods with consistent results. The upper limit of the classical initial rate method should be high enough so that data after reaction of about 5.0 min for enzyme activity at such an upper limit are suitable for kinetic analysis of reaction curve. The integration strategy gives an approximated linear range from the lower limit of linear response by the classical initial rate method to the upper limit of linear response by kinetic analysis of LDH-coupled ALT reaction curve (Yang, et al., 2010).

2.4.2 New integration strategy for enzyme substrate assay

Analysis of a biochemical as the substrate of a typical tool enzyme, *i.e.*, enzymatic analysis of substrate in biological samples, is important in biomedicine (Bergmeyer, 1983; Dilena, 1986; Guilbault, 1976; Moss, 1980). In general, there are the kinetic method and the end-point method for enzyme substrate assay; the end-point method is called the equilibrium method, and it determines the difference between the initial signal for a reaction system before enzyme action and the last signal after the completion of enzyme reaction; such differences proportional to S_0 can serve as an index of substrate concentration (Dilena, et al., 1986; Guilbault, 1976; Moss, 1980; Zhao, et al., 2009). For better analysis efficiency and lower cost on tool enzymes, kinetic methods for enzyme substrate assay are preferred. Among available kinetic methods, the initial rate method based on the response of initial rates of an enzyme at a fixed quantity to substrate concentrations is conventional; however, it tolerates sensitivity to any factor affecting enzyme activities, requires tool enzymes of high K_m, and has narrow linear ranges. Kinetic analysis of reaction curve with a differential rate equation to estimate S_0 is proposed with favourable resistance to variations in enzyme activities and has upper limit of linear response over K_m, but it suffers from high sensitivity to background and has unfavourable lower limit of linear response (Dilena, et al., 1986; Hamilton & Pardue, 1982; Moss, 1980). Hence, new kinetic methods for enzyme substrate assay are still desired.

For enzymatic analysis of substrate, the equilibrium method can still be preferable as long as it has desirable analysis efficiency and favourable cost on tool enzyme. In theory, the last signal for the stable product or the background in the equilibrium method can be estimated by kinetic analysis of reaction curve with data far before the completion of reaction. This process can be a new kinetic method for enzyme substrate assay and is distinguished from the equilibrium method and other kinetic methods by its prediction of the last signal after the completion of enzyme reaction (Liao, 2005; Liao, et al., 2003, 2005a, 2006; Zhao, L.N., et al., 2006; Zhao, Y.S., et al., 2006, 2009). This new kinetic method should have resistance to factors affecting enzyme activities and upper limit of linear response higher than K_m besides all advantages of common kinetic methods.

An enzyme reaction curve can be monitored by absorbance of a stable product or the substrate itself (Figure 2). The initial absorbance before enzyme action (A_0) thus is the background (A_b) when absorbance for a stable product is quantified, or is the absorbance of the substrate plus background when absorbance of the substrate is quantified. The last absorbance after the completion of enzyme reaction, which is predicted by kinetic analysis of reaction curve, is the maximum absorbance of the stable product plus the background (A_m) or A_b itself. There is strong covariance between the initial signal and the last signal for the same reaction system; this assertion enhances precision of this kinetic method for substrate assay (Baywenton, 1986; Liao, et al, 2005b; Zhao, Y.S., et al., 2009).

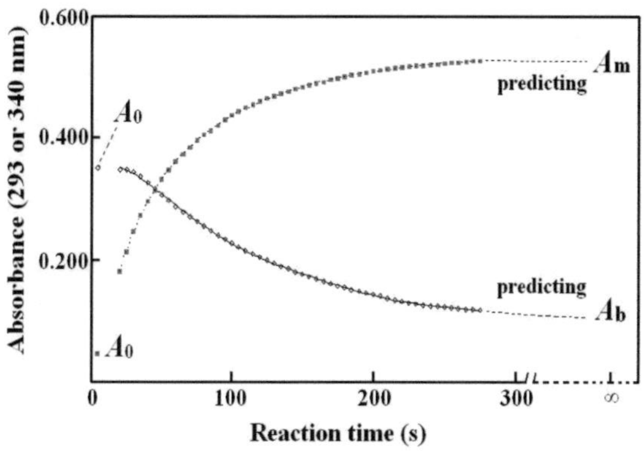

Fig. 2. Demonstration of reaction curves of uricase (293 nm) and glutathione-S-tranferase (340 nm), and the prediction of the last absorbance after infinite reaction time

However, this new kinetic method for substrate assay can by no means concomitantly have wider linear ranges and desirable analysis efficiency. Due to the prerequisites of the quality of data for kinetic analysis of reaction curve, the activity of a tool enzyme for enzymatic analysis of substrate should be reasonably high for higher upper limit of linear response, but it should be reasonably low for favourable lower limit of linear response. On the other hand, the duration to monitor reaction curves should be long enough to have higher upper limit at reasonable cost on a tool enzyme, but should be as short as possible for favourable analysis efficiency. Thus, this new kinetic method alone requires tough optimizations of conditions. Moreover, there are the inevitable random noises from any instrument to record an enzyme reaction curve; when there is much small difference between the initial signal before enzyme action and the last signal recorded after the preset reaction duration, this kinetic method for substrate assay always has unsatisfactory precision. Therefore, this new kinetic method itself is still much beyond satisfaction for substrate assay.

To concomitantly have wider linear ranges, desirable analysis efficiency and favourable precision for enzyme substrate assay, the integration of kinetic analysis of reaction curve with the equilibrium method can be used. The indexes of substrate quantities by the two methods have exactly the same physical meanings, and thus the integration strategy can be easily realized for enzyme substrate assay. By this integration strategy, there should still be an overlapped range of concentrations of the substrate measurable consistently by both methods, besides a switch threshold within such an overlapped region to change from the equilibrium method to kinetic analysis of reaction curve. Additionally, this overlapped region of substrate concentration measurable by both methods with consistent results should localize in a range of substrate concentration high enough for reasonable precision of substrate assay based on kinetic analysis of enzyme reaction curve. These requirements can be met as described below. (a) The upper limit of linear response by the equilibrium method should be optimized to be high enough, so that the difference between the initial signal before enzyme action and the last recorded signal for about 80% of this upper limit is 50 times higher than the random noise of an instrument to record enzyme reaction curves; such a difference can be used as the switch threshold. (b) The activity of a tool enzyme and the duration to monitor reaction curve as experimental conditions should be optimized; kinetic parameters except V_m for kinetic analysis of reaction curve are optimized as well. The resistance of the predicted last signal to reasonable variations in data ranges for analysis can be a criterion to judge the optimized set of preset parameters. For favourable analysis efficiency in clinical laboratories, reaction duration can be about 5.0 min. This reaction duration results in a minimum activity of the tool enzyme for the integration strategy so that the upper limit of linear response by the equilibrium method can be high enough to switch to kinetic analysis of reaction curve. This integration strategy after optimizations can simultaneously have wider linear ranges, higher analysis efficiency and lower cost, better precision and stronger resistance to factors affecting enzyme activities.

Similarly, with the integration strategy for enzyme substrate assay, we also use twice the lower limit of the equilibrium method as the lower limit by the integration strategy if the standard error of estimate is much larger; or else, three times the standard error of estimate by the integration strategy is taken as the lower limit of linear response.

In general, the following steps are required to realize this integration strategy for enzyme substrate assay: (a) to work out the integrated rate equation with the predictor variable of reaction time; (b) to optimize individually the (kinetic) parameters preset as constants for kinetic analysis of reaction curve; (c) to optimize the activity of the tool enzyme so that data for the upper limit of linear response by the equilibrium method within about 5.0-min reaction are suitable for kinetic analysis of reaction curve. As demonstrated later, this integration strategy is applicable to enzymes suffering from strong product inhibition.

2.5 Applications of new methods to some typical enzymes

We investigated kinetic analysis of reaction curve with arylesterase (Liao, et al., 2001, 2003a, 2007b), alcohol dehydrogenase (ADH) (Liao, et al., 2007a), gama-glutamyltransfease (Li, et al., 2011), uricase (Liao, 2005; Liao, et al., 2005a, 2005b, 2006; Liu, et al., 2009; Zhao, Y.S., et al., 2006, 2009), glutathione-S-tranferase (GST) (Liao, et al., 2003b; Zhao, L.N., et al., 2006), butylcholineasterase (Liao, et al., 2009; Yang, et al., 2011), LDH (Cheng, et al., 2008) and LDH-coupled ALT reaction systems (Yang, et al., 2010). Uricase of simple kinetics is a good example to study new methods for kinetic analysis of reaction curve; reactions of GGT and ADH suffer product inhibition and kinetic analyses of their reaction curves are complicated because they require unreported parameters. Hence, our new methods for kinetic analysis of reaction curve and the integration strategies for quantifying enzyme substrates and initial rates are demonstrated with uricase, GST and ADH as examples.

2.5.1 Uricase reaction

Uricase follows simple Michaelis-Menten kinetics on single substrate in air-saturated buffers, and suffers neither reversible reaction nor product inhibition (Liao, 2005; Liao, et al., 2005a, 2005b; Zhao, Y.S., et al., 2006). Uricase reaction curve can be monitored by absorbance at 293 nm. The potential interference from the intermediate 5-hydroxylisourate with uric acid absorbance at 293 nm can be alleviated by analyzing data of steady-state reaction in borate buffer at high reaction pH (Kahn & Tipton, 1998; Priest & Pitts, 1972). The integrated rate equation for uricase reaction with the predictor variable of reaction time is Equ.(4). Uricases from different sources have different K_m (Liao, et al., 2005a, 2006; Zhang, et al., 2010; Zhao, Y.S., et al., 2006). Using Equ.(4), K_m of *Candidate* utilis is estimated with reasonable reliability (Liao, et al., 2005a). Using Equ.(9) to estimate the ratio of V_m to K_m, uricase mutants of better catalytic capacity and their sensitivity to xanthine are routinely characterized (data unpublished). Thus, we used uricases of different K_m as models to test the two integration strategies for enzyme substrate assay and initial rate assay, respectively.

Uricase from *Bacillus* fastidiosus A.T.C.C. 29604 has high K_m to facilitate predicting A_b (Zhang, et al., 2010; Zhao, Y.S., et al., 2006, 2009). Reaction curves at low levels of uric acid with this uricase at 40 U/L are demonstrated in Fig. 3. Steady-state reaction is not reached within 30 s since reaction initiation; it is difficult to get more than 5 data with absorbance changes over 0.003 for kinetic analysis of reaction curve at uric acid levels below 3.0 μmol/L. At 40 U/L of this uricase, the absorbance after reaction for 5.0 min has negligible difference from that after reaction for 30 min for uric acid below 5.0 μmol/L. To quantify the difference between A_0 and A_b after reaction for 5.0 min, the equilibrium method has an upper limit of about 5.0 μmol/L, while kinetic analysis of reaction curve with K_m as a constant is feasible for S_0 of about 5.0 μmol/L. Thus, the change of absorbance over 0.050 between A_0 and the absorbance after reaction for 5.0 min can be the switch threshold to change from the equilibrium method to kinetic analysis of reaction curve.

This integration strategy for enzyme substrate assay gives the linear response from about 1.5 μmol/L up to 60 μmol/L uric acid at 40 U/L uricase (Fig.4, unpublished), and shows resistance to the action of xanthine at 30 μmol/L in reaction solutions (this level of xanthine always caused negative interference with all available kits commercialized for serum uric acid assay). Therefore, the integration strategy for uric acid assay is clearly superior to any other uricase method reported.

Uricases from *Candida* sp. with K_m of 6.6 μmol/L (Sigma U0880) and *Bacillus* fastidious uricase from A.T.C.C. 29604 with K_m of 0.22 mmol/L are used to test the integration strategy for initial rate assay. The use of uric acid at S_0 of 25 μmol/L to monitor reaction curves

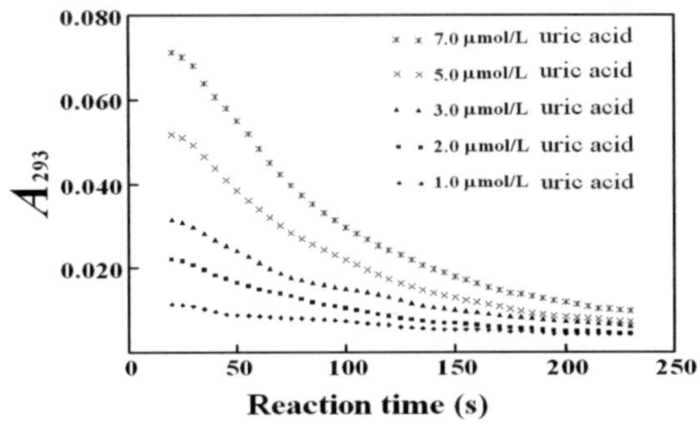

Fig. 3. Reaction curves (absorbance at 293 nm) at low levels of uric acid and 40 U/L uricase (recombinant uricase in *E.* Coli BL21 was as reported before (Zhang, et al., 2010)).

within 8.0 min or at S_0 of 75 μmol/L to monitor reaction curves within 5.0 min, the integration strategy to measure initial rates of both uricases is feasible; the use of PSC of 93% S_0 to convert V_m into initial rates gives the linear range of about two magnitudes (Liu, et al., 2009). Therefore, the integration strategy for enzyme initial rate assay is also advantageous.

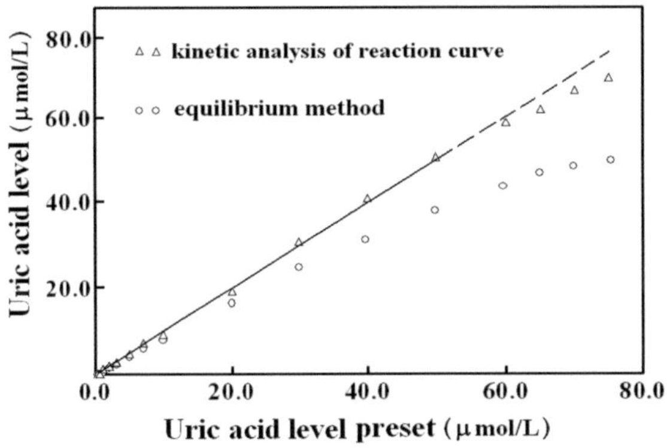

Fig. 4. Response of absorbance change at 293 nm to preset uric acid levels at 40 U/L uricase.

2.5.2 Glutathione-S-transferase reaction

Using purified alkaline GST isozyme from porcine liver as model on glutathione (GSH) and 2,4,-dinitrochlorobenzene (CDNB) as substrates, GST reaction curves are monitored by absorbance at 340 nm (Kunze, 1997; Pabst, et al, 1974; Zhao, L.N., et al., 2006). To promote reaction on single substrate, CDNB is fixed at 1.0 mmol/L while GSH concentrations are kept below 0.10 mmol/L (Zhao, L.N., et al., 2006). Because the concentration of product is calculated from absorbance at 340 nm, the background absorbance before GST reaction is adjusted to zero so that there is no need to treat A_b as a parameter. This treatment of background absorbance eliminate the estimation of A_b and thus confronts with no problem of covariance between A_b and A_m for NLSF. However, GST reaction is more complicated than uricase because it suffers strong product inhibition with an unreported inhibition constant (Kunze, 1997; Pabst, et al, 1974). Thus, the effectiveness of the two integration strategies is tested for measuring initial rate and GSH levels after the inhibition constant of the product is optimized for kinetic analysis of GST reaction curve.

The following symbols are assigned: C to instantaneous concentration of CDNB, B to instantaneous concentration of GSH, Q to instantaneous concentration of the product, K_{ma} to K_m of GST for CSNB, K_{mb} to K_m of GST for GSH, K_{ia} to the dissociation constant of GSH, K_{iq} to the dissociation constant of the product, A for instantaneous absorbance, A_m for the maximal absorbance of the product, ε to difference in absorptivity of product and CDNB, V_m for the maximal reaction rate of GST. The differential rate equation for GST reaction is Equ.(13). After the definition of M_1, M_2 and M_3, the integrated rate equation with the predictor variable of reaction time is Equ.(19) if GST reaction is irreversible and a process similar to that for Equ.(4) is employed (Zhao, L.N., et al., 2006).

$$\frac{1}{V} = \left(K_{mb}/V_m\right) \times \left[1 + K_{ib} \times K_{ma} \times Q/\left(K_{iq} \times K_{mb} \times C\right)\right]/B + \left[1 + K_{ma} \times \left(1 + Q/K_{iq}\right)/C\right]/V_m \tag{13}$$

$$M_1 = K_{ma}/\left(\varepsilon \times K_{iq}\right) \tag{14}$$

$$M_2 = K_{ma} - K_{ib} \times K_{ma}/K_{iq} - A_m \times K_{ma}/\left(\varepsilon \times K_{iq}\right) + C - A_0 \times K_{ma}/\left(\varepsilon \times K_{iq}\right) \tag{15}$$

$$M_3 = K_{ma} \times A_m + \varepsilon \times K_{mb} \times C + C \times A_m - K_{ib} \times K_{mb} \times A_0/K_{iq} - K_{ma} \times A_m \times A_0/\left(K_{iq} \times \varepsilon\right) \tag{16}$$

$$\frac{M_1 \times A^2 + M_2 \times A - M_3}{A - A_m} \times dA = C \times \varepsilon \times V_m \times dt \tag{17}$$

$$Y = M_1 \times \left(A - A_m\right)^2/2 + \left(2 \times M_1 \times A_m + M_2\right) \times \left(A - A_m\right) + \left(M_1 \times A_m^2 + M_2 \times A_m - M_3\right) \times Ln\left|A - A_m\right| \tag{18}$$

$$Y = C \times \varepsilon \times V_m \times \left(t - T_{lag}\right) = a + b \times t \tag{19}$$

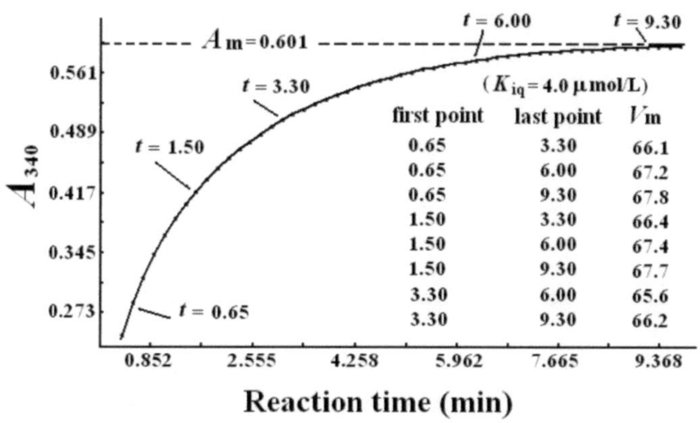

Fig. 5. Estimated V_m to changes in data ranges for analyses with 60 μmol/L GSH.

As demonstrated in the definition of M_1, M_2 and M_3, kinetic parameters preset as constants for kinetic analysis of GST reaction curve should have strong covariance. Except K_{iq} as an unknown kinetic parameter for optimization, other kinetic parameters are those reported (Kunze, 1997; Pabst, et al, 1974). To optimize K_{iq}, two criteria are used. The first is the consistency of predicted A_m at a series of GSH concentrations using data of 6.0-min reaction with that by the equilibrium method after 40 min reaction (GST activity is optimized to complete the reaction within 40 min). The second is the resistance of V_m to reasonable changes in data ranges for analyses. After stepwise optimization, K_{iq} is fixed at 4.0 μmol/L; A_m predicted for GSH from 5.0 μmol/L to 50 μmol/L is consistent with that by the equilibrium method (Zhao, L.N., et al. 2006); the estimation of V_m is resistant to changes of data ranges (Fig. 5). Therefore, K_{iq} is optimized and fixed as a constant at 4.0 μmol/L.

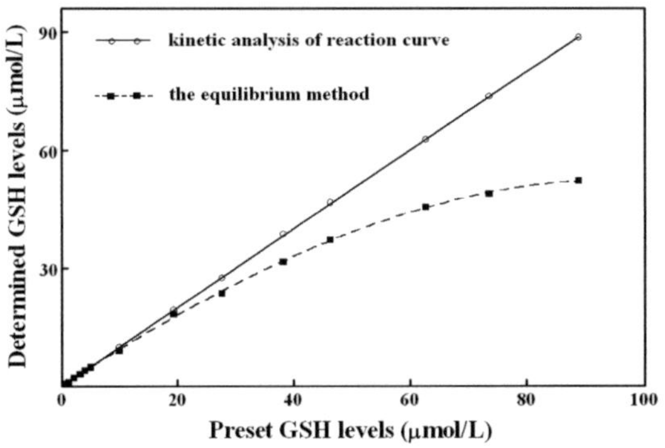

Fig. 6. Response of GSH concentration determined to preset GSH concentrations (the equilibrium method uses data with 6.0 min reaction).

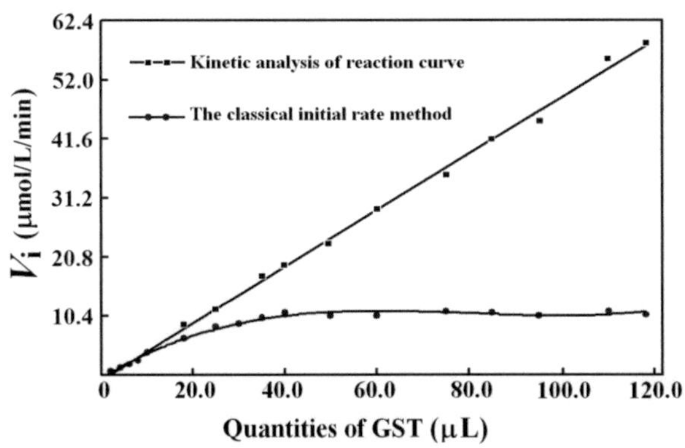

Fig. 7. Response of initial rates to quantities of purified porcine alkaline GST.

Kinetic analysis of GST reaction curve can predict A_m for GSH over 4.0 μmol/L, but there are no sufficient data for analyses at GSH below 3.0 μmol/L; after optimization of GST activity for complete conversion of GSH at 5.0 μmol/L within 6.0 min, reaction curve within 5.0 min for GSH at 5.0 μmol/L can be used for kinetic analysis of reaction curve to predict A_m. With the optimized GST activity for reaction within 5.0 min, the linear range for GSH assay is from 1.5 μmol/L to over 90.0 μmol/L by the integration strategy while it is from 4.0

μmol/L to over 90.0 μmol/L by kinetic analysis of reaction curve alone (Fig. 6, unpublished). By the equilibrium method alone for reaction within 5.0 min, the assay of 80.0 μmol/L GSH requires GST activity that is 50 folds higher due to the inhibition of GST by the accumulated product. Therefore, the integration strategy for GSH assay is obviously advantageous.

The integration strategy for measuring GST initial rates is tested. For convenience, S_0 of the final GSH is fixed at 50 μmol/L and the duration to monitor reaction curve is optimized. After the analyses of reaction curves recorded within 10 min, it is found that reaction for 6.0 min is sufficient to provide the required overlapped region of GST activities measurable by both methods. By using K_{iq} fixed at 4.0 μmol/L as a constant, the reaction duration of 6.0 min and PSC at 48 μmol/L to convert V_m to initial rates, the integration strategy gives a linear range from 2.0 U/L to 60 U/L; kinetic analysis of reaction curve alone gives the linear range from 5.0 U/L to 60 U/L while the classical initial rate method alone gives a linear range from 1.0 U/L to 5.0 U/L (Fig. 7, unpublished). Clearly, with enzyme suffering strong product inhibition, the integration strategy for enzyme initial rate assay is advantageous.

2.5.3 Alcohol dehydrogenase reaction

ADH is widely used for serum ethanol assay. ADH kinetics is sophisticated due to the reversibility of reaction and the inhibition by both acetaldehyde and NADH as products. To simplify ADH kinetics, some special approaches are employed to make ADH reaction apparently irreversible on single substrate (alcohol). Thus, reaction pH is optimized to 9.2 to scavenge hydrogen ion; semicabarzide at final 75 mmol/L is used to remove acetaldehyde as completely as possible; final nicotinamide adenine dinucleotide (NAD^+) is 3.0 mmol/L; final ADH is about 50 U/L (Liao, et al., 2007a). By assigning the maximal absorbance at 340 nm for reduced nicotinamide adenine dinucleotide (NADH) by the equilibrium method to A_{me} and that by kinetic analysis of reaction curve to A_{mk}, kinetic analysis of ADH reaction curve should predict A_{mk} consistent with A_{me}, but requires some special efforts.

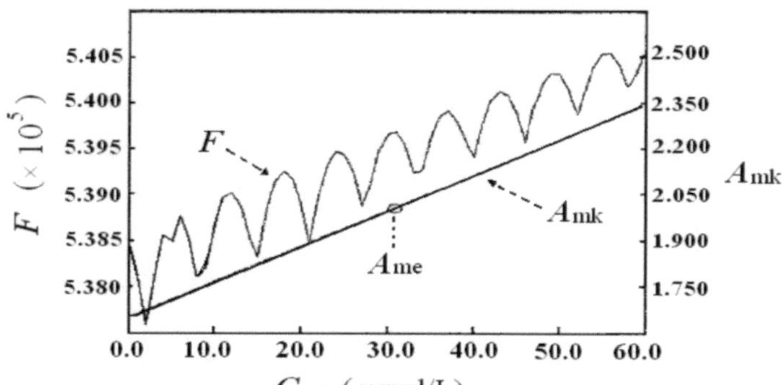

Fig. 8. Response of F values to preset C_{ald} for kinetic analysis of reaction curve for 0.31 mmol/L ethanol (reproduced with permission from Liao, et al, 2007a).

The use of semicabarzide reduces concentrations of acetaldehyde (C_{ald}) to unknown levels, and thus complicates the treatment of acetaldehyde inhibition on ADH. The integration rate equation with the predictor variable of reaction time can be worked out for ADH (Liao, et al., 2007a). All kinetic parameters and NAD^+ concentrations are preset as those used or reported (Ganzhorn, et al. 1987). However, there are multiple maxima of the goodness of fit with the continuous increase in steady-state C_{ald} for kinetic analysis of reaction curve (Fig. 8). Thus, C_{ald} can not be concomitantly estimated by kinetic analysis of reaction curve, and a special approach is used to approximate steady-state C_{ald} for predicting A_{mk}.

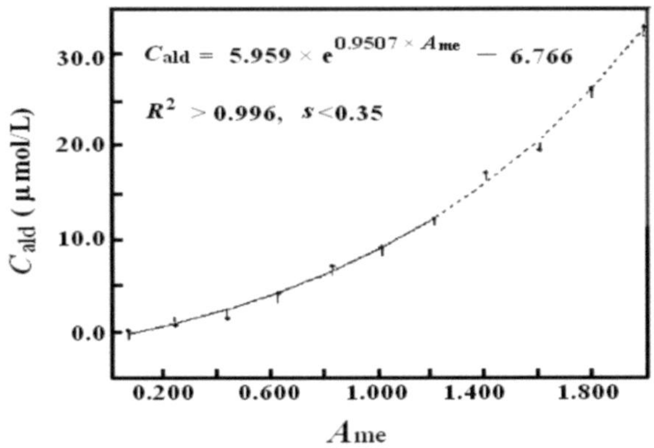

Fig. 9. Correlation function of the best steady-state C_{ald} with A_{me} (reproduced with permission from Liao, et al, 2007a).

Under the same reaction conditions, the equilibrium method can determine A_{me} for ethanol below 0.20 mmol/L after reaction for 50 min. For kinetic analyses of such reaction curves, the lag time for steady-state reaction is estimated to be over 40 s and is used to select data of steady-state reaction for analysis. Using the equilibrium method as the reference method, the best steady-state C_{ald} for data of 6.0-min reaction is obtained for consistency of A_{mk} with A_{me} at each tested ethanol level from 10 µmol/L to 0.17 mmol/L. After dilution and determination by the equilibrium method, A_{me} for each tested ethanol level from 0.17 mmol/L to 0.30 mmol/L is also available. Consequently, an exponential additive function is obtained to approximate the correlation of the best C_{ald} for predicting A_{mk} consistent with A_{me} (Fig. 9). This special correlation function for C_{ald} and A_{mk} is used as a restriction function to iteratively adjust C_{ald} for predicting A_{mk}; namely, iterative kinetic analysis of reaction curve with C_{ald} predicted from the restriction function using previous A_{mk} finally gives the desired A_{mk}. Such an artificial intelligence approach to the steady-state C_{ald} for kinetic analysis of reaction curve can hardly be found in publications.

To start kinetic analysis of an ADH reaction curve, the highest absorbance under analysis is taken as A_{mk} to predict the best C_{ald} for the current run of kinetic analysis of reaction curve. The estimated A_{mk} is then used to predict the second C_{ald} for the second run of kinetic analysis of reaction curve (Fig. 10). Such an iterative kinetic analysis of reaction curve can predict A_{mk} consistent with A_{me} for 0.31 mmol/L ethanol when reaction duration is just 6.0 min and the convergence criterion is set for absorbance change below 0.0015 in A_{mk}. Usually convergence is achieved with 7 runs of the iterative kinetic analysis of reaction. Moreover, it is resistant to the change of ADH activities by 50% and coefficients of variation (CV) are below 5% for final ethanol levels from 20 µmol/L to 310 µmol/L in reaction solutions.

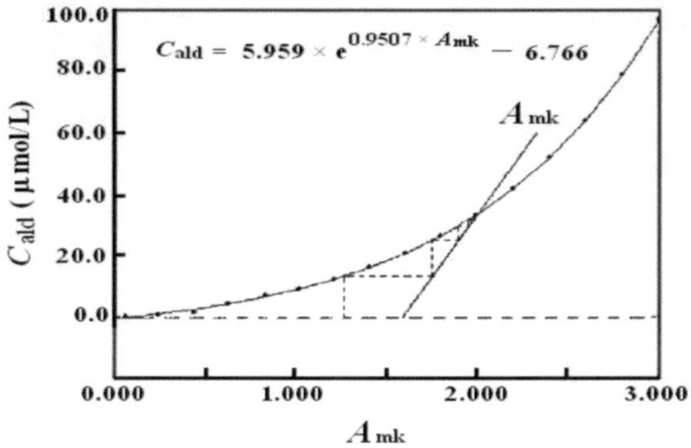

Fig. 10. Iterative adjustment of C_{ald} to predict A_{mk} for 0.31 mmol/L ethanol at 50 U/L ADH (reproduced with permission from Liao, et al, 2007a).

Obviously, by this special approach for kinetic analysis of ADH reaction curve, the upper limit of linear response is excellent, but the lower limit of linear response is over 5.0 μmol ethanol. Under the stated reaction conditions, the equilibrium method after reaction for 8.0 min is effective to quantify ethanol up to final 6.0 μmol. Thus, the equilibrium method with reaction duration of 8.0 min can be integrated with iterative kinetic analysis of reaction curve for quantifying ethanol; this integration strategy gives the linear range from about final 2.0 μmol to about 0.30 mmol/L ethanol in reaction solutions; it has CVs below 8% for ethanol below 10 μmol/L, and CVs below 5% for ethanol over 20 μmol/L (Liao, et al., 2007a). These results clearly supported the advantage of the new integration strategy for substrate assay and the importance of chemometrics in kinetic enzymatic analysis of substrate.

2.6 Programming for kinetic analysis of enzyme reaction curve

Most software package like Origin, SAS, MATLAB can perform kinetic analysis of reaction curve, but they are usually ineffective to implicit functions for kinetic analysis of reaction curve. For convenience and the use of some complicated methods for kinetic analysis of reaction curve in widow-aided mode, self-programming is still favourable.

For simplicity in programming, we used Visual Basic 6.0 to write the source code and working windows (Liu, et al., 2011). The executable program has the main window to perform kinetic analysis of reaction curve (Fig. 11). Original data for each reaction curve is stored as a text file, and keywords are used to indicate specific information related to the reaction curve including sample numbering, the enzyme used, the quantification method, some necessary kinetic parameters, and usually initial signal before enzyme action. Such information is read into memory by the software for kinetic analysis of reaction curve.

On the main window to perform kinetic analysis of reaction curve, original data are listed and plotted for eyesight-checking of data for steady-state reaction. Text boxes are used to input some common parameters like K_m, and most parameters are read from the text file for the reaction curve. Subprogram for an enzyme reaction system is called for running; results are displayed on the main window and may be saved in text file for further analysis.

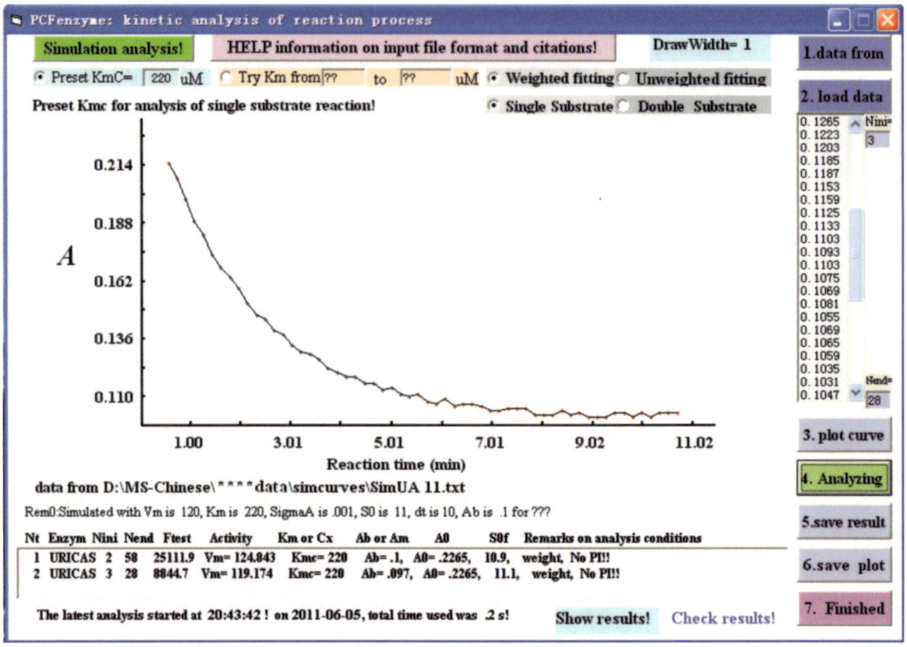

Fig. 11. Main window for the executable PCFenzyme

We called the software PCFenzyme. An old version of the executable PCFenzyme can be downloaded at http://dx.doi.org/10.1016/j.clinbiochem.2008.11.016. The latest version of the executable PCFenzyme with new methods included is available upon request by e-mail.

3. Conclusion

The following conclusions can be drawn. (a) Kinetic analysis of reaction curve can give the initial substrate concentration before enzyme action, the maximal reaction rate, Michaelis-Menten constant and other related parameters of an enzyme reaction system; for reliability, however, it is better to just estimate the initial substrate concentration before enzyme action and the maximal reaction rate with Michaelis-Menten constant and other parameters fixed as constants after optimization. (b) For an enzyme whose integrated rate equation with the predictor variable of reaction time is accessible, kinetic analysis of reaction curve can estimate parameters *via* nonlinear-least-square-fitting after transformation of data from the reaction curve under analysis. (c) For an enzyme reaction system whose kinetics is described by a set of differential rate equations or is difficult to be integrated with the predictor of reaction time, iterative numerical integration of the differential rate equation(s) with a series of preset parameters can produce serial calculated reaction curves; such calculated reaction curves can be fit to the reaction curve under analysis for estimating parameters based on nonlinear-least-square-fitting. This approach is applicable to enzyme-coupled reaction systems of sophisticated kinetics. (d) The integration of kinetic analysis of reaction curve with the equilibrium method can quantify enzyme substrates with expanded linear ranges, favourable analysis efficiency, low cost on tool enzyme, desirable resistance to factors affecting enzyme activities and enhanced precision; it can be applied to enzyme reaction suffering strong product inhibition. (e) The integration of kinetic analysis of reaction curve with the classical initial rate method can measure enzyme initial rates with wide linear ranges, favourable analysis efficiency and practical levels of substrates; it can be applicable to enzyme-coupled reaction curve or enzyme reaction suffering product inhibition.

Taken together, kinetic analysis of enzyme reaction curves under optimized conditions can screen common reversible inhibitors and enzyme mutants; the integration strategy for measuring enzyme activities can quantify serum enzymes and enzyme labels in enzyme-immunoassays to expand the quantifiable ranges, and can be applied to quantify irreversible inhibitors as environmental pollutants; the integration strategy to quantify enzyme substrate can be the second-generation approaches and potentially find wide applications in clinical laboratory medicine. Therefore, these new methodologies for enzymatic analyses based on chemometrics can potentially find their important applications in biomedical sciences.

4. Acknowledgment

This work is supported by the program for New Century Excellent Talent in University (NCET-09), high-technology-program "863" of China (2011AA02A108), National Natural Science Foundation of China (nos. 30200266, 30672009, 81071427), Chongqing Municipal Commission of Sciences and Technology (CQ CSTC2011BA5039), and Chongqing Education Commission (KJ100313).

5. References

Atkins, G.L. & Nimmo, I.A. (1973). The reliability of Michaelis–Menten constants and maximum velocities estimated by using the integrated Michaelis–Menten equation. *The Biochemical Journal*, vol. 135, no.4, (December 1973), pp. 779–784, ISSN 0264-6021

Baywenton, P. R. (1986). *Data process and error analysis* (Translated into Chinese by Weili Qiu, Genxin Xu, Enguang Zhao, and Shengzhong Chen), ISBN 13214.84, Knowledge Press, Beijing, China

Bergmeyer, H.U. (1983). Methods of Enzymatic Analysis, Vol. *I. Fundamentals (3rd Ed.)*, ISBN 978-3527260416, Wiley VCH, Weinheim, Germany

Burden, R.L. & Faires, J.D. (2001). *Numerical Analysis* (7th ed.) , ISBN 978-0534382162, Academic Internet Publishers, Ventura, Carliforlia, USA

Burguillo, J., Wright, A.J. & Bardsley, W. G. (1983). Use of the F test for determining the degree of enzyme-kinetic and ligand-binding data. *The Biochemical Journal*, vol. 211, no.1, (April 1983), pp. 23–34, ISSN0264-6021

Cheng, Y.C. & Prusoff, W.H. (1973). Relationship between the inhibition constant (KI) and the concentration of inhibitor which causes 50 per cent inhibition (I50) of an enzymatic reaction. Biochemical Pharmacology, vol.22, no. 23, (December 1973), pp. 3099-3108, ISSN 0006-2952.

Cheng, Z.L., Chen, H., Zhao, Y.S., Yang, X.L., Lu, W., Liao, H., Yu, M.A., & Liao, F. (2008). The measurement of the activity of rabbit muscle lactic dehydrogenase by integrating the classical initial rate method with an integrated method. *2nd International Conference on Bioinformatics and Biomedical Engineering, iCBBE 2008*, pp.1209-1212, ISBN 978-1-4244-1748-3, Shanghai, China, May 26-28, 2008

Claro, E. (2000). Understanding initial velocity after the derivatives of progress curves. *Biochemistry and Molecular Biology Education*, Vol.28, no.6, (November 2000), pp. 304-306, ISSN 1470-8175

Cornish-Bowden, A. (1975). The use of the direct linear plot for determining initial velocities. *The Biochemical Journal*, vol. 149, no.2, (August 1975), pp. 305-312, ISSN *0264-6021*

Cornish-Bowden, A. (1995). *Analysis of enzyme kinetic data*, ISBN 978-0198548775, Oxford University Press, London, UK

Dagys, R., Tumas, S., Zvirblis, S. & Pauliukonis, A. (1990) Determination of first and second derivatives of progress curves in the case of unknown experimental error. *Computers and Biomedical Research*, Vol.23, no. 5, (October 1990), pp. 490-498, ISSN 0010-4809

Dagys, R., Pauliukonis, A., Kazlauskas, D., Mankevicius, M. & Simutis, R. (1986). Determination of initial velocities of enzymic reactions from progress curves. *The Biochemical Journal*, vol.237, no.3, (August 1986), pp. 821-825, ISSN 0264-6021

del Rio F.J., Riu, J. & Rius, F. X. (2001). Robust linear regression taking into account errors in the predictor and response variables. *Analyst*, vol. 126, no. , (July 2001), pp. 1113–1117, ISSN 0003-2654

Dilena, B.A., Peake, M.J., Pardue, H.L., Skoug, J.W. (1986). Direct ultraviolet method for enzymatic determination of uric acid, with equilibrium and kinetic data-processing options. *Clinical Chemistry*, vol. 32, no.3, (May 1986), pp. 486-491, ISSN 0009-9147

Dixon, M.C. & Webb, EC. (1979). *Enzymes* (3rd ed.), ISBN 0122183584, Academic Press, New York, USA

Draper, N.R. & Smith, H. (1998). *Applied regression analysis* (3rd ed.), ISBN 978-0471170822, Wiley-Interscience; New York, USA

Duggleby, R. G. (1983). Determination of the kinetic properties of enzymes catalysing coupled reaction sequences. *Biochimica et Biophysica Acta (BBA) - Protein Structure and Molecular Enzymology*, Vol.744, no. 3, (May 1983), pp. 249-259, ISSN 0167-4838

Duggleby, R.G. (1985). Estimation of the initial velocity of enzyme-catalysed reactions by non-linear regression analysis of progress curves. *The Biochemical Journal*, vol. 228, no.1, (May 1985), pp. 55-60, ISSN 0264-6021

Duggleby, R. G. (1994). Analysis of progress curves for enzyme-catalyzed reactions: application to unstable enzymes, coupled reactions, and transient state kinetics. *Biochimica et Biophysica Acta (BBA) – General subjects*, vol. 1205, no.2, (April 1994), pp. 268-274, ISSN 0304-4165

Fresht, A. (1985). *Enzyme structure and Mechanism* (2nd Ed.), ISBN 978-0716716143, Freeman WH, New York, USA

Ganzhorn, A.J., Green, D.W., Hershey, A.D., Gould, R.M., Plapp, B.V. (1987). Kinetic characterization of yeast alcohol dehydrogenases. Amino acid residue 294 and substrate specificity. The Journal of Biological chemistry,vol.262, no.8, (March 1987), pp. 3754-3761, ISSN 0021-9258

Guilbault, G. G. (1976). *Handbook of enzymatic methods of analysis*, ISBN 978-0824764258, Marcel Dekker, New York, USA

Gutierrez, O.A. & Danielson, U. H. (2006). Sensitivity analysis and error structure of progress curves. Analytical Biochemistry, vol.358, no.1, (August 2006), pp.1-10, ISSN 0003-2697

Hamilton, S. D. & Pardue, H. L. (1982). Kinetic method having a linear range for substrate concentration that exceed Michaelis–Menten constants. *Clinical Chemistry*, vol. 28, no.12, (December 1982), pp.2359–2365, ISSN 0009-9147

Hasinoff, B. B. (1985). A convenient analysis of Michaelis enzyme kinetic progress curves based on second derivatives. *Biochimica et Biophysica Acta (BBA) - General Subjects*, Vol. 838, no. 2, (February 1985), pp. 290-292, ISSN 0304-4165

Kahn, K. & Tipton, P.A. (1998). Spectroscopic characterization of intermediates in the urate oxidase reaction. *Biochemistry*, vol. 37, no. (August 1998), pp. 11651-11659, ISSN 0006-2960.

Koerber, S. C. & Fink, A. L. (1987). The analysis of enzyme progress curves by numerical differentiation, including competitive product inhibition and enzyme reactivation. *Analytical Biochemistry*, vol. 165, no.1, (December 2004), pp. 75-87, ISSN 0003-2697

Li, Z.R., Liu,Y., Yang, X.Y., Pu,J., Liu, B.Z., Yuan, Y.H., Xie, Y.L. & Liao, F. (2011). Kinetic analysis of gamma-glutamyltransferase reaction process for measuring activity via an integration strategy at low concentrations of gamma-glutamyl p-nitroaniline. *Journal of Zhejiang University Sciecnce B*, vol. 12, no.3, (March 2011), pp. 180-188, ISSN 1673-1581

Liao, F. (2005). *The method for quantitative enzymatic analysis of uric acid in body fluids by predicting the background absorbance*. China patent: ZL O3135649.4, 2005-08-31

Liao, F., Li, J.C., Kang, G.F., Zeng, Z.C., Zuo, Y.P. (2003a). Measurement of mouse liver glutathione-S- transferase activity by the integrated method. *Journal of Medical Colleges of PLA*, vol. 18, no.5, (October 2003), pp. 295-300, ISSN 1000-1948

Liao, F., Liu, W.L., Zhou, Q.X., Zeng, Z.C., Zuo, Y.P. (2001). Assay of serum arylesterase activity by fitting to the reaction curve with an integrated rate equation. *Clinica Chimica Acta*, vol. 314, no.1-2, (December 2001), pp.67-76, ISSN 0009-8981

Liao, F., Tian, K.C., Yang, X., Zhou, Q.X., Zeng, Z.C., Zuo, Y.P. (2003b). Kinetic substrate quantification by fitting to the integrated Michaelis-Menten equation. *Analytical Bioanalytical Chemistry*, vol. 375, no. 6, (Febrary 2003), pp. 756-762, ISSN1618-2642

Liao, F., Yang, D.Y., Tang, J.Q., Yang, X.L., Liu, B.Z., Zhao, Y.S., Zhao, L.N., Liao, H. & Yu, M.A. (2009). The measurement of serum cholinesterase activities by an integration strategy with expanded linear ranges and negligible substrate-activation. *Clinical Biochemistry*, vol.42, no.6, (December 2008), pp.926-928. ISSN 0009-9120

Liao, F., Zhao, L.N., Zhao, Y.S., Tao, J., Zuo, Y.P. (2007a). Integrated rate equation considering product inhibition and its application to kinetic assay of serum ethanol. *Analytical Sciences*, vol. 23, no.4, (April 2007), pp. 439-444 , ISSN 0910-6340

Liao, F., Zhao, Y.S., Zhao, L.N., Tao, J., Zhu, X.Y., Liu, L. (2006). The evaluation of a direct kinetic method for serum uric acid assay by predicting the background absorbance of uricase reaction solution with an integrated method. *Journal of Zhejiang University Science B*, vol. 7, no.6, pp. 497-502, ISSN 1673-1581

Liao, F., Zhao, Y.S., Zhao, L.N., Tao, J., Zhu, X.Y., Wang, Y.M., Zuo, Y.P. (2005b). Kinetic method for enzymatic analysis by predicting background with uricase reaction as model. *Journal of Medical Colleges of PLA*, vol.20, no.6, (Deember 2005), pp. 338-344, ISSN 1000-1948

Liao, F., Zhu, X.Y., Wang, Y.M., Zhao, Y.S, Zhu, L.P., Zuo, Y.P. (2007b). Correlation of serum arylesterase activity on phenylacetate estimated by the integrated method to common classical biochemical indexes of liver damage. *Journal of Zhejiang University Science B*, vol. 8, no.4, (April 2007), pp.237-241, ISSN 1673-1581

Liao, F., Zhu,X.Y., Wang, Y.M., Zuo, Y.P. (2005a). The comparison on the estimation of kinetic parameters by fitting enzyme reaction curve to the integrated rate equation of different predictor variables. *Journal of Biochemical Biophysical Methods*, vol. 62, no.1, (January 2005), pp. 13-24, ISSN 0165-022X

Liu, B.Z., Zhao, Y.S., Zhao, L.N., Xie, Y.L., Zhu,S., Li,Z.R., Liu,Y., Lu,W., Yang,X.L., Xie, G.M., Zhong, H.S., Yu, M.A., Liao,H. & Liao, F. (2009). An integration strategy to estimate the initial rates of enzyme reactions with much expanded linear ranges using uricases as models. *Analytica Chimica Acta*, vol.631, no.1, (October 2008), pp. 22-28. ISSN 0003-2670

Liu, M., Yang, X.L., Yuan, Y.H., Tao, J. & Liao, F. (2011). PCFenzyme for kinetic analyses of enzyme reaction processes. *Procedia Environmental Sciences*, vol. 8, (December 2011), pp.582-587, ISSN 1878-0296

Lu, W.P. & Fei, L. (2003). A logarithmic approximation to initial rates of enzyme reactions. *Analytical Biochemistry*, vol. 316, no. 1, (May 2003), pp.58-65, ISSN 0003-2697

Marangoni, A. G. (2003). *Enzyme kinetics:a modern approach*, ISBN 978-0471159858, Wiley-Interscience, New York, USA

Mey1er-Almes, F.J. & Auer, M. (2000). Enzyme inhibition assay using fluorescence correlation spectroscopy: a new algorithm for the derivation of Kcat/KM.and Ki values at substrate concentration much lower than the Michaelis constant. *Biochemistry*, vol. 39, no.43 (October 2000), pp. 13261–13268, ISSN 0006-2960

Miller, J. C. & Miller, J. N. (1993). *Statistics for analytical chemistry* (3rd), ISBN 978-0130309907, Ellis Horwood, Chichester, New York, USA

Morishita, Y., Iinuma, Y., Nakashima, N., Majima, K., Mizuguchi, K. & Kawamura, Y. (2000). Total and pancreatic amylase measured with 2-chloro-4-nitrophenyl-4-*O*-ß-D-galactopyranosylmaltoside. *Clinical Chemistry*, vol. 46, no.7, (July 2000), pp. 928-933, ISSN 0009-9147

Moruno-Davila, M.A., Solo, C.G., Garcla-Moreno, M., Garcla-CAnovas, F. & Varon, R. (2001). Kinetic analysis of enzyme systems with suicide substrate in the presence of a reversible, uncompetitive inhibitor. *Biosystems*, vol. 61, no.1, (June 2001), pp.5-14, ISSN 0303-2647

Moss, D.W. (1980). Methodological principles in the enzymatic determination of substrates illustrated by the measurement of uric acid. *Clinica Chimica Acta*, Vol. 105, no. 3, (August 1980), pp. 351-360, ISSN 0009-8981

Newman, P.F.J., Atkins, G.L. & Nimmo, I. A. (1974). The effects of systematic error on the accuracy of Michaelis constant and maximum velocities estimated by using the integrated Michaelis–Menten equation. *The Biochemical Journal*, vol. 143, no. 3, (December 1974), pp. 779–781. ISSN 0264-6021

Northrop, D. B. (1983). Fitting enzyme-kinetic data to V/K. *Analytical Biochemistry*, vol. 132, No. 2, (July 1983), pp. 457–61, ISSN 0003-2697

Orsi, B.A. & Tipton, K. F. (1979). Kinetic analysis of progress curves. In: *Methods in Enzymology*, vol. 63, D. L. Purich, (Ed.), 159-183, Academic Press, ISBN 978-0-12-181963-7, New York, USA

Priest, D.G. & Pitts, O.M. (1972). Reaction intermediate effects on the spectrophotometric uricase assay. *Analytical Biochemistry*, vol.50, no.1, (November 1972), pp. 195-205, ISSN 0003-2697

Stromme,J.H. & Theodorsen, L. (1976). Gamma-glutamyltransferase: Substrate inhibition, kinetic mechanism, and assay conditions. *Clinical Chemistry*, vol. 22, no.4, (April 1976), pp. 417-421, ISSN 0009-9147

Varon, R., Garrido-del Solo, C., Garcia-Moreno, M., Garcoa-Canovas, F., Moya-Garcia, G., Vidal de Labra, J., Havsteen BH. (1998). Kinetics of enzyme systems with unstable suicide substrates. *Biosystems*, vol. 47, no.3, (August 1998), pp.177-192, ISSN 0303-2647

Walsh, R., Martin, E., Darvesh, S. (2010). A method to describe enzyme-catalyzed reactions by combining steady state and time course enzyme kinetic parameters. *Biochimica et Biophysica Acta-General Subjects*, vol.1800, no.1, (October 2009), pp1-5, ISSN 0304-4165.

Yang, D., Tang, J., Yang, X., Deng, P., Zhao, Y., Zhu, S., Xie, Y., Dai, X., Liao, H., Yu, M., Liao, J. & Liao, F. (2011). An integration strategy to measure enzyme activities for detecting irreversible inhibitors with dimethoate on butyrylcholinesterase as model. *International Journal of Environmental Analytical Chemistry*, vol.91, no.5, (March 2011), pp.431-439, ISSN 0306-7319

Yang, X. L., Liu, B.Z., Sang,Y., Yuan, Y.H., Pu, J., Liu,Y., Li, Z.R., Feng,J., Xie, Y.L, Tang, R. K., Yuan, H.D. & Liao, F. (2010). Kinetic analysis of lactate-dehydrogenase-coupled reaction process and measurement of alanine transaminase by an integration strategy. *Analytical Sciences*, vol.26, no. 11, (November 2010), pp. 1193-1198, ISSN0910-6340

Zhang, C., Yang, X.L., Feng,J., Yuan, Y.H., Li,X., Bu, Y.Q., Xie, Y.L., Yuan, H.D. & Liao, F. (2010). Effects of modification of amino groups with poly(ethylene glycol) on a recombinant uricase from Bacillus fastidiosus. *Bioscience Biotechnology Biochemistry*, 2010; vol.74, no.6, (June 2010), pp. 1298-1301, 0916-8451. ISSN 0916-8451.

Zhao, L.N., Tao, J., Zhao, Y.S., Liao, F. (2006). Quantification of reduced glutathione by analyzing glutathine-S-transferase reaction process taking into account of product inhibition. Journal of Xi'an Jiaotong University (Medical Sciences), vol. 27, no.3, (June 2006), pp.300-303, ISSN 1671-8259

Zhao,Y.S., Yang, X.Y., Lu,W., Liao,H. & Liao, F. (2009). Uricase based method for determination of uric acid in serum. *Microchimica Acta*, vol. 164, no.1, (May 2008), pp.1-6, ISSN 0026-3672

Zhao,Y.S., Zhao,L.N., Yang,G.Q., Tao, J., Bu, Y.Q. & Liao, F. (2006). Characterization of an intracellular uricase from Bacillus fastidious ATCC 26904 and its application to serum uric acid assay by a patented kinetic method. *Biotechnology Applied Biochemistry*, vol. 45, no.2, (September 2006), pp. 75-80, ISSN 0885-4513

Zou, G.L. & Zhu, R.F. (1997). *Enzymology*, Wuhan University Press, ISBN 7-307-02271-0/Q, Wuhan, China

8

Chemometric Study on Molecules with Anticancer Properties

João Elias Vidueira Ferreira[1], Antonio Florêncio de Figueiredo[2], Jardel Pinto Barbosa[3] and José Ciríaco Pinheiro[3]

[1]*Universidade do Estado do Pará*

[2]*Instituto Federal de Educação, Ciência e Tecnologia do Pará*

[3]*Laboratório de Química Teórica e Computacional, Universidade Federal do Pará*

Brasil

1. Introduction

Cancer is a class of diseases characterized by uncontrolled growth of abnormal cells of an organism. All over the world millions of people die every year owing to one of the different types of cancer. Unfortunately cancer chemotherapy finds a serious limitation since treatment with drugs is followed by drug resistance in the tumorous cells and side effects (Efferth, 2005). So researches have been directed to make chemotherapy treatment more efficient.

In the late years literature has reported the research on natural products as a good strategy to discover new chemotherapy agents. One of the plants that have shown anticancer properties is *Artemisia annua L.* (*qinghao*). It has the active ingredient artemisinin, which is used as antimalarial. Artemisinin and derivatives have excellent efficacy against multidrug-resistant strains of *P. falciparum* and they are very well tolerated (Price et al., 1998). Recently the sensibility to artemisinin has been evaluated in some tumorous cells. Studies suggest that artemisinin is more toxic to cancerous cells than to normal cells, so giving a new perspective in cancer therapy (Lai et al, 2009).

However ... This book is on chemometrics and what has chemometrics to do with cancer chemotherapy? Well... understanding how these two different areas can be related to one another is the purpose of this chapter. You just must keep on reading this chapter and you will see the many ways chemometrics can be employed to investigate the "behavior" molecules exhibit considering anticancer activity and to make predictions about drugs that were not tested yet. The potential application of chemometrics to analytical data arising from problems in biology and medicine is enormous and, in fact, the applications of chemometrics have diversified substantially over the last few years (Brereton, 2007; 2009). At the end of the chapter you will note that, as in many areas of research, chemometrics plays an important role in medicinal chemistry, fortunately.

Firstly it is necessary to remember that producing a drug is something that takes time and money, so the process must be rationalized! However, in the past, drugs were discovered by synthesizing a lot of molecules, rather without rigorous criteria, and testing experimentally all of them to evaluate their capacity of cure of the disease or at least to control it. But in process of time this methodology became more and more inadequate, for the more new compounds are studied the less a new compound may be discovered to be potent against a disease. It has long been desired to design active structures on the basis of logic and calculations, not relying on chance or trial-and-error (Fujita, 1995).

Nowadays, in science, there is a basic assumpion that molecular properties and structural characteristics are closely connected to biological functions of the compounds. It is often assumed that compounds with similar properties and structures also display similar biological responses. Chemical structure encodes a large amount of information explaining why a certain molecule is active, toxic or insoluble (Rajarshi, 2008). Thus to understand the mechanism of action of a drug it is necessary to interpret the role played by its molecular and structural properties.

In the last decades, much scientific research has focused on how to capture and convert the information encoded in a molecular structure into one or more numbers used to establish quantitative relationships between structures and properties, biological activities or other experimental properties (Puzyn et al., 2010). Quantitative structure-activity relationship (QSAR) studies have been of great value in medicinal chemistry. Statistical tools can be used for the prediction of the biological activities of new compounds based only on the knowledge of their chemical structures, i.e., not depending on experimental data, which are unknown. Such a strategy gives very useful information for the understanding of the mechanisms of the action of drugs and proposals for syntheses, in this way rationalizing drug discovery. QSAR is alive and well (Doweyko, 2008), that is, QSAR has been used with success and so it is still of relevance today.

Moreover advances in computation brought software that made possible to get many different types of information (descriptors) about the molecules. Consequently data gathered through experiments and computers can produce a huge matrix whose elements are information related to molecules. But it seems that analyzing all them will require infinite patience!

What to do?

Chemometics has the solution!

That is true because chemometrics is the art of extracting chemically relevant information from data produced in chemical experiments (Wold, 1995). Most people only think of statistics when faced with a lot of quantitative information to process (Bruns et al., 2006). In this text we show a common and efficient methodology used in medicinal chemistry to rationalize the process of producing a new drug by employing chemometric methods. It is presented a molecular modeling and a chemometric study of 25 artemisinins, which involves artemisinin and derivatives (training set, Fig. 1) with different degrees of cytotoxicities against human hepatocellular carcinoma HepG2 (Liu et al, 2005), since among the malignant tumors in the liver, the hepatocellular carcinoma is very commom. Literature has showed the application of the methodology here described to investigate biological properties (antimalarial and anticancer) of artemisinin and derivatives (Barbosa et al., 2011); (Cardoso et al., 2008); (Pinheiro et al., 2003).

2. Methodology

Any chemometric study requires data. In this study data are obtained from molecular descriptors calculated through computation. The start point is the molecular modeling step, which consists on the construction of the structures and the complete optimization of their geometries through a quantum chemistry approach implemented in computer. This is necessary to represent molecules as real as possible and thus to compute their molecular descriptors. The B3LYP/6-31G** method (Levine, 1991) as implemented in the Gaussian 98 program was employed (Frisch et al., 1998), considering this strategy is suitable for optimizing well all structures since a good description of the geometrical parameters of artemisinin is achieved.

The 25 compounds investigated include artemisinin, amides, esters, alcohols, ketones, and five-membered ring derivatives. All compounds have been associated to their in vitro bioactivity against a human hepatocellular carcinoma cell line, HepG2, and were labeled previously into two classes according with their activities: (-) less active (those with $IC_{50} \geqslant 97$ μM) and (+) more active (those with $IC_{50} < 97$ μM) derivatives. The criteria for choosing this value of IC_{50} are rather subjective. Nevertheless it is convenient to say that 97 μM is the IC_{50} for artemisinin and the higher IC_{50} the less active is the compound.

After molecular modeling, 1700 descriptors (independent variables) were computed for each molecule in the training set. They represent different source of chemical information (features) regarding the molecules and include geometric, electronic, quantum-chemical, physical-chemical, topological descriptors and others. They are assumed to be important to understand molecular characteristics such as bioactivity against cancer. In fact one of the purposes of a research like this is to find which descriptors of the molecules are better related to the disease under study, in this example cancer. The software used to compute these descriptors were e-Dragon (Virtual Computational Laboratory , 2010), a product from the Virtual Computational Laboratory and Gaussian 98 (Frisch et al., 1998).

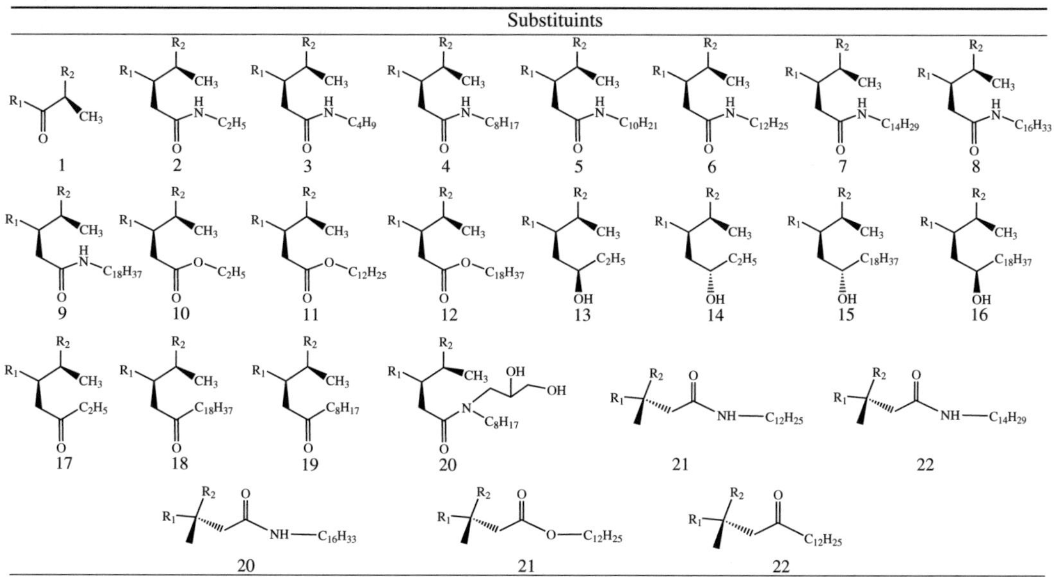

Fig. 1. Artemisinin and derivatives (training set) with different degrees of cytotoxicities against human hepatocellular carcinoma HepG2

However, a crucial point to be considered in any data analysis is preprocessing. The original data matrix usually does not have optimal value distribution for the analysis (for example it has different units and variances in variables), which requires some pretreatment prior to multivariate analysis. In general, the autoscale preprocessing, which results in scaled variables with zero mean and unit variance, is used (Ferreira, 2002). Then, all variables were auto-scaled as a preprocessing so that they could be standardized and this way could have the same importance regarding the scale.

Then the next step consists on application of multivariate statistical methods to find key features involving molecules, descriptors and anticancer activity. The methods include principal component analysis (PCA), hiererchical cluster analysis (HCA), K-nearest neighbor method (KNN), soft independent modeling of class analogy method (SIMCA) and stepwise discriminant analysis (SDA). The analyses were performed on a data matrix with dimension 25 lines (molecules) x 1700 columns (descriptors), not shown for convenience. For a further study of the methodology applied there are standard books available such as (Varmuza & Filzmoser, 2009) and (Manly, 2004).

2.1 PCA

Suppose that in your study, like in the example exhibited in this chapter, you have a large set of data, certainly it will not be a simple task to analyze so many variables and extract useful information from them. It would be a "revolution" in your research if you could confidently interpret all data in a simpler way. Fortunately, with the aid of PCA technique, this "revolution" can happen. Through PCA you can reduce the total number of variables to a smaller set while maintaining as much of the original information as is possible. No matter your area of research this is a great advantage.

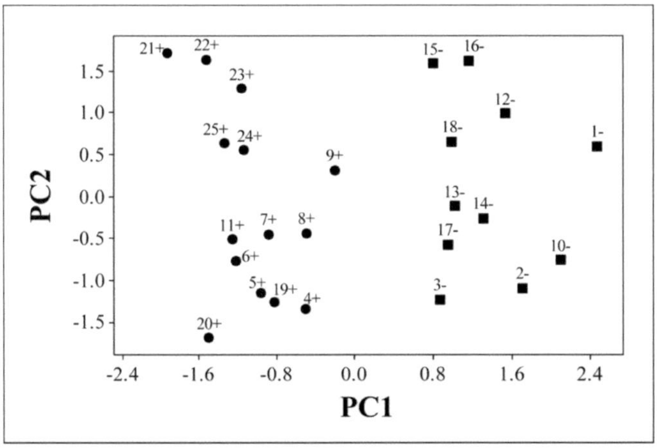

Fig. 2. Plot of PC1-PC2 scores for artemisinin and derivates (training set) with activity against human hepatocellular carcinoma HepG2. More active compounds displayed on the left side (plus sign) while less active ones on the right side (minus sign)

Now considering our data matrix, PCA was employed looking for a small group of descriptors so that they alone were responsible for classifying all 25 samples into two distinct classes: more active and less active. Besides it is desirable to choose uncorrelated descriptors that could be easier to interpret and analyze, trying to associate them to cytotoxicities against human hepatocellular carcinoma HepG2.

Furthermore, given the large quantity of multivariate data available, it was necessary to reduce the number of variables. Thus, if two any descriptors had a high Pearson correlation coefficient (r > 0.8), one of the two was randomly excluded from the matrix, since theoretically they describe the same property to be modeled (biological response). Therefore it is sufficient to use only one of them as an independent variable in a predictive model (Ferreira, 2002). Moreover those descriptors that showed the same values for most of the samples were eliminated too.

Compound	*IC5*	*Mor29m*	*O1*	*MlogP*	**Activity**
1	4.862	-0.305	-0.246	2.845	97
2	5.253	-0.308	-0.200	2.630	>100
3	5.389	-0.372	-0.202	3.080	>100
4	5.628	-0.445	-0.194	4.845	9.5
5	5.684	-0.474	-0.205	5.250	2.8
6	5.624	-0.525	-0.214	5.644	1.2
7	5.501	-0.514	-0.211	6.027	0.46
8	5.364	-0.518	-0.191	6.400	0.79
9	5.225	-0.501	-0.210	6.765	4.2
10	5.217	-0.236	-0.205	3.036	>100
11	5.597	-0.526	-0.218	6.050	0.72
12	5.197	-0.179	-0.225	7.171	>100
13	5.253	-0.364	-0.246	3.141	>100
14	5.253	-0.322	-0.237	3.141	>100
15	5.159	-0.294	-0.259	7.095	>100
16	5.159	-0.232	-0.258	7.095	>100
17	5.180	-0.443	-0.219	2.996	>100
18	5.168	-0.307	-0.209	7.131	>100
19	5.624	-0.485	-0.186	5.644	1.8
20	5.856	-0.518	-0.218	3.941	3.5
21	5.543	-0.562	-0.344	5.449	1.3
22	5.419	-0.560	-0.320	5.837	0.77
23	5.280	-0.591	-0.281	6.215	0.74
24	5.516	-0.498	-0.269	5.855	3.7
25	5.488	-0.545	-0.273	5.815	0.47
Mean	5.378	-0.425	-0.234	5.164	
Stardard Deviation	0.225	0.121	0.040	1.570	

Table 1. Values of the four descriptors selected through PCA for compounds from the training set

After this step, PCA was performed in order to continue reducing the dimensionality of the data, find descriptors that could be useful in characterizing the behavior of the compounds acting against cancer and look for natural clustering in the data and outlier samples. While processing PCA, several attempts to obtain a good classification of the compounds are made. At each attempt, one or more variables are removed, PCA is run and the score and loading plots are analyzed.

The score plot gives information about the compounds (similarities and differences). The loading plot gives information about the variables (how they are connected to each other and which are the best to describe the variance in the original data). Depending on the results displayed by the plots, variables remain removed or included in the data matrix. If a removal of a variable contributes to separate compounds showed by the score plot into two classes (more and less active), then in the next attempt PCA is run without this variable. But if no improvement is achieved, then the variable removed is inserted in the data matrix, another variable is selected to be removed and PCA is run again. The loadings plot gives good clues on which variables must be excluded. Variables that are very close to one another indicate they are correlated and, as stated before, only one of them needs to remain.

This methodology comprises part of the art of variable selection: patience and intuition are the fundamental tools here. It is not necessary to mention that the more you know about the system you are investigating (samples and variables and how they are connected), the more you can have success in the process of finding variables that really are important to your investigation. Variable selection does not occur like magic, at least, not always!

The descriptors selected in PCA were *IC5*, *Mor29m*, *O1* and *MlogP*, which represent four distinct types of interactions related to the molecules, especially between the molecules and the biological receptor. These descriptors are classified as steric (*IC5*), 3D-morse (*Mor29m*), electronic (*O1*) and molecular (*MlogP*). The main properties of a drug that appear to influence its activity are its lipophilicity, the electronic effects within the molecule and the size and shape of the molecule (steric effects) (Gareth, 2003).

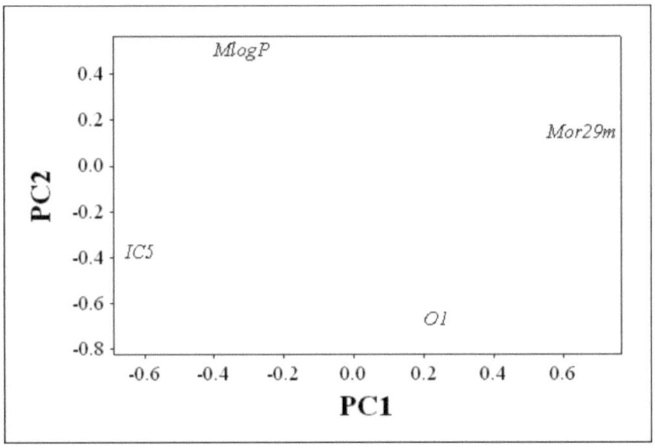

Fig. 3. Plot of the PC1-PC2 loadings for the four descriptors selected through PCA

The PCA results show the score plot (Fig. 2) relative to the first and second principal components. In PC1, there is a distinct separation of the compounds into two classes. More active compounds are on the left side, while less active are on the right side. They were chosen among all data set (1700 descriptors) and they are assumed to be very important to investigate anticancer mechanism involving artemisinins. Table 1 displays the values computed for these four descriptors. This step was crucial since a matrix with 1700 columns was reduced to only 4 columns. No doubt it is more appropriate to deal with a smaller matrix. The first three principal components, PC1, PC2 and PC3 explained 43.6%, 28.7% and 20.9% of the total variance, respectively. The Pearson correlation coefficient between the variables is in general low (less than 0.25, in absolute values); exception occurs between *Mor29m* and *IC5*, which is -0.65).

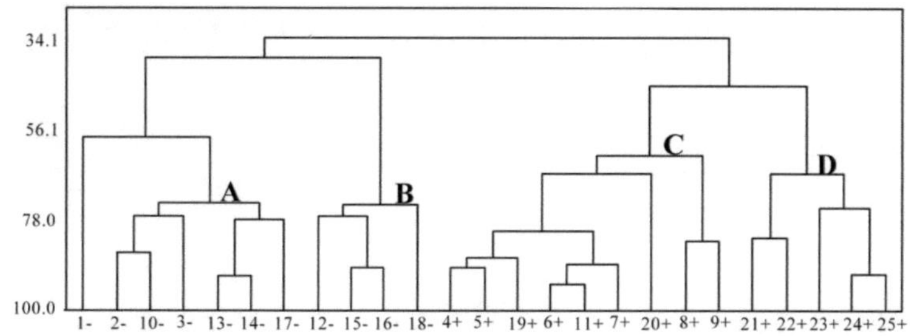

Fig. 4. HCA dendogram for artemisinin and derivatives (training set) with biological activity against human hepatocellular carcinoma HepG2. Plus sign for more active compounds while minus sign for less active ones

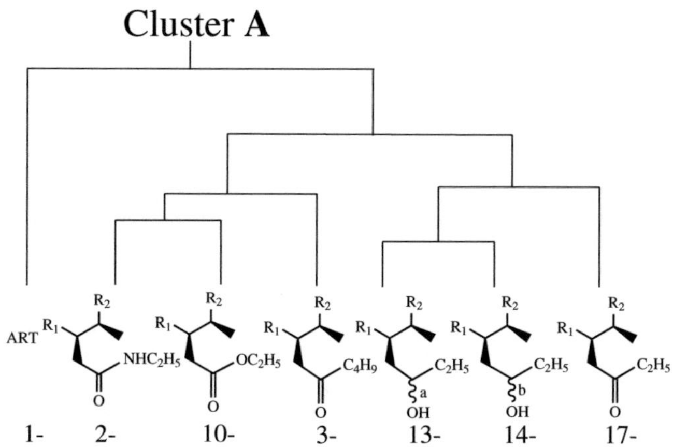

Fig. 5. Cluster **A**

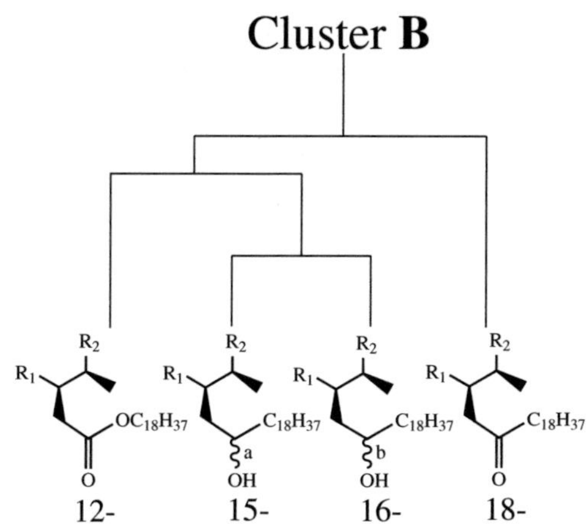

Fig. 6. Cluster **B**

The loading plot relative to the first and second principal components can be seen in Fig. 3. PC1 and PC2 are expressed in Equations 1 and 2, respectively, as a function of the four selected descriptors. They represent quantitative variables that provide the overall predictive ability

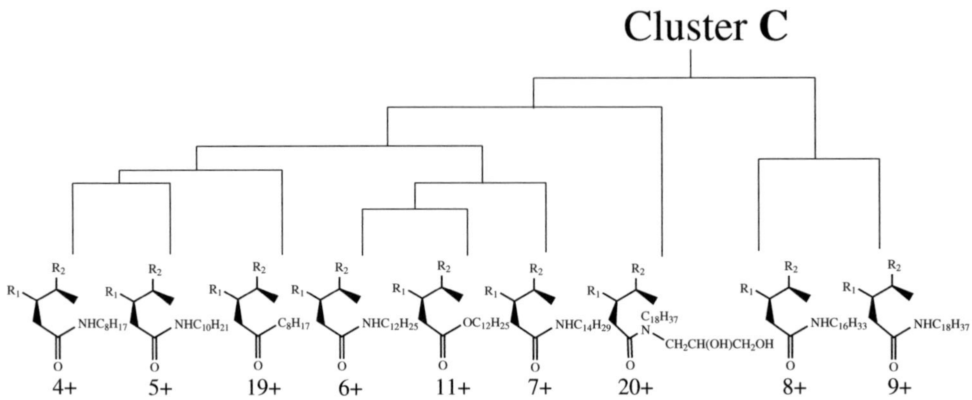

Fig. 7. Cluster **C**

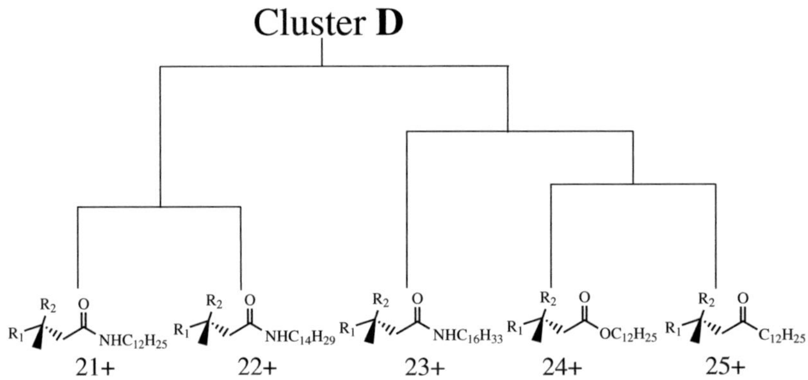

Fig. 8. Cluster **D**

of the different sets of molecular descriptors selected. In Equation 1 the loadings of *IC5* and *MlogP* are negative whereas they are positive for *Mor29m* and *O1*. Among all of them *IC5* and *Mor29m* are the most important to PC1 due to the magnitude of their coefficients (-0.613 and 0.687, respectively) in comparison to *O1* and *MlogP* (0.234 and -0.313, respectively). For a compound to be more active against cancer, it must generally be connected to negative values for PC1, that is, it must present high values for *IC5* and *MlogP*, but more negative values for *Mor29m* and *O1*.

$$PC1 = -0.613IC5 + 0.687Mor29m + 0.234O1 - 0.313MlogP \quad (1)$$

$$PC2 = -0.445IC5 + 0.081Mor29m - 0.743O1 + 0.493MlogP \quad (2)$$

2.2 HCA

Considering the necessity of grouping molecules of similar kind into respective categories (more and less active ones), HCA is suitable for this purpose since it is possible to visualize the disposition of molecules with respect to their similarities and so make suppositions of how they may act against the disease. When performing HCA many approaches are available. Each one differs basically by the way samples are grouped.

Compound	K1	K2	K3	K4	K5	K6
1	-	-	-	-	-	-
2	-	-	-	-	-	-
3	-	-	-	-	-	-
4	+	+	+	+	+	+
5	+	+	+	+	+	+
6	+	+	+	+	+	+
7	+	+	+	+	+	+
8	+	+	+	+	+	+
9	+	+	+	+	+	+
10	-	-	-	-	-	-
11	+	+	+	+	+	+
12	-	-	-	-	-	-
13	-	-	-	-	-	-
14	-	-	-	-	-	-
15	-	-	-	-	-	-
16	-	-	-	-	-	-
17	-	-	-	-	-	-
18	-	-	-	-	-	-
19	+	+	+	+	+	+
20	+	+	+	+	+	+
21	+	+	+	+	+	+
22	+	+	+	+	+	+
23	+	+	+	+	+	+
24	+	+	+	+	+	+
25	+	+	+	+	+	+

Table 2. Classification of compounds from the training set according to KNN method

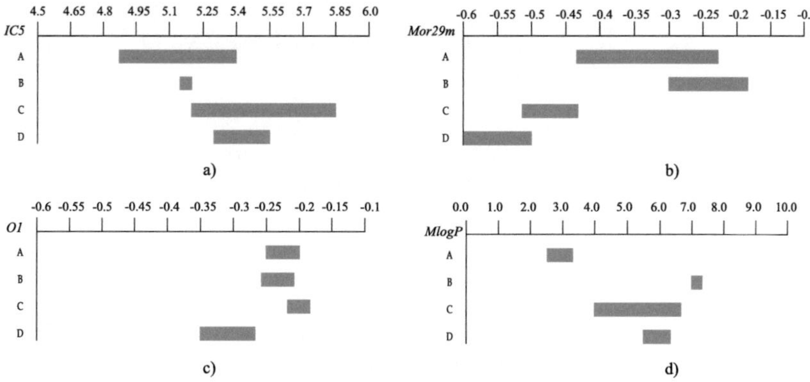

Fig. 9. Variations in descriptors: a) Variations in *IC5* for each cluster; b) Variations in *Mor29m* for each cluster; c) Variations in *O1* for each cluster; d) Variations in *MlogP* for each cluster

In this work, classification through HCA was based on the Euclidean distance and the average group method. This method established links between samples/cluster. The distance between two clusters was computed as the distance between the average values (the mean vector or centroids) of the two clusters. The descriptors employed in HCA were the same selected in

PCA, that is, *IC5, Mor29m, O1* and *MlogP*. The representation of clustering results is shown by the dendogram in Fig. 4, which depicts the similarity of samples. The branches on the bottom of the dendogram represent single samples. The length of the branches linking two clusters is related to their similarity. Long branches are related to low similarity while short branches mean high similarity. On the scale of similarity, a value of 100 is assigned to identical samples and a value of 0 to the most dissimilar samples. For a better interpretation of the dendogram, the clusters are also analyzed alone (Figs. 5, 6, 7 and 8), and variations in descriptors in each cluster are presented in Fig. 9. The scale above each figure is associated to the property considered and the letters indicate the cluster in the dendogram. It is easily recognized that descriptors in clusters in general have different pattern of variations, a characteristic supported by the fact that clusters have different groups of molecules.

Group or class	Number of Compounds	Compounds wrongly classified					
		K1	K2	K3	K4	K5	K6
Less active	11	0	0	0	0	0	0
More active	14	0	0	0	0	0	0
%Correct information	25	100	100	100	100	100	100

Table 3. Classification matrix obtained by using KNN

The dendogram shows compounds classified into two different classes according to their activities with no sample incorrectly classified. Less active compounds are on the left side and are divided into clusters **A** (Fig. 5) and **B** (Fig. 6). In cluster **A** substituints have either C_2H_5 (**2, 10, 13, 14** and **17**) or C_4H_9 (**3**). Here the lowest values for *IC5* (Fig. 9a) and *MlogP* are found (Fig. 9d). In cluster **B** (**12, 15, 16** and **18**) all substituints have $C_{18}H_{37}$ and are present the highest values for *MlogP* (Fig. 9d). Considering more active samples, right side of the figure, in cluster **C** (Fig. 7) compounds have amide group (exception is **11**, ester, and **19**, ketone) and attached to this group there is an alkyl chain of 8 to 18 carbon atoms. Here the descriptor *IC5* displays the highest values (Fig. 9a). In Cluster **D** (Fig. 8) substituints have an alkyl chain of 12 to 16 carbon atoms and the six-membered ring molecules with oxygen O_{11} are replaced by five-membered ring molecules. Compounds display the lowest values for *Mor29m* (Fig. 9b) and *O1* (Fig. 9c).

Besides these two methods of classification (PCA and HCA), others (KNN, SIMCA and SDA) were applied to data. They are important to construct reliable models useful to classify new compounds (test set) regarding their ability to face cancer. This is certainly the ultimate purpose of many researches in planning a new drug.

2.3 KNN

This method categorizes an unknown object based on its proximity to samples already placed in categories. After built the model, compounds from the test set are classified and their classes predicted taking into account the multivariate distance of the compound with respect to K samples in the training set. The model built for KNN in this example employs leave one out method, has 6 (six) as a maximum k value and autoscaled data. Table 2 shows classification for each sample at each k value. Column number corresponds to k setting so that the first column of this matrix holds the class for each training set sample when only one neighbor (the nearest) is polled whereas the last column holds the class for the samples when the kmax nearest neighbors are polled. Tables 2 and 3 summarizes the results for KNN analysis. All 6-nearest neighbors classified samples correctly.

2.4 SIMCA

The SIMCA method develops principal component models for each training set category. The main goal is the reliable classification of new samples. When a prediction is made in SIMCA, new samples insufficiently close to the PC space of a class are considered non-members. Table 4 shows classification for compounds from the training set. Here sample **9** was classified incorrectly since its activity is 4.2 (more active) but it is classified by SIMCA as less active.

	Compound												
	1	2	3	4	5	6	7	8	9	10	11	12	13
Class	-	-	-	+	+	+	+	+	-	-	+	-	-
	Compound												
	14	15	16	17	18	19	20	21	22	23	24	25	
Class	-	-	-	-	-	+	+	+	+	+	+	+	

Table 4. Classification of compounds from the training set according to SIMCA method

Probably the reason for this misclassification lies in the fact that compound **9** may not be "well grouped" into one of the two classes. In fact when you analyze Fig. 2 you note that **9** is the compound classified as more active that is closer to compounds classified as less active.

Group or Class	Number of Compounds	True group	
		More active	Less active
Less active	11	0	11
More active	14	14	0
Total	25		
%Correct information		100	100

Table 5. Classification matrix obtained by using SDA

2.5 SDA

SDA is also a multivariate method that attempts to maximize the probability of correct allocation. The main objectives of SDA are to separate objects from distinct populations and to allocate new objects into populations previously defined.

The discrimination functions for less active and more active classes are, respectively, Equations 3 and 4, given below:

$$Y_{LESS} = -5.728 - 2.825MlogP - 0.682O1 - 3.243IC5 + 7.745Mor29m \quad (3)$$

$$Y_{MORE} = -3.536 + 2.220MlogP + 0.536O1 + 2.548IC5 - 6.086Mor29m \quad (4)$$

The way the method is used is based on the following steps:

(a) Initially, for each molecule, the values for descriptors (*IC5, Mor29m, O1* and *MlogP*) are computed;

(b) The values from (a) are inserted in the two discrimination functions (Equation 3 and Equation 4). However, since these equations were obtained from autoscaled values from Table 1 (training set), it is necessary that values from Table 7 (test set) are autoscaled before inserted into the equations;

(c) The two values computed from (b) are compared. In case the value calculated from Equation 3 is higher than that from Equation 4, then the molecule is classified as less active. Otherwise, the molecule is classified as more active.

Group or Class	Number of Compounds	True group	
		More active	Less active
Less active	11	0	11
More active	14	14	0
Total	25		
%Correct information		100	100

Table 6. Classification matrix obtained by using SDA with Cross Validation

Through SDA all compounds of the training set were classified as presented in Table 5. The classification error rate was 0% resulting in a satisfactory separation between more and less active compounds.

The reliability of the model is determined by carrying out a cross-validation test, which uses the leave-one-out technique. In this procedure, one compound is omitted of the data set and the classification functions are built based on the remaining compounds. Afterwards, the omitted compound is classified according to the classification functions generated. In the next step, the omitted compound is included and a new compound is removed, and the procedure goes on until the last compound is removed. The obtained results with the cross-validation methodology are summarized in Table 6. Since the total of correct information was 100%, the model can be believed as being a good model.

Compound	*IC5*	*Mor29m*	*O1*	*MlogP*
26	5.371	-0.437	-0.238	2.461
27	5.526	-0.544	-0.249	2.496
28	5.402	-0.516	-0.241	1.649
29	5.336	-0.481	-0.239	2.461
30	5.572	-0.553	-0.238	2.305
31	5.464	-0.411	-0.226	3.117
32	5.584	-0.323	-0.244	3.328
33	5.282	-0.496	-0.226	3.225
34	5.483	-0.570	-0.345	2.090
35	5.583	-0.667	-0.262	2.922

Table 7. Values of the four descriptors for the compounds from the test set

2.6 Classification of unknown compounds

The models built from compounds from the training set through PCA, HCA, KNN, SIMCA and SDA now can be used to classify others compounds (test set, Fig. 10) whose anticancer activities are unknown. So ten compounds were proposed here to verify if they must be classified as less active or more active against a human hepatocellular carcinoma cell line, HepG2. In fact, they were not selected from any literature, so it is supposed that they have not been tested against this carcinoma. These compounds were selected so that they have substituitions at the same positions as those for the training set (R1 and R2) and the same type of atoms. It is important to keep the main characteristics of the compounds that generated the models. This way good predictions can be achieved. The classification of the test set was based on the four descriptors used in the models: *IC5, Mor29m, O1* and *MlogP*, according to Table 7.

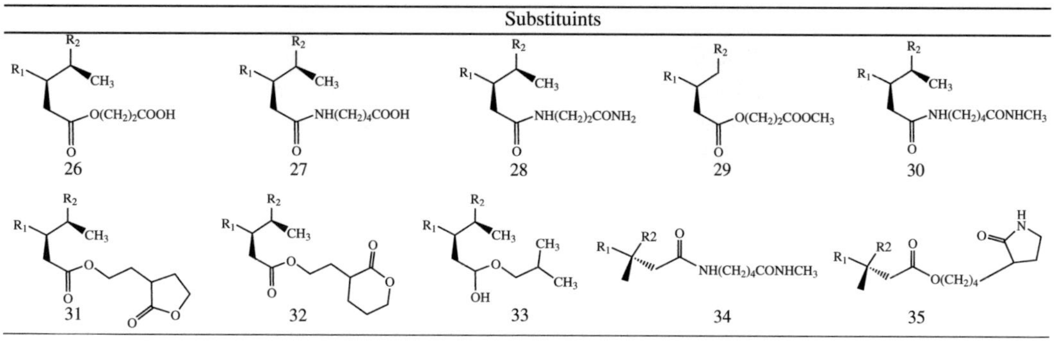

Fig. 10. Compounds from the test set which must be classified as either less active or more active

Compound	PCA	HCA	KNN	SIMCA	SDA
26	-	-	-	-	-
27	+	+	+	+	+
28	-	-	-	-	-
29	-	-	-	-	-
30	+	+	+	+	+
31	-	-	-	-	-
32	-	-	-	-	-
33	-	-	-	-	-
34	+	+	+	+	+
35	+	+	+	+	+

Table 8. Predicted classification for unknown compounds from the test set through different methods. Minus sign (-) for a compound classified as less active while plus sign (+) for a compound classified as more active

The result presented in Table 8 reveal that all samples (test set) receive the same classification by the four methods. Compounds **26, 28, 29, 31, 32** and **33** were classified as less active while compounds **27, 30, 34** and **35** were classified as more active. If you look for an explanation for such a pattern you will note that **26** and **27** present carboxylic acid group at the end of the chain, but only **27** is classified as more active. So it is possible that the change of an ester group by an amide group causes increase in activity. However when two amide groups are considered as occurs in **28** and **30** more carbon atoms in substituent means more active. Now comparing **26, 29, 31** and **32,** all of them have ester group associated with another different group and they all are classified as less active. The presence of the second group seams not to modify activity too much. The same effect is found in **34** and **35,** both more active.

3. Conclusion

All multivariate statistical methods (PCA, HCA, KNN, SIMCA and SDA) classified the 25 compounds from the training set into two distinct classes: more active and less active according to their degree of anticancer HepG2 activity. This classification was based on *IC5,*

Mor29m, O1 and *MlogP* descriptors. They represent four distinct classes of interactions related to the molecules, especially between the molecules and the biological receptor: steric (*IC5*), 3D-morse (*Mor29m*), electronic (*O1*) and molecular (*MlogP*).

A test set with ten molecules with unknown anticancer activity has its molecules classified, according to their biological response, into more active or less active compound. The results reveal in which classes they are grouped. In general molecules classified as more active must be seen as more efficient in cancer treatment than those classified as less active. Then the developed studies with PCA, HCA, KNN, SIMCA and SDA can provide valuable insight into the experimental process of syntheses and biological evaluation of the new artemisinin derivatives with activity against cancer HepG2. Without chemometrics no model and, consequently, no classification could be possible unless you are a prophet!

The interfacioal location of chemometrics, falling between measurements on the one side and statistical and computational theory and methods on the other, poses a challenge to the new practioner (Brow et al., 2009). The future of chemometrics lies in the development of innovative solutions to interesting problems. Some of the most exciting opportunities for innovation and new developments in the field of chemometrics lie at the interface between chemical and biological sciences. These opportunities are made possible by the exciting new scientific advances and discoveries of the past decade (Gemperline, 2006).

Finally, after reading this chapter you certainly must have noticed that chemometrics is a useful tool in medicinal chemistry, mainly when the great diversity of data is taken into account, because a lot of conclusions can be achieved. A study like this one here presented, where different methods are employed, is one of the examples of how chemometrics is important in drug design. Thus applications of statistics in chemical data analysis looking for the discovery of more efficacious drugs against diseases must continue and will certainly help researches.

4. References

Barbosa, J.; Ferreira, J.; Figueiredo, A.; Almeida, R.; Silva, O.; Carvalho, J.; Cristino, M.; Ciriaco-Pinheiro, J.; Vieira, J. & Serra, R. (2011). Molecular Modeling and Chemometric Study of Anticancer Derivatives of Artemisini. *Journal of the Serbian Chemical Society*, Vol. 76, No. 9, (September 2011), pp. 1263-1282, ISSN 0352-5139

Brereton, R. (2009). *Chemometrics for Pattern Recognition*, John Wiley & Sons, Ltd, ISBN 978-0-470-74646-2, West Sussex,England

Brereton, R. (2007). *Applied Chemometrics for Scientists*, John Wiley & Sons, Ltd, ISBN 978-0-470-01686-2, West Sussex, England

Brown, S.; Tauler, R. & Walczak, B. (Ed(s)) (2009). *Compreensive Chemometrics: Chemical and Biochemical Data Analysis*, Vol. 1, Elsevier, ISBN 978-0-444-52702-8, Amsterdan, The Netherlands

Bruns, R.; Scarminio, I. & Barrros Neto, B. (2006) *Statistical Design - Chemometrics*, Elsevier, ISBN 978-0-444-52181-1, Amsterdan, The Netherlands

Cardoso, F.; Figueiredo, A.; Lobato, M.; Miranda, R.; Almeida, R. & Pinheiro, J. (2008). A Study on Antimalarial Artemisinin Derivatives Using MEP Maps and Multivariate QSAR. *Journal of Molecular Modeling*, Vol. 14, No. 1, (January 2008), pp. 39-48, ISSN 0948-5023

Doweyko, A. (2008). QSAR: Dead or Alive? *Journal of Computer-Aided Molecular Design*, Vol. 22, No. 2, (February 2008), pp. 81-89, ISSN 1573-4951

Efferth, T. (2005). Mechanistic Perspectives for 1,2,4-trioxanes in Anti-cancer Therapy. *Drug Resistance. Updat*, Vol. 8, No.1-2, (February 2005), pp. 85-97, ISSN 1368-7646

Ferreira, M. (2002). Multivariate QSAR. *Journal of the Brazilian Chemical Society*, Vol.13, No. 6, (November/December 2002), pp. 742-753, ISSN 1678-4790

Fujita, T. (1995). *QSAR and Drug Design: New Developments and Applications*, Elsevier, ISBN 0-444-88615-X, Amsterdan, The Netherlands

Gareth, T. (2003). *Fundamental of Medicinal Chemistry*, John Wiley & Sons, Ltd, ISBN 0-470-84307-1, West Sussex, England

Frisch, M. J.; Trucks, G. W.; Schlegel, H. B.; Scuseria, G. E.; Robb, M. A.; Cheeseman, J. R.; Zakrzewski, V. G.; Montgomery, Jr J. A.; Stratmann, R. E.; Burant, J. C.; Dapprich, S.; Millam, J. M.; Daniels, A. D.; Kudin, K.N.; Strain, M. C.; Farkas, O.; Tomasi, J.; Barone, V.; Cossi, M.; Cammi, R.; Mennucci, B.; Pomelli, C.; Adamo, C.; Clifford, S.; Ochterski, J.; Petersson, G. A.; Ayala, P. Y.; Cui, Q.; Morokuma, K.; Salvador, P.; Dannenberg, J. J.; Malick, D. K.; Rabuck, A. D.; Raghavachari, K. J.; Foresman, B.; Cioslowski, J.; Ortiz, J. V.; Baboul, A. G.; Stefanov, B. B.; Liu, G.; Liashenko, A.; Piskorz, P.; Komaromi, I.; Gomperts, R.; Martin, R. L.; Fox, D. J.; Keith, T.; Al-Laham, M. A.; Peng, C. Y.; Nanayakkara, A.; Challacombe, M.; Gill, P. M. W.; Johnson, B.; Chen, W.; Wong, M. W.; Andres, J. L.; Gonzalez, C.; Head-Gordon, M.; Replogle, E. S. & Pople, J. A. (1998) *Gaussian, Inc.*, Gaussian 98 Revision A.7, Pittsburgh PA

Gemperline, P. (2006). *Practical Guide to Chemometrics* (2nd), CRC Press, ISBN 1-57444-783-1, Florida, USA

Varmuza, K. & Filzmoser, P. (2009). *Introduction to Multivariate Statistical Analysis in Chemometrics*, CRC Press, ISBN 9781420059472, Florida, USA

Lai, H.; Nakasi, I.; Lacoste, E.; Singh, N. & Sasaki (2009). T. Artemisinin-Transferrin Conjugate Retards Growth of Breast Tumors in the Rat. *Anticancer Research*, Vol. 29, No. 10, (October 2009), pp. 3807-3810, ISSN 1791-7530

Levine, I. (1991). *Quantum Chemistry* (4th), Prentice Hall, ISBN 0-205-12770-3, New Jersey, USA

Liu,Y.; Wong, V.; Ko, B.; Wong, M. & Che, C. (2005). Synthesis and Cytotoxicity Studies of Artemisinin Derivatives Containing Lipophilic Alkyl Carbon Chains. *Organic Letters*, Vol. 7, No. 8, (March 2005), pp. 1561-1564. ISSN 1523-7052

Pinheiro, J.; Kiralj, R.; & Ferreira, M. (2003). Artemisinin Derivatives with Antimalarial Activity against Plasmodium falciparum Designed with the Aid of Quantum Chemical and Partial Least Squares Methods. *QSAR & Combinatorial Science*, Vol. 22, No. 8, (November 2003), pp. 830-842, ISSN 1611-0218

Manly, B. (2004). *Multivariate Statistical Methods: A Primer* (3), Chapman and Hall/CRC, ISBN 9781584884149, London, England

Price, R.; van Vugt, M.; Nosten, F.; Luxemburger, C.; Brockman, A.; Phaipun, L.; Chongsuphajaisiddhi, T. & White, N. (1998). Artesunate versus Artemether for the Treatment of Recrudescent Multidrug-resistant Falciparum Malaria. *The American Journal of Tropical Medicine and Hygiene*, Vol. 59, No. 6, (December 1998), pp. 883-888, ISSN 0002-9637

Puzyn, T.; Leszczynski, J. & Cronin, M. (Ed(s)). (2010). *Recent Advances in QSAR Studies: Methods and Applications*, Springer, ISBN 978-1-4020-9783-6, New York, USA

Rajarshi, G. (2008). On the interpretation and interpretability of quantitative structure-activity relationship models. *Journal of Computer-Aided Molecular Design*, Vol. 22, No. 12, (December 2008), pp. 857-871, ISSN 1573-4951

Wold, S. (1995). *Chemometrics, what do we mean with it, and what do we want from it? Chemometrics and Intelligent Laboratory Systems*, Vol. 30, No. 1, (November 1995), pp. 109-115, ISSN 0169-7439

Virtual Computational Laboratory, VCCLAB In: e-Dragon, 13.05.2010, Available from http://www.vcclab.org

9

Electronic Nose Integrated with Chemometrics for Rapid Identification of Foodborne Pathogen

Yong Xin Yu and Yong Zhao*
College of Food Science and Technology, Shanghai Ocean University, Shanghai, China

1. Introduction

Diseases caused by foodborne pathogens have been a serious threat to public health and food safety for decades and remain one of the major concerns of our society. There are hundreds of diseases caused by different foodborne pathogenic microorganisms, including pathogenic viruses, bacteria, fungi, parasites, marine phytoplankton, and cyanobacteria, etc (Hui, 2001). Among these, bacteria such as *Salmonella* spp., *Shigella* spp., *Escherichia coli, Staphylococcus aureus, Campylobacter jejuni, Campylobacter coli, Bacillus cereus, Vibrio parahaemolyticus* and *Listeria monocytogenes* are the most common foodborne pathogens (McClure, 2002), which can spread easily and rapidly under requiring food, moisture and a favorable temperature (Bhunia, 2008).

Identification and detection pathogens in clinical, environmental or food samples usually involves time-consuming growth in selective media, subsequent isolation and laborious biochemical and molecular diagnostic procedures (Gates, 2011). Many of these techniques are also expensive or not sensitive enough for the early detection of bacterial activity (Adley, 2006). The development of alternative analytical techniques that are rapid and simple has become increasingly important to reduce sample preparation time investment and to conduct real time analyses.

It is well known that microorganisms can produce species-specific microbial volatile organic compounds (MVOCs), or odor compounds, which characterize as odor fingerprinting (Turner & Magan, 2004). Early in this research area, the question arose as to can we use odor fingerprinting like DNA fingerprinting to identify or detect microbe in pure culture or in food samples. To date it is still a very interesting scientific question. Many studies (Bjurman, 1999, Kim et al., 2007, Korpi et al., 1998, Pasanen et al., 1996, Wilkins et al., 2003), especially those using analytical tools such as gas chromatography (GC) or gas chromatography coupled with mass spectrometry (GC–MS) for headspace analysis, have shown that microorganisms produce many MVOCs, including alcohols, aliphatic acids and terpenes, some of which have characteristic odors (Schnürer et al., 1999).

* Corresponding Author, mail address: yzhao@shou.edu.cn

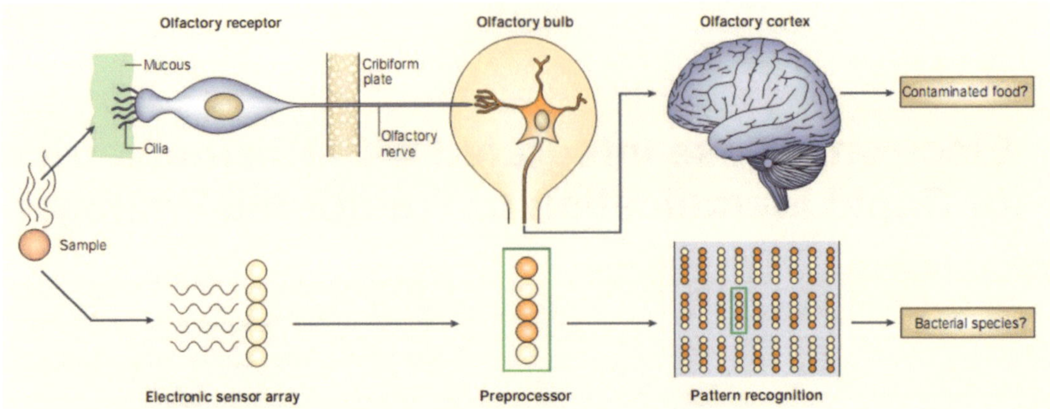

Fig. 1. Electronic nose devices mimic the human olfactory system.

The electronic devices simulate the different stages of the human olfactory system, resulting in volatile odor recognition, which can now be used to discriminate between different bacterial infections. (Turner & Magan, 2004)

During the past three decades there has been significant research interest in the development of electronic nose (E-nose) technology for food, agricultural and environmental applications (Buratti et al., 2004, Pasanen et al., 1996, Romain et al., 2000, Wilkins et al., 2003). The term E-nose describes a machine olfaction system, which successfully mimics human olfaction and intelligently integrates of multitudes of technologies like sensing technology, chemometrics, microelectronics and advanced soft computing (see Fig. 1). Basically, this device is used to detect and distinguish complex odor at low cost. Typically, an electronic nose consists of three parts: a sensor array which is exposed to the volatiles, conversion of the sensor signals to a readable format, and software analysis of the data to produce characteristic outputs related to the odor encountered. The output from the sensor array may be interpreted via a variety of chemometrics methods (Capone et al., 2001, Evans et al., 2000, Haugen & Kvaal, 1998) such as principal component analysis (PCA), discriminant function analysis (DFA), cluster analysis (CA), soft independent modelling by class analogy (SIMCA), partial least squares (PLS) and artificial neural networks (ANN) to discriminate between different samples. The data obtained from the sensor array are comparative and generally not quantitative or qualitative in any way. It has the potential to be a sensitive, fast, one-step method to characterize a wide array of different volatile chemicals. Since the first model of an intelligent electronic gas sensing model was described, a significant amount of gas sensing research has been focused on several industrial applications.

Recently, some novel microbiological applications of E-nose have been reported, such as the characterization of fungi (Keshri et al., 1998, Pasanen et al., 1996, Schnürer et al., 1999), bacteria (Dutta et al., 2005, Pavlou et al., 2002a) and the diagnosis of disease (Gardner et al., 2000, Pavlou et al., 2002b, Zhang et al., 2000). It is more and more clear that E-nose techniques coupled with different chemometrics analyses of the odor fingerprinting offer a wide range of applications for food microbiology, including identification of foodborne pathogen.

2. Detection strategies

Several E-nose devices have been developed, all of which comprise three basic building blocks: a volatile gas odor passes over a sensor array, the conductance of the sensors changes owing to the level of binding and results in a set of sensor signals, which are coupled to data-analysis software to produce an output (Turner & Magan, 2004).

The main strategy of foodborne pathogen identification based on E-nose, which is composed of three steps: headspace sampling, gas sensor detection and chemometrics analysis (see Fig. 2).

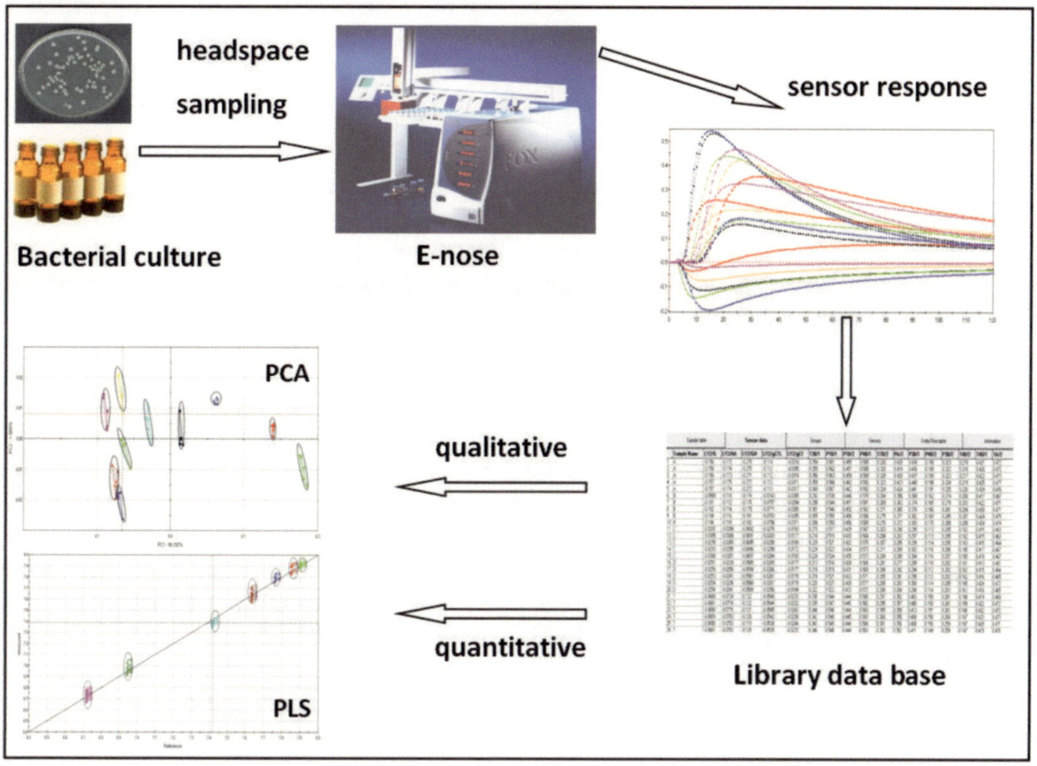

Fig. 2. Electronic nose and chemometrics for the identification of foodborne pathogen. The main strategy of foodborne pathogen identification based on E-nose.

2.1 Headspace sampling

Before analysis, the bacterial cultures should be transferred into standard 20 ml headspace vials and sealed with PTFE-lined Teflon caps to equilibrate the headspace. Sample handling is a critical step affecting the analysis by E-nose. The quality of the analysis can be improved by adopting an appropriate sampling technique. To introduce the volatile compounds present in the headspace (HS) of the sample into the E-nose's detection system, several headspace sampling techniques have been used in E-nose. Typically, the methods of headspace sampling (Ayoko, 2004) include static headspace (SHS) technique, purge and trap (P&T) technique, stir bar sorptive extraction (SBSE) technique, inside-needle dynamic extraction (INDEX) technique, membrane introduction mass spectrometry (MIMS) technique and solid phase micro extraction (SPME) technique.

Unlike the other techniques, SPME has a considerable concentration capacity and is very simple because it does not require especial equipment. The principle involves exposing a silica fibre covered with a thin layer of adsorbent in the HS of the sample in order to trap the volatile components onto the fibre. The adsorbed compounds are desorbed by heating and introduced into the detection system. A SPME sampler consists of a fused silica fiber that is coated by a suitable polymer (e.g. PDMS, PDMS/divinylbenzene, carboxen/PDMS) and housed inside a needle. The fiber is exposed to headspace volatile and after sampling is complete, it is retracted into the needle. Apart from the nature of the adsorbent deposited on the fiber, the main parameters to optimize are the equilibration time, the sample temperature and the duration of extraction. Compared with other sampling methods, SPME is simple to use and reasonably sensitive, so it is a user-friendly pre-concentration method.

In our studies, the headspace sampling method of E-nose was optimized for MVOCs analysis. The samples were placed in the HS100 auto-sampler in arbitrary order. The automatic injection unit heated the samples to 37°C with an incubation time of 600 seconds. The temperature of the injection syringe was 47°C. The delay time between two injections was 300 seconds. Then the adsorbed compounds are desorbed by heating and introduced into the detection system (Yu Y. X., 2010a, Yu Y. X., 2010b).

2.2 Gas sensor detection

The most complicated part of electronic olfaction process is odor capture and sensor technology to be deployed for such capturing. Once the volatile compounds of samples are introduced into the gas sensor detection system, the sensor array is exposed to the volatile compounds and then the odor fingerprint of samples is generated from sensor respond. By chemical interaction between the volatile compounds and the gas sensors, the state of the sensors is altered giving rise to electrical signals that are registered by the instrument of E-nose. In this way the signals from the individual sensor represent a pattern that is unique for the gas mixture measured and those data based on sensors is transformed to a matrix. The ideal sensors to be integrated in an electronic nose should fulfill the following criteria (Barsan & Weimar, 2001, James et al., 2005): high sensitivity toward the volatile chemical compounds, that is, the chemicals to be detected may be present in the concentration range of ppm or ppb, and the sensor should be sufficiently sensitive to small concentration level of gaseous species within a volatile mixture, similar to that of the human nose (down to 10^{-12} g/ml); low sensitivity toward humidity and temperature; medium selectivity, that is, they must respond to a range of different compounds present in the headspace of the sample; high stability; high reproducibility and reliability; high speed of response, short reaction and recovery time, that is, in order to be used for online measurements, the response time of the sensor should be in the range of seconds; reversibility, that is, the sensor should be able to recover after exposure to gas; robust and durable; easy calibration; easily processable data output; and small dimensions.

The E-nose used in our studies is a commercial equipment (FOX4000, Alpha M.O.S., Toulouse, France), with 18 metal oxide sensors (LY2/AA, LY2/G, LY2/gCT, LY2/gCTl, LY2/Gh, LY2/LG, P10/1, P10/2, P30/1, P30/2, P40/1, P40/2, PA2, T30/1, T40/2, T70/2,

T40/1, TA2), and this sensor array system is used for monitoring the volatile compounds produced by microorganism, and so on. The descriptors associated with the sensors are shown in Table 1. FOX4000 E-nose assay measurements showed signal with maximum intensities changing with the type of samples, which indicate that discrimination is obtained.

Sensors	Volatile description	Sensors	Volatile description
LY2/LG	Fluoride, chloride, oxynitride, sulphide	P30 /1	Hydrocarbons, ammonia, ethanol
LY2 /G	Ammonia, amines, Carbon oxygen compounds	T70 /2	Toluene, xylene, carbon monoxide
LY2 /AA	Alcohol, acetone, ammonia	T40 /1	Fluorine
LY2 /GH	Ammonia, amines compounds	P40 /1	Fluorine, chlorine
P40 /2	Chlorine, hydrogen sulfide, fluoride	LY2 /gCTL	hydrogen sulfide
P30 /2	Hydrogen sulphide, ketone	LY2 /gCT	Propane, butane
T30 /1	Polar compound, hydrogen chloride	T40 /2	chlorine
P10 /1	Nonpolar compound: hydrocarbon, Ammonia, chlorine	PA /2	Ethanol, ammonia, amine compounds
P10 /2	Nonpolar compound: Methane, ethane	TA /2	ethanol

Table 1. Sensor types and volatile descriptors of FOX4000 E-nose.

Each sensor element changes its electrical resistance (R_{max}) when exposed to volatile compounds. In order to produce consistent data for the classification, the sensor response is presented with a volatile chemical relative to the baseline electrical resistance in fresh air, which is the maximum change in the sensor electrical resistance divided by the initial electrical resistance, as follows:

$$\text{Relative electrical resistance change} = (R_{max} - R_0) / R_0$$

where R_0 is the initial baseline electrical resistance of the sensor and $R_{max} - R_0$ is the maximum change of the sensor electrical resistance. The baseline of the sensors was acquired in a synthetic air saturated steam at fixed temperature. The relative electrical resistance change value was used for data evaluation because it gives the most stable result, and is more robust against sensor baseline variation (Siripatrawan, 2008).

Data of the relative electrical resistance changes from the 18 sensors can combine with every sample to form a matrix (see Fig. 2: The library data base) and the data is without preprocessing prior to chemometrics analysis. The sensor response is stored in the computer through data acquisition card and these data sets are analyzed to extract information.

2.3 Chemometrics analysis

The matrix of signal is interpreted by multivariate chemometrics techniques like the PCA, PLS, ANN, and so on. Samples with similar odor fingerprinting generally give rise to similar sensor response patterns, while samples with different odor fingerprinting show differences in their patterns. The sensors of an E-nose can respond to both odorous and odorless volatile compounds.

These various chemometrics methods are used in those works, according to the aim of the studies. Generally speaking, the chemometrics methods can be divided into two types: unsupervised and supervised methods(Mariey et al., 2001). The objective of unsupervised methods is to extrapolate the odor fingerprinting data without a prior knowledge about the bacteria studied. Principal component analysis (PCA) and Hierarchical cluster analysis (HCA) are major examples of unsupervised methods. Supervised methods, on the other hand, require prior knowledge of the sample identity. With a set of well-characterized samples, a model can be trained so that it can predict the identity of unknown samples. Discriminant analysis (DA) and artificial neural network (ANN) analysis are major examples of supervised methods.

PCA is used to reduce the multidimensionality of the data set into its most dominant components or scores while maintaining the relevant variation between the data points. PCA identifies the natural clusters in the data set with the first principal component (PC) expressing the largest amount of variation, followed by the second PC which conveys the second most important factor of the remaining analysis, and so forth(Di et al., 2009, Huang et al., 2009, Ivosev et al., 2008). Score plots can be used to interpret the similarities and differences between bacteria. The closer the samples are within a score plot, the more similar they are with respect to the principal component score evaluated(Mariey et al., 2001). In our studies, each sample data of 18 sensors is then compared to the others in order to make homogeneous groups. A scatter plot can then be drawn to visualize the results, each sample being represented by a plot.

3. Application of E-nose and chemometrics for bacteria identification

With the success of the above applications of the E-nose have been published, the authors were interested in determining whether or not an E-nose would be able to identify bacteria. A series of experiments were designed to determine this. In this part, bacteria identification at different levels (genus, species, strains) was cited as an example to illustrate using this integrated technology to foodborne bacteria effective identification.

3.1 At genus level

In this study, three bacteria, *Listeria monocytogenes, Staphylococcus lentus* and *Bacillus cereus,* which from three different genus, were investigated for the odor fingerprint by E-nose. The result of PCA (Fig.3a) shows that, the fingerprints give a good difference between the blank culture and the bacterial culture, and the three bacteria can be classified from each other by the odor fingerprints. Using the cluster analysis to represent the sensor responses (Fig. 3b), it is also possible to obtain a clear separation between the blank control and culture inoculated with bacteria. And the CA result also reveals that successful discrimination between the bacteria at different genus is possible(Yu Y. X., 2010a).

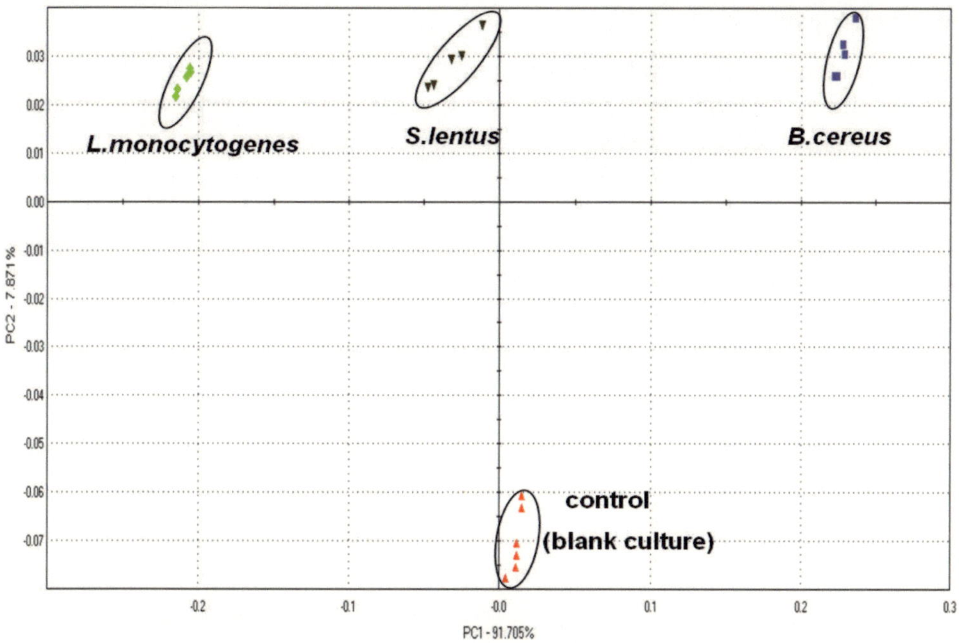

Fig. 3(a). Principal components analysis (PCA) for the discrimination of three bacteria from different genus on the basis of E-nose. The plot displays clear discrimination between the four groups, accounting for nearly 99% of the variance within the dataset.

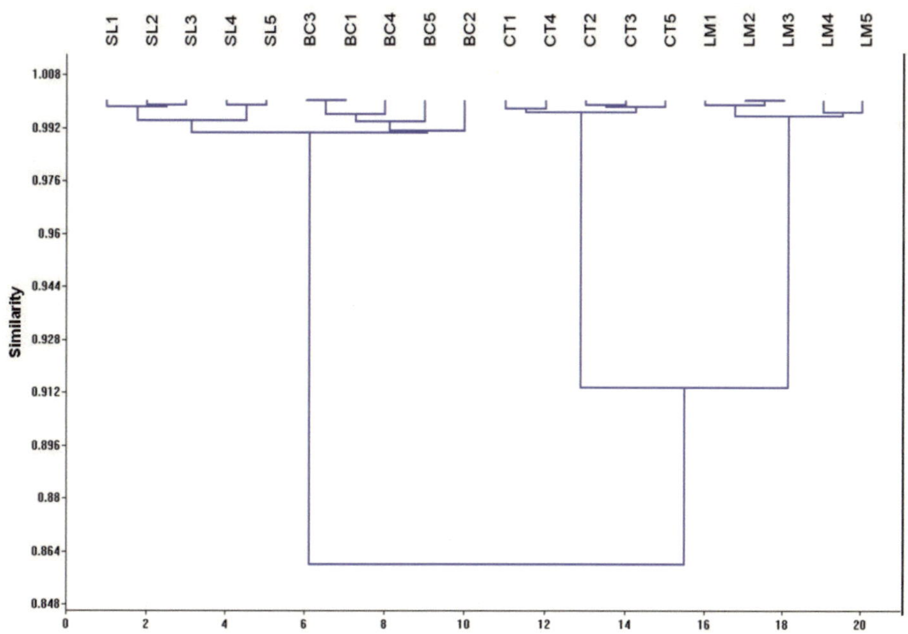

Fig. 3(b). Cluster analysis (CA) for the discrimination of three bacteria from different genus on the basis of E-nose. (*S. lentus*: SL1-SL5, *B. cereus*: BC1-BC5, *L. monocytogenes*: LM1-LM5, control blank culture: CT1-CT5).

3.2 At species level

In this study, using the same collection methodology, the E-nose was tested for its ability to distinguish among bacterial pathogens at species levels. Four species bacteria selected from *Pseudomonas* sp, named *Pseudomonas fragi, Pseudomonas fluorescens, Pseudomonas putida* and *Pseudomonas aeruginosa,* were investigated for the odor fingerprint by E-nose. It is clear that the E-nose was able to distinguish amongst all specimens tested. The PCA result in Fig.4(a) shows a representative experiment, where individual species of bacteria clustered in individual groups, separate from each other and the bacteria *Pseudomonas fragi* is given a great difference form the three other bacteria by the odor fingerprints. The result of cluster analysis in Fig. 4(b) also reveals that successful discrimination between the different bacteria at strains level is possible.

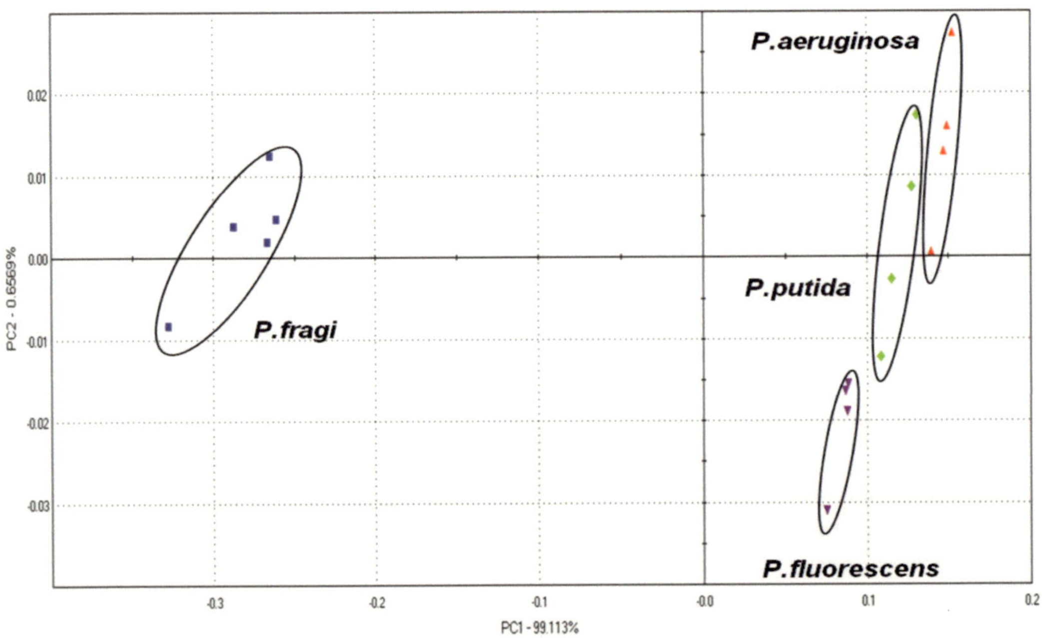

Fig. 4(a). Principal components analysis (PCA) for the discrimination of four different species of Pseudomonas sp on the basis of E-nose. The plot displays clear discrimination between the four groups, accounting for nearly 99% of the variance within the dataset.

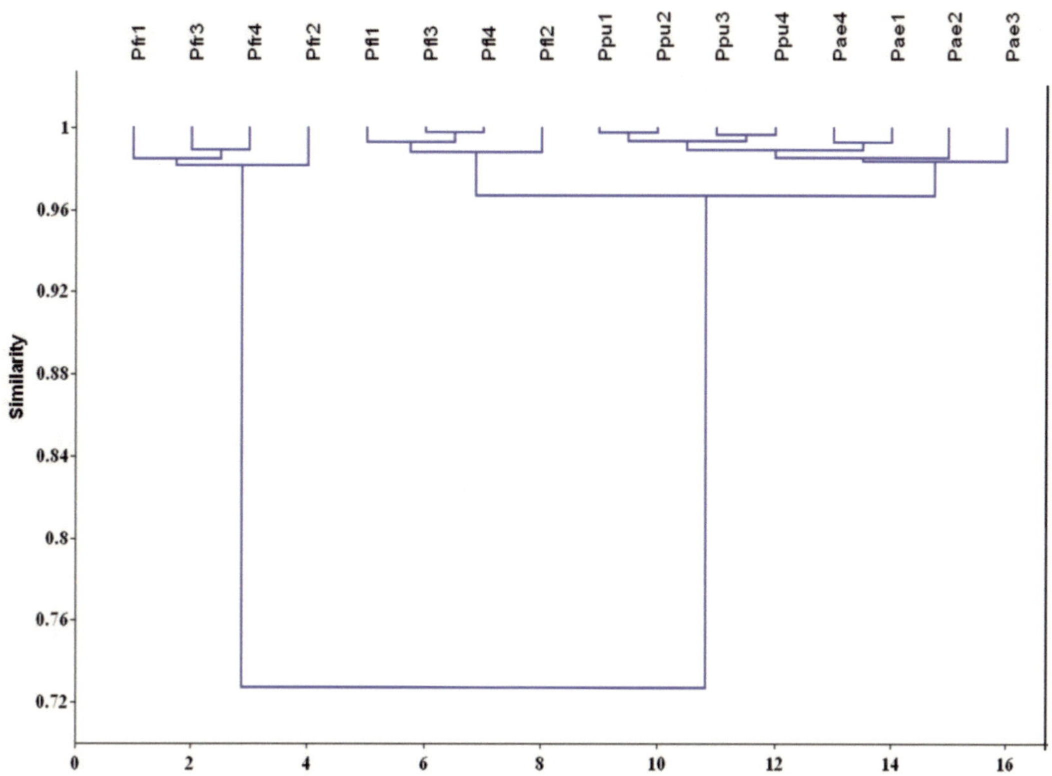

(*P. fragi*: Pfr1-Pfr4, *P. fluorescens*: Pfl1-Pfl4, *P. putida*: Ppu1-Ppu4, *P. aeruginosa*: Pae1-Pae4).

Fig. 4(b). Cluster analysis (CA) for the discrimination of four different species of Pseudomonas sp on the basis of E-nose.

3.3 At strains level

The next set of experiments involved testing the integrated method to see whether it could correctly differentiate bacteria samples as different strains. In this study, four strains of *Vibrio parahaemolyticus*, named *V. parahaemolyticus* F01, *V. parahaemolyticus* F13, *V. parahaemolyticus* F38 and *V. parahaemolyticus* F54, were compared with the odor fingerprint by E-nose. As shown in a representative data set in Fig. 5(a), the four strains of *V. parahaemolyticus* are separated from each other. However, the result from cluster analysis in Fig. 5(b) shows that some overlap appeared between *V. parahaemolyticus* F01 and *V. parahaemolyticus* F13, and it indicate that the odor fingerprints of these two strains may be too similar to identify by this method.

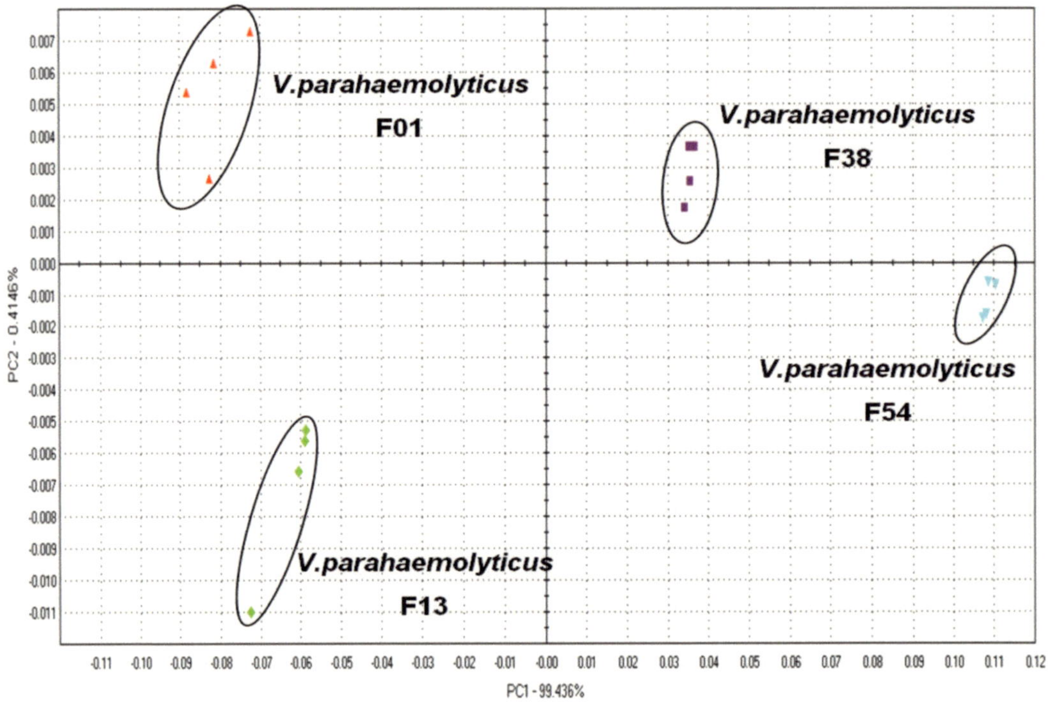

Fig. 5(a). Principal components analysis (PCA) for the discrimination of four different strains of *V. parahaemolyticus* on the basis of E-nose. The plot displays clear discrimination between the four groups, accounting for nearly 99% of the variance within the dataset.

4. Future perspectives

Electronic nose technology is relatively new and holds great promise as a detection tool in food safety area because it is portable, rapid and has potential applicability in foodborne pathogen identification or detection. On the basis of the work described above, we have demonstrated that the E-nose integrated with chemometrics can be used to identify pathogen bacteria at genus, species and strains levels.

As to know, bacteria respond to environmental triggers by switching to different physiological states. If such changes can be detected in the odor fingerprints, then E-nose analysis can produce information that can be very useful in determining virulence,

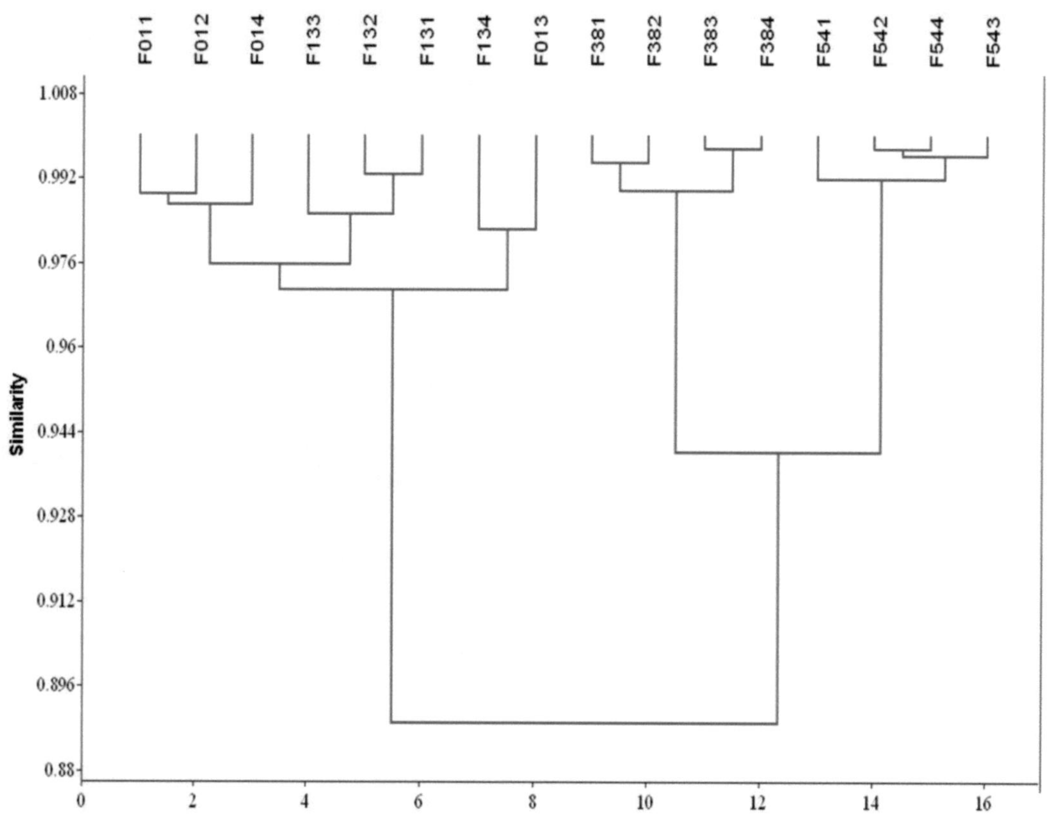

(*V.p* F01: F011-F014, *V.p* F13: F131-F134, *V.p* F38: F381-F384, *V.p* F54: F541-F544).

Fig. 5(b). Cluster analysis (CA) for the discrimination of four different strains of *V. parahaemolyticus* on the basis of E-nose.

conducting epidemiological studies, or determining the source of a food poisoning outbreak. Of course the ability to produce information on the physiological state of a microorganism offers many potential benefits. Nevertheless, a variety of different fingerprints, produced under a variety of growth conditions, must be developed for each pathogen, for inclusion in the reference database. To avoid this complication, we should culture the pathogens under controlled conditions. Otherwise, the identification algorithm must be capable of sorting through them all, to find a single, reliable, positive identification for the unknown.

Recently developed chemometrics algorithms are particularly suited to the rapid analysis and depiction of this data. Chemometrics is one approach that may offer novel insights into our understanding of the difference of microbiology. Adopting appropriate chemometrics methods will improve the quality of analysis.

Odor fingerprinting method based on E-nose is still in its infancy. Many recent technological advances, which are outside the scope of this chapter, can be used to transform the odor fingerprinting concept into user-friendly, automated systems for high-throughput analyses. The introduction of smaller, faster and smarter instrumentation of E-nose to the market could also depend much on the embedding of chemometrics. In addition, more and more classification techniques based on odor fingerprinting may be developed to classify the pathogens into exact levels such as genus, species and stains. Further investigation may contribute to make a distinction between the pathogen and non-pathogen bacterial.

In short, E-nose integrated with chemometrics is a reliable, rapid, and economic technique which could be explored as a routine diagnostic tool for microbial analysis.

5. Acknowledgments

The authors acknowledge the financial support of the project of Shanghai Youth Science and Technology Development (Project No: 07QA14047), the Leading Academic Discipline Project of Shanghai Municipal Education Commission (Project No: J50704), Shanghai Municipal Science, Technology Key Project of Agriculture Flourishing plan (Grant No: 2006, 10-5; 2009, 6-1), Public Science and Technology Research Funds Projects of Ocean (Project No: 201105007), Project of Science and Technology Commission of Shanghai Municipality (Project No: 11310501100), and Shanghai Ocean University youth teacher Fund (Project No: A-2501-10-011506).

6. References

Adley C, 2006. Food-borne pathogens: methods and protocols. *Humana Pr Inc.*

Ayoko GA, 2004. Volatile organic compounds in indoor environments. *Air Pollution*, 1-35.

Barsan N, Weimar U, 2001. Conduction model of metal oxide gas sensors. *Journal of Electroceramics* 7, 143-67.

Bhunia AK, 2008. Foodborne microbial pathogens: mechanisms and pathogenesis. *Springer Verlag.*

Bjurman J, 1999. Release of MVOCs from microorganisms. *Organic Indoor Air Pollutants*, 259-73.

Buratti S, Benedetti S, Scampicchio M, Pangerod E, 2004. Characterization and classification of Italian Barbera wines by using an electronic nose and an amperometric electronic tongue. *Anal Chim Acta* 525, 133-9.

Capone S, Epifani M, Quaranta F, Siciliano P, Taurino A, Vasanelli L, 2001. Monitoring of rancidity of milk by means of an electronic nose and a dynamic PCA analysis. *Sensors and Actuators B: Chemical* 78, 174-9.

Di CZ, Crainiceanu CM, Caffo BS, Punjabi NM, 2009. Multilevel functional principal component analysis. *Annals of Applied Statistics* 3, 458-88.

Dutta R, Morgan D, Baker N, Gardner JW, Hines EL, 2005. Identification of Staphylococcus aureus infections in hospital environment: electronic nose based approach. *Sensors and Actuators B: Chemical* 109, 355-62.

Evans P, Persaud KC, Mcneish AS, Sneath RW, Hobson N, Magan N, 2000. Evaluation of a radial basis function neural network for the determination of wheat quality from electronic nose data. *Sensors and Actuators B: Chemical* 69, 348-58.

Gardner JW, Shin HW, Hines EL, 2000. An electronic nose system to diagnose illness. *Sensors and Actuators B: Chemical* 70, 19-24.

Gates KW, 2011. Rapid Detection and Characterization of Foodborne Pathogens by Molecular Techniques. *Journal of Aquatic Food Product Technology* 20, 108-13.

Haugen JE, Kvaal K, 1998. Electronic nose and artificial neural network. *Meat Sci* 49, S273-S86.

Huang SY, Yeh YR, Eguchi S, 2009. Robust kernel principal component analysis. *Neural Comput* 21, 3179-213.

Hui YH, 2001. Foodborne Disease Handbook: Plant Toxicants. *CRC*.

Ivosev G, Burton L, Bonner R, 2008. Dimensionality reduction and visualization in principal component analysis. *Anal Chem* 80, 4933-44.

James D, Scott SM, Ali Z, O'hare WT, 2005. Chemical sensors for electronic nose systems. *Microchimica Acta* 149, 1-17.

Keshri G, Magan N, Voysey P, 1998. Use of an electronic nose for the early detection and differentiation between spoilage fungi. *Lett Appl Microbiol* 27, 261-4.

Kim JL, Elfman L, Mi Y, Wieslander G, Smedje G, Norbäck D, 2007. Indoor molds, bacteria, microbial volatile organic compounds and plasticizers in schools–associations with asthma and respiratory symptoms in pupils. *Indoor Air* 17, 153-63.

Korpi A, Pasanen AL, Pasanen P, 1998. Volatile compounds originating from mixed microbial cultures on building materials under various humidity conditions. *Appl Environ Microbiol* 64, 2914.

Mariey L, Signolle J, Amiel C, Travert J, 2001. Discrimination, classification, identification of microorganisms using FTIR spectroscopy and chemometrics. *Vibrational Spectroscopy* 26, 151-9.

Mcclure PJ, 2002. Foodborne pathogens: hazards, risk analysis, and control. *Woodhead Pub Ltd.*

Pasanen AL, Lappalainen S, Pasanen P, 1996. Volatile organic metabolites associated with some toxic fungi and their mycotoxins. *Analyst* 121, 1949-53.

Pavlou A, Turner A, Magan N, 2002a. Recognition of anaerobic bacterial isolates in vitro using electronic nose technology. *Lett Appl Microbiol* 35, 366-9.

Pavlou AK, Magan N, Mcnulty C, *et al.*, 2002b. Use of an electronic nose system for diagnoses of urinary tract infections. *Biosensors and Bioelectronics* 17, 893-9.

Romain AC, Nicolas J, Wiertz V, Maternova J, Andre P, 2000. Use of a simple tin oxide sensor array to identify five malodours collected in the field. *Sensors and Actuators B: Chemical* 62, 73-9.

Schnürer J, Olsson J, Börjesson T, 1999. Fungal volatiles as indicators of food and feeds spoilage. *Fungal Genetics and Biology* 27, 209-17.

Siripatrawan U, 2008. Rapid differentiation between E. coli and Salmonella Typhimurium using metal oxide sensors integrated with pattern recognition. *Sensors and Actuators B: Chemical* 133, 414-9.

Turner APF, Magan N, 2004. Electronic noses and disease diagnostics. *Nature Reviews Microbiology* 2, 161-6.

Wilkins K, Larsen K, Simkus M, 2003. Volatile metabolites from indoor molds grown on media containing wood constituents. *Environmental Science and Pollution Research* 10, 206-8.

Yu Y. X., Liu Y., Sun X. H., Pan Y. J., Zhao Y., 2010a. Recognition of Three Pathogens Using Electronic Nose Technology. *Chinese Journal of Sensors and Actuators* 23, 10-3.

Yu Y. X., Sun X. H., Pan Y. J., Zhao Y., 2010b. Research on Food-borne Pathogen Detection Based on Electronic Nose. *chemistry online (in Chinese)*, 154-9.

Zhang Q, Wang P, Li J, Gao X, 2000. Diagnosis of diabetes by image detection of breath using gas-sensitive laps. *Biosensors and Bioelectronics* 15, 249-56.

Part 3

Technology

10

Chemometrics in Food Technology

Riccardo Guidetti, Roberto Beghi and Valentina Giovenzana
Department of Agricultural Engineering,
Università degli Studi di Milano, Milano,
Italy

1. Introduction

The food sector is one of the most important voices in the economic field as it fulfills one of the main needs of man. The changes in the society in recent years have radically modified the food industry by combining the concept of globalization with the revaluation of local production. Besides the production needs to be global, in fact, there are always strong forces that tend to re-evaluate the expression of the deep local production like social history and centuries-old tradition.

The increase in productivity, in ever-expanding market, has prompted a reorganization of control systems to maximize product standardization, ensuring a high level of food security, promote greater compliance among all batches produced. The protection of large quantities of production, however, necessarily passes through systems to highlight possible fraud present throughout the production chain: from the raw materials (controlled by the producer) to the finished products (controlled by large sales organizations). The fraud also concern the protection of local productions: the products of guaranteed origin must be characterized in such a way to identify specific properties easily and detectable by objective means.

The laboratories employ analytical techniques that are often inadequate because they require many samples, a long time to get the response, staff with high analytical ability. In a context where the speed is an imperative, technology solutions must require fewer samples or, at least no one (non-destructive techniques); they have to provide quick answers, if not immediate, in order to allow the operator to decide quickly about further steps to control or release the product to market; they must be easy to use, to promote their use throughout the production chain where it is not always possible to have analytical laboratories. The technologies must therefore be adapted to this new approach to production: the sensors and the necessary related data modeling, which allows the "measure", are evolving to meet the needs of the agri-food sector. The trial involves, often, Research Institutions on the side of Companies, a sign of a great interest and a high level of expectations. The manufacturers of technologies, often, provide devices that require calibration phases not always easy to perform, but that are often the subject of actual researches. These are particularly complex when the modeling approach must be based on chemometrics.

This chapter is essentially divided into two parts: the first part analyzes the theoretical principles of the most important technologies, currently used in the food industry, that used a chemometric approach for the analysis of data (spectrophotometry Vis/NIR (Visible and Near InfraRed) and NIR (Near InfraRed), Image Analysis with particular regard to Hyperspectral Image Analysis and Electronic Nose); the second part will present some case studies of particular interest related to the same technologies (fruit and vegetables, wine, meat, fish, dairy, olive, coffee, baked goods, etc.) (Frank & Todeschini, 1994; Massart et al., 1997 and 1998; Basilevsk, 1994; Jackson, 1991).

2. Technologies used in the food sector combined with chemometrics

2.1 NIR and Vis/NIR spectroscopy

Among the non-destructive techniques has met a significant development in the last 20 years the optical analysis in the region of near infrared (NIR) and visible-near infrared (Vis/NIR), based on the use of information arising from the interaction between the structure of food and light.

2.1.1 Electromagnetic radiation

Spectroscopic analysis is a group of techniques allowing to get information on the structure of matter through its interaction with electromagnetic radiation.

Radiation is characterized by (Fessenden & Fessenden, 1993):

- a wavelength (λ), which is the distance between two adjacent maxima and is measured in nm;
- a frequency (v), representing the number of oscillations described by the wave per unit of time and is measured in hertz (cycles/s);
- a wave number (n), which represents the number of cycles per centimeter and is measured in cm^{-1}.

The entire electromagnetic spectrum is divided into several regions, each characterized by a range of wavelengths (Fig.1)

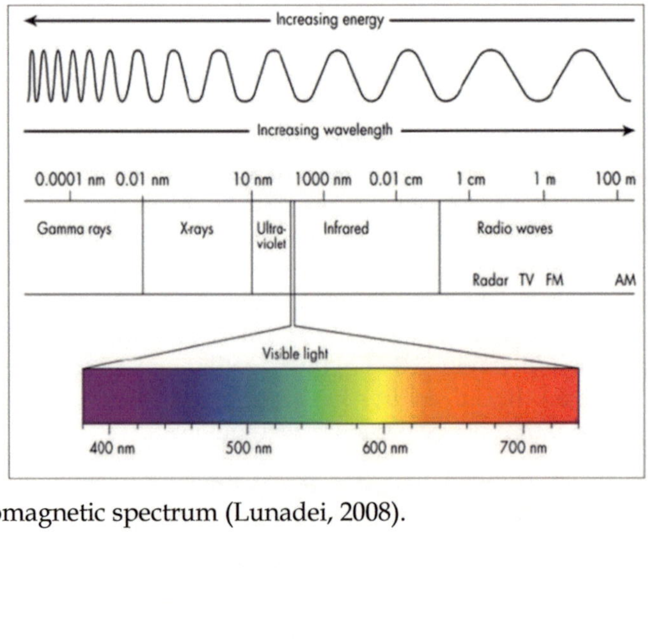

Fig. 1. The electromagnetic spectrum (Lunadei, 2008).

2.1.2 Transitions in the near infrared region (NIR)

The radiation from the infrared region is able to promote transitions at vibrational level. The infrared spectroscopy is used to acquire information about the nature of the functional groups present in a molecule. The infrared region is conventionally divided into three sub-regions: near (750-2500 nm), medium (2500-50000 nm) and far infrared (50-1000 μm).

Fundamental vibrational transitions, namely between the ground state and first excited state, take place in the mid-infrared, while in the region of near-infrared absorption bands are due to transitions between the ground state and the second or the third excited state. This type of transitions are called overtones and their absorption bands are generally very weak. The absorption bands associated with overtones can be identified and correlated to the corresponding absorption bands arising from the fundamental vibrational transitions because they fall at multiple wavelengths of these.

Following the process of absorption of photons by molecules the intensity of the radiation undergoes a decrease. The law that governs the absorption process is known as the Beer-Lambert Law:

$$A = \log (I_0/I) = \log (1/T) = \varepsilon \cdot l \cdot c \qquad (1)$$

where:

A = absorbance [log (incident light intensity/transmitted beam intensity)];
T = transmittance [beam intensity transmitted/incident light intensity];
I_0 = radiation intensity before interacting with the sample;
I = radiation intensity after interaction with the sample;
ε = molar extinction coefficient characteristic of each molecule ($l \cdot mol^{-1} \cdot cm^{-1}$);
l = optical path length crossed by radiation (cm);
c = sample concentration (mol/l).

The spectrum is a graph where in the abscissa is reported a magnitude related to the nature of radiation such as the wavelength (λ) or the wave number (n) and in the Y-axis a quantity related to the change in the intensity of radiation as absorbance (A) or transmittance (T).

2.1.3 Instruments

Since '70s producers developed analysis instruments specifically for NIR analysis trying to simplify them to fit also less skillful users, thanks to integrated statistical software and to partial automation of analysis.

Instruments built in this period can be divided in three groups: desk instruments, compact portable instruments and on-line compatible devices.

Devices evolved over the years also for the systems employed to select wavelength. First instruments used filter devices able to select only some wavelength (Fig. 2). These devices are efficient when specific wavelength are needed. Since the second half of '80s instruments capable to acquire simultaneously the sample spectrum in a specific interval of wavelength were introduced, recording the average spectrum of a single defined sample area (diode array systems and FT-NIR instruments) (Stark & Luchter, 2003). At the same time, chemometric data analysis growth helped to diffuse NIR analysis.

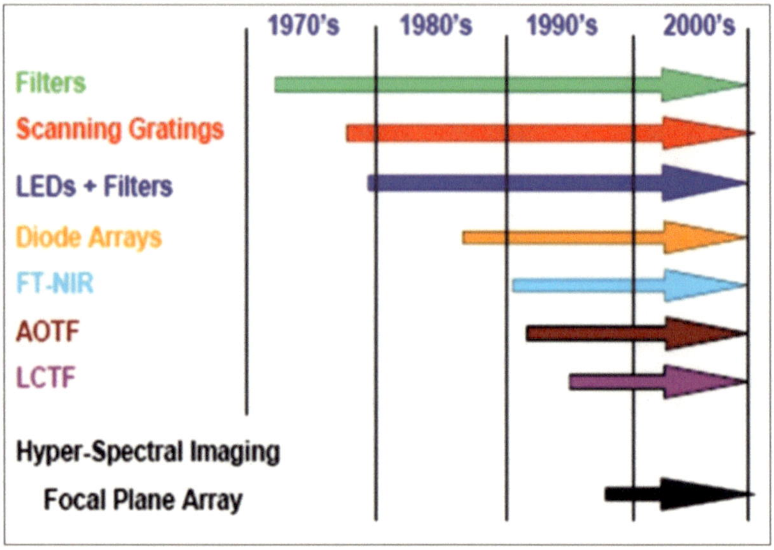

Fig. 2. Development of the different analysis technologies scheme (Stark & Luchter, 2003).

Particularly, food sector showed interest towards NIR and Vis/NIR instruments, both mobile and on-line. Devices based on diode array spectrophotometers and FT-NIR desk systems proved to be the best for this sector.

Both in the case of portable and stationary instruments, the fundamental components of these systems are common and are four:

- Light source;
- Light radiation transport system;
- Sample compartment and measurement zone;
- Spectrophotometer and Personal Computers.

Light source

Tungsten filament halogen lamps are chosen as the light source by most of the instruments. This is due to a good compromise between good performance and relatively low cost. This type of lamps are particularly suitable for use in low voltage. A little drawback may be represented by sensitivity to vibration of the filament.

Halogen bulbs are filled with halogen gas to extend their lives by using the return of evaporated tungsten to the filament. The life of the lamp depends on the design of the filament and the temperature of use, on average ranges from a minimum of 50 hours and a maximum of 10000 hours at rated voltage. The lamp should be chosen according to the use conditions and the spectral region of interest. An increase in the voltage of the lamp may cause a shift of the peaks of the emission spectrum towards the visible region but can also lead to a reduction of 30% of its useful life. On the contrary, use of lower voltages can increase the lamp life together, however, with an intensity reduction of light radiation, especially in the visible region. Emission spectrum of the tungsten filament changes as a function of temperature and emissivity of the tungsten filament. The spectrum shows high intensity in the VNIR region (NIR region close to the area of the visible).

Even if less common, alternative light sources are available. For example, LED light sources and ad laser sources could be used. LED sources (light emitting diodes) are certainly interesting sources thank to their efficiency and their small size. They meet, however, a limited distribution due to limited availability of LEDs emitting at wavelengths in the NIR region. Technology to produce LEDs to cover most of the NIR region already exists, but demand for this type of light sources is currently too low and the development of commercial product of this type is still in an early stage.

The use of laser sources guarantees very intense emission in a narrow band. But the reduced spectral range covered by each specific laser source can cause problems in some applications. In any case the complexity and high cost of these devices have limited very much their use so far, mostly restricted to the world of research.

Light radiation transport system

Light source must be very close to the sample to light it up with good intensity. This is not always possible, so systems able to convey light on the samples are needed. Thanks to optic fibers this problem was solved, allowing the development of different shapes devices.

The use of fiber optics allows to separate the area of placement of the instrument from the measuring proper area. There are indeed numerous circumstances on products sorting line in which environmental conditions do not fulfill direct installation of measure instruments. For example, high temperature, excessive vibrations or lack of space are restricting factors to the use of on-line NIR devices. In all these situations optic fibers are the solution to the problem of conveying light. They transmit light from lamp to sample and from sample to spectrophotometer. They allow to have an immediate measure on a localized sample area, thanks to their small dimensions, reaching areas difficult to access. Furthermore, they are made of a dielectric material that protects from electric and electromagnetic interferences. 'Optic fibers' means fibers optically transparent, purposely studied to transmit light thanks to total internal reflection phenomenon. Internal reflection is said to be total because it is highly efficient, in fact more than 99,999% radiation energy is transmitted in every reflection. This means that radiation can be reflected thousands of times during the way without suffer an appreciable attenuation of intensity (Osborne et al., 1993).

Optic fiber consists of an inner core, a covering zone and of an external protection cover. The core is usually made of pure silica, but can also be used plastics or special glasses. The cladding area consists of material with a lower refractive index, while the exterior is only to protect the fiber from mechanical, thermal and chemical stress.

In figure 3 are shown the inner core and the cladding of an optical fiber. Index of refraction of inner core have to be bigger than cladding one. Each ray of light that penetrates inside the fiber with an angle $\leq \theta_{max}$ (acceptance angle) is totally reflected with high efficiency within the fiber.

Sample compartment and measurement zone

Samples compartment and measurement zone are highly influenced by the technique of acquisition of spectra. Different techniques are employed, depending on type of samples, solid or liquid, small or large, to be measured in plan or in line, that influence the geometry of the measurement zone.

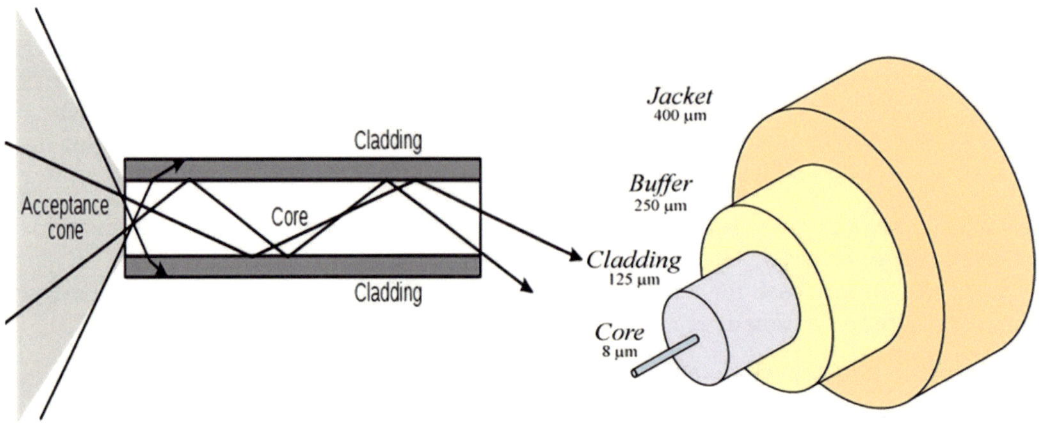

Fig. 3. Scheme of an optical fiber. The acceptance cone is determined by the critical angle for incoming light radiation. Buffer absorbs the radiation not reflected by the cladding. Jacket has a protective function.

The techniques to acquire spectra are four: transmittance, reflectance, transflectance and interactance. They are different mainly for the different positioning of the light source and of the measurement sensor around the sample (Fig. 4).

a. **Transmittance** - The transmittance measurements are based on the acquisition of spectral information by measuring the light that goes through the whole sample (Lu & Ariana, 2002). The use of analysis in transmittance can explore much of the internal structure of the product. This showed that is a technique particularly well suited to detect internal defects. To achieve significant results with this technique is required a high intensity light source and a high sensitivity measuring device. This because intensity of light able to cross the product is often very low. The transmittance measurements generally require a particular geometry of the measuring chamber, which can greatly influence the design of the instrument.
b. **Reflectance** - This technique measures the component of radiation reflected from the sample. The radiation is not reflected on the surface but penetrates into the sample a few millimeters, radiation is partly absorbed and partly reflected back again. Measuring this component of reflected radiation after interacting with the sample is possible to establish a relationship of proportionality between reflectance and analyte concentration in the sample. The reflectance measurement technique is well suited to the analysis of solid matrices because the levels of intensity of light radiation after the interaction with the sample are high.

   This technique also allows to put in a limited space inside a tip the bundle of fibers that illuminate the sample and the fibers leading to the spectrophotometer the radiation after the interaction with the product. Therefore the use of this type of acquisition technique is particularly versatile and is suitable for compact, portable instruments, designed for use in field or on the process line. The major drawback using this technique is related to the possibility to investigate only the outer area of the sample without having the chance to go deep inside.

c. **Transflectance** – This technique is used in case it is preferable to have a single point of measurement, as in the case of acquisitions in reflectance. In this case, however, the incident light passes through the whole sample, is reflected by a special reflective surface, recross the sample and strikes the sensor located near the area of illumination. The incident light so makes a double passage through the sample. Obviously this type of technique can be used only in the case of samples very permeable to light radiation such as partially transparent fluid. It is therefore not applicable to solid samples.
d. **Interactance** - This technique is considered a hybrid between transmittance and reflectance, as it uses characteristics of both techniques previously seen. In this case the light source and sensor are located in areas near the sample but between them physically separated. So the radiation reaches the measurement sensor after interacting with part of internal structure of the sample. This technique is mainly used in the analysis of big solid samples, for example, a whole fruit. Interactance is thus a compromise between reflectance and transmittance and has good ability to detect internal defects of the product combined with a good intensity of light radiation. This analysis is widely used on static equipment where, through the use of special holders, is easily obtained the separation between the areas of incidence of light radiation and the area to which the sensor is placed. It is instead difficult to use this configuration on-line because is complicated to place a barrier between incident and returning light to the sensor directly on the process line.

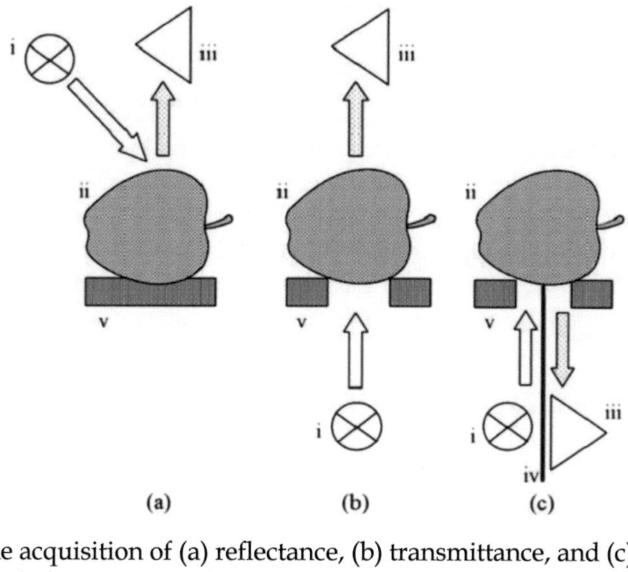

Fig. 4. Setup for the acquisition of (a) reflectance, (b) transmittance, and (c) interactance spectra, with (i) the light source, (ii) fruit, (iii) monochromator/detector, (iv) light barrier, and (v) support. In interactance mode, light due to specular reflection is physically prevented from entering the monochromator by means of a light barrier (Nicolai et al., 2007).

Spectrophotometer and Personal Computers

Spectrophotometer can be considered the heart of an instrument for NIR analysis. The employed technology for the wavelengths selection greatly influences the performance of the instrument. For example, the use of filters allows instruments to record the signal of a single wavelength at a time. Modern instruments (diode array instruments and interferometers) allow to record the spectrum of the entire wavelengths range.

Instruments equipped with a diode array spectrophotometer are those who have met the increased use for portable and online applications in food sector. This is due to their compact size, versatility and robustness, thanks to the lack of moving parts during operation and also thanks to a relatively low cost.

As seen before, fiber optic sensor collects the portion of the electromagnetic radiation after interaction with the internal structure of the sample and transfers it to the spectrophotometer. The optical fiber is connected to the optical bench of the instrument. The optical bench allows to decompose the electromagnetic radiation and recording the intensity at different wavelengths.

Optical bench of this type of instrument generally consists of five components:

a. Optical fiber connector: connects the fiber optic with the optical bench of the instrument.
b. First spherical mirror (collimating mirror), has the function of collimating the light and send it to the diffraction grating.
c. Diffraction grating: in this area of the instrument, the light is split into different wavelengths and sent to the second spherical mirror.
d. Second spherical mirror (focussing mirror), collects diffracted radiation from the grating and sends them to the CCD sensor.
e. Matrix CCD sensor (diode array): records the signal intensity at each wavelength.

High sensitivity of the CCD matrix sensor compensate the low intensity of light radiation input due to the reduced diameter of the optical fibers used. Sensors used are generally Si-diode array or InGaAs-diode array. The first ones are certainly the most common and cheap and allow the acquisition of the spectrum in the range between 400 and 1100 nm, so are used for Vis/NIR analysis. InGaAs sensors, more expensive, are used in applications requiring the acquisition of spectra at longer wavelengths, their use should range from 900 to 2300 nm.

Recorded signal by the CCD sensor is digitized and acquired by a PC using the software management tool of the instruments. Software records and allows to display graphically the spectrum of the analyzed sample. The management software also allows to interface with the spectrophotometer enabling to change some parameters during the acquisition of spectra.

2.2 Image analysis

In the food industry, since some time, there is a growing interest in image analysis techniques, since the appearance of a food contains a variety of information directly related to the quality of the product itself and this characteristics are difficult to measure through use of classical methods of analysis. In addition, image analysis techniques: provide information much more accurate than human vision, are objective and continuous over time and offer the great advantage of being non-destructive. These features, enable vision systems to be used in real time on the process lines, allowing on-line control and automation of sorting and classification within the production cycle (Guanasekaran & Ding, 1994).

The objective of the application of image analysis techniques, in the food sector, is the quantification of geometric and densitometric characteristic of image, acquired in a form that represents meaningful information (at macro and microscopic level) of appearance of an object (Diezak, 1988). The evolution of these techniques and their implementation in the vision machine in form of hardware and specialized software, allows a wide flexibility of applications, a high capacity of calculation and a rigorous statistical approach.

The benefits of image analysis techniques (Brosnan & Sun, 2004) that rely on the use of machine vision systems can be summarized as follows:

a. are non-destructive techniques;
b. techniques are use-friendly, rapid, precise, accurate and efficient;
c. generate objective data that can be recorded for analysis deferred;
d. allow a complete analysis of the lots and not just a single sample of the lot;
e. reduce the involvement of human personnel in performing tedious tasks and allow the automation of various functions that would require intensive work shifts;
f. are reasonably affordable cost.

These suggest the reason that drives scientific research, of the agro-food sector, to devote to the study and analysis of machine vision systems to analyze the internal and external quality characteristics of food, valued according to the optical properties of the products. With a suitable light source, it is possible to extract information about color, shape, size and texture. From these features, it is possible to know many objective aspects of the sample, to be able to correlate, through statistical analysis, the characteristics defined by quality parameters (degree of maturation, the presence of external or internal mechanical defects, class, etc.) (Du & Sun, 2006, Zheng et al., 2006).

The image analysis may have several applications in the food industry: as a descriptor or as gastronomic and technologic parameter. Vision machine can also be used to know size, structure, color in order to quantify the macro and microscopic surface defects of a product or for the characterization and identification of foods or to monitor the shelf life (Riva, 1999). "Image analysis" is a wide designation that include, in addition to classical studies in grayscale and RGB images, the analysis of images collected by mean multiple spectral channels (multispectral) or, more recently, hyperspectral images, technique exploited for its full extension in the spectral direction.

The hyperspectral image (Chemical and Spectroscopic Imaging) is an emerging technology, non-destructive, which complements the conventional imaging with spectroscopy in order to obtain, from an object, information, both spectral and spatial. The hyperspectral images are digital images in which each element (pixel) is not a single number or a set of three numbers, like the color pictures (RGB), but a whole spectrum associated with that point. They are three-dimensional blocks of data. Their main advantage is that they provide spatial information necessary for the study of non-homogeneous samples. The advantage of this technique is the ability to detect in a foodstuff even minor constituents, isolated spatially.

To support image analysis, chemometric techniques are necessary to process and to model, data sets, in order to extract the highest possible information content. Methods of classification, modeling, multivariate regression, similarity analysis, principal components analysis, methods of experimental design and optimization, must be applied on the basis of each different condition and needs.

2.2.1 The vision system

The machine vision systems, appeared in the early sixties and then spread over time, in many fields of application, are composed of a lighting system, a data acquisition system connected to a computer, via a capture card, which digitizes (converts the signal into numerical form) and stores the analogic electrical signal, at the output from the camera sensor (Russ et al., 1988). The scanned image is then "converted" into a numerical matrix. Captured images are elaborated by appropriate processing softwares in order to acquire the useful information. In figure 5 shows an example of vision machine.

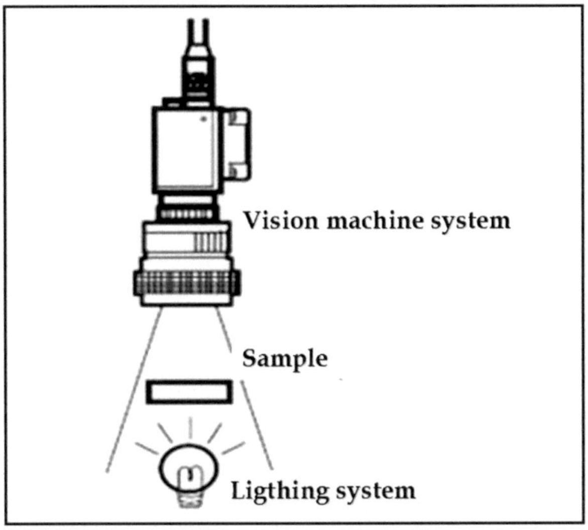

Fig. 5. Example of image acquisition of an inspected sample.

It is important to choose the localization of light source, but also the type of light source (incandescent, halogen, fluorescent, etc.) influences the performance of the analysis. In fact, although the light sources emitting electromagnetic radiation corresponding to the visible (VIS, 400-700 nm), ultraviolet (UV 10-400 nm) and near infrared (NIR, 700-1400 nm) are the most widely used, to create a digital image can also be used other types of light sources in order to emit different radiations, depending on the purpose of analysis. For example, to determine the internal structure of objects and/or identify any internal defects, it's possible to use an X-ray source, even if, although this type of source gives good results, its application is much more widespread in the medical field than in the agro-food, this is due to high costs of equipment and low speed of operating.

2.2.2 The digital image

A digital image is generated from the conversion of an analogic video signal produced by a digital camera into an electronic signal (scanning), then stored in the memory of a PC in the form of binary information. Any digital image can be considered as an array of points, the pixels, that make up the smallest element of an image. Each pixel contains a double set of information: its position in space, identified by the values (x, y) and the value of its intensity. Digital images can be represented using only two colors, typically black and white (binary image), or shades of gray (monochrome image) or a range of colors (multichannel image). The value of light intensity will be different depending on the type of image.

The pixels, in the binary images, can have, as the intensity value, or 0 (equivalent to black) or 1 (white). The value of intensity in monochrome images, will be within a range, defined gray scale, from 0 to L, which usually corresponds to the interval from 0 (or 1) to 255 (or 256), where a value of 0 corresponds to black, a value of 255 corresponds to white, and intermediate values to the various shades of gray.

Finally, in multi-channel images, the color of each pixel will be identified by three or four values, depending on the reference color model. For example, in RGB color space, each pixel will be characterized by a three values, each between 0 and 255, respectively, corresponding to the intensity in the red, green and blue. When all three values are 0, the color of object is black, and when all three have maximum value, the object will be white, while, when there are equal levels of R, G and B, the gray color is generated. The images of this type, in fact, may be considered as a three-dimensional matrix, consisting of three overlapping matrices having the same number of rows and columns, where the elements of the first matrix represent the pixel intensity in the red channel, those in the second matrix, the green channel and those of third matrix, in the blue channel.

2.2.3 Multispectral and hyperspectral images

RGB images, represented by three overlapping monochrome images, are the simplest example of multichannel images. In medical applications, in geotechnical, in the analysis of materials and of remote sensing, instead, are often used sensors capable of acquiring multispectral and hyperspectral images, two particular types of multi-channel images. The multispectral images are typically acquired in three/ten spectral bands including in the range of Vis, but also in the field of IR, fairly spaced (Aleixos et al., 2002). In this way it's possible to extract a larger number of information from the images respect those normally obtained from the analysis of RGB images.

An example of bands normally used in this type of analysis, are the band of blue (430-490 nm), green (491-560 nm), red (620-700 nm), NIR (700-1400 nm), MIR (1400-1750 nm). The spectral combinations can be different depending on the purpose of analysis. The combination of NIR-RG (near infrared, red, green) is often used to identify green areas in satellite images, because the green color reflects a lot in the NIR wavelength. The combination of NIR-R-B (near infrared, red, blue) is very useful to verify the ripening stage of fruit, this is due to the chlorophyll that shows a peak of adsorption in the wavelength of the red. Finally, the combination of NIR-MIR-blue (NIR, MIR and blue) is useful to observe the sea depth, the green areas in remote sensing images.

Hyperspectral imaging (HSI) combines spectroscopy and the traditional imaging to form a three-dimensional structure of multivariate data (hypercube). The hyperspectral images are consist of many spectral bands acquired in a narrow and contiguous way, allowing to analyze each pixel in the multiple wavelengths simultaneously and, therefore, to obtain a spectrum associated with a single pixels. The set of data constituting an hyperspectral image can be thought as a kind of data cube, with two spatial directions, ideally resting on the surface observed, and a spectral dimension. Extracting a horizontal plane from the cube it is possible to get a monochrome image, while the set of values, corresponding to a fixed position in the plane (x, y), is the spectrum of a pixel of the image (Fig. 6).

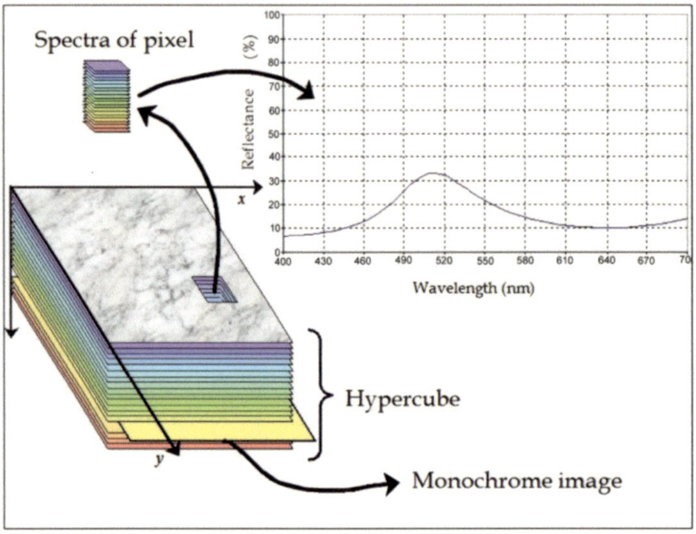

Fig. 6. Example of hyperspectral image.

With the hyperspectral imaging, you can acquire the spectra in reflectance, in transmission and fluorescence as a function of the different kind of sample to analysis, even if the most of the scientific works, present in the literature, using spectral images acquired in reflectance, transmission and emission.

The significant time savings that can be made to the industrial production processes, encourage the use of this instrumentation. The hyperspectral image analysis has many advantages, but still has some defects. The advantages of using hyperspectral analysis for what concerns the agro-food sector can be summarized as follows:

a. does not necessary to prepare the test sample;
b. it is a non-invasive, non-destructive methodology, it avoids the sample loss that can be used for other purposes or analysis;
c. can be regarded as an economic tool that it allows a saving of time, labor, reagents, and a strong cost-saving for the waste treatment;
d. for each pixel of the analyzed sample is possible to have the full spectrum and not a only absorbance value for few wavelength;
e. many constituents can be determined simultaneously within a sample, such as color and morphological characteristics;
f. due to its high spectral resolution, it is possible to estimate both qualitative than quantitative information;
g. it is also possible to select a single region of interest of the sample, and save it in a spectral library.

As mentioned previously, one of the advantages HSI is the large volume of data available in each hypercube, with which to create the calibration and validation set. But, the information derived from the analysis, contain also redundant information. This abundance of data has two drawback, one due to the high computational load of heavy data size and the second is due to the long acquisition times, given the size of the data being collected (Firtha et al. 2008). Therefore, it is desirable to reduce the load to manageable levels, especially if the goal is the application of HSI techniques in real time, on-line on production lines. In fact, in many cases, the large amount of data acquired from the spectral image, is appropriately reduced (with chemometric processing) so as to select only those wavelengths interesting for the intended purpose. Once the spectral bands of interest were identified, a multispectral system, with only selected wavelengths, can be engineered a system for industrial application. Another negative aspect is that the spectral image analysis is an indirect method to which it is necessary to apply appropriate chemometric techniques and a procedure of data transfer.

The spectral image is not suitable for liquid and homogeneous samples. In fact, the value of this type of image is evident when applied to heterogeneous samples, and many foods are an excellent heterogeneous matrix. Despite the novelty of applying HSI in the food sector, many jobs are already present in the literature.

The traditional image analysis, based on a computer system, has had a strong development in the food sector with the aim of replacing the human eye on saving costs and improving efficiency, speed and accuracy. But the computer vision technology is not able to select between objects of similar colors, to make complex classifications, to predict quality characteristics (e.g. chemical composition) or detect internal defects. Since the quality of a food is not an individual attribute but it contains a number of inherent characteristics of the food itself, to measure the optical properties of food products has been one of the most studied non-destructive techniques for the simultaneous detection of different quality parameters. In fact, the light reflected from the food contains information about constituents near and at the surface of the foodstuff. Near-infrared spectroscopy technology (NIRS) is rapid, non-destructive, easy to apply on-line and off-line. With this technology, it is possible to obtain spectroscopic information about the components of analyzed sample, but it is not possible to know the position of the component.

The only characteristic of appearance (color, shape, etc.) however, are easily detectable with conventional image analysis. The combination of image analysis technology and spectroscopy is the chemical imaging spectroscopy that allows to get spatial and spectral information for each pixel of the foodstuff. This technology allowing to know the location of each chemical component in the scanned image. Table 1 summarizes the main differences between the three analytical technologies: imaging, spectroscopy and hyperspectral imaging.

2.3 Electronic nose (e-nose)

"An instrument which comprises an array of electronic chemical sensors with partial specificity and appropriate pattern recognition system, capable of recognizing simple or complex odors" is the term of "electronic nose" coined in 1988 by Gardner and Bartlett (Gardner and Bartlett, 1994).

Features	Imaging	Spectroscopy	Hyperspectral Imaging
Spatial information	√	x	√
Spectral information	x	√	√
Multi-costituent information	x	√	√
Building chimica images	x	x	√
Flexibility of spectral information extraction	x	x	√

Table 1. Main differences among imaging, spectroscopy and hyperspectral imaging techniques (ElMarsy & Sun, 2010).

Scientific interest in the use of electronic noses was formalised, the first time, in a workshop on chemosensory information processing during a session of the North Atlantic Treaty Organization (NATO) that was entirely dedicated to the topic of artificial olfaction. Since 1991, interest in biological sensors technology has grown considerably as is evident by numerous scientific articles. Moreover, commercial efforts to improve sensor technologies and to develop tools of greater sophistication and improved capabilities, with diverse sensitivities, are with ever-expanding (Wilson & Baietto, 2009).

Electronic noses are emerging as an innovative analytical-sensorial tool to characterize the sensory comparison of food in terms of freshness, determination of geographical origin, seasoning. The first electronic nose goes back to the '80s, when Persaud and Dodd of the University of Warwick (UK) tried to model and simulate the operation of the olfactory system of mammals with solid state sensors. Since then, artificial olfactory systems are designed closer to the natural one.

The electronic nose is a technology that tends to replace/complement the human olfactory system. The tool does not analyze the chemical composition of the volatile fraction, but it identifies the olfactory fingerprint.

Currently, these electronic devices are characterized by complex architecture, where it is possible to try to reproduce the functioning of the olfactory system of mammals. The tool is a biomimetic system that is designed to mimic the functioning of the olfactory systems that we find in nature, specifically human olfactory system. Typically, an electronic nose collects information through an array of sensors, able to respond in a selective mode and reversible to the presence of chemicals, generating electrical signals as a function of their concentration. Currently, the sensors that have reached the highest level of development are made from metal oxides semiconductor (MOS). The sensors are usually characterized by fast response, low energy consumption, small size, high sensitivity, reliability, stability and reproducibility. In addition to semiconductor of metal, the sensor can be made of transistors, plated with metal semiconductor (MOSFETs), or conductive polymers. The MOS sensors are inorganic, typically made of tin oxide, zinc oxide, titanium oxide, tungsten oxide. The absorption of gas by them change their conductivity. These sensors operate at high temperatures, between 200 and 500 °C and are relatively cheap. In figure 7 is represented the main parts of a typical sensor.

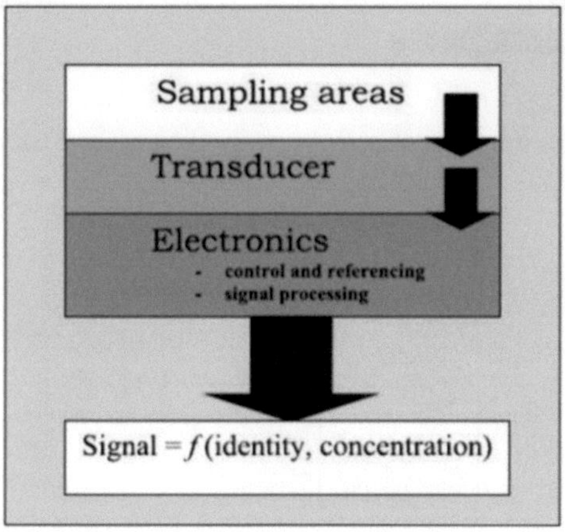

Fig. 7. The main parts of a typical sensor (Deisingh et al. 2004)

Below, ten possible MOS sensors for reading specific molecules (Table 2):

Sensor	Molecules detected
W1C	Aromatic compounds
W5S	Oxides of nitrogen, low specificity
W3C	Ammonium compounds, aromatic
W6S	Hydrogen
W5C	Alkanes, aromatic compounds, less polar compounds
W1S	Methane, low specificity
W1W	Sulfur compounds, terpenes, limonene, pyrazines
W2S	Alcohol, partially aromatic compounds, low specificity
W2W	Aromatic compounds, organic sulfur compounds
W3S	Methane

Table 2. Example of sensors in the electronic nose, with their categories of compounds that can to determine

The information is initially encoded as electrical unit, but are immediately captured and digitized in order to be numerically translated by a computer system. In practice, an odorant is described by the electronic nose, based on the responses of individual sensors, as a point or a region of a multidimensional space.

Thanks to special algorithms, derived from the discipline called pattern recognition, the system is able to build an olfactory map in order to allow a qualitative and quantitative analysis, discriminating a foodstuff simply by its olfactory fingerprint.

The architecture of an electronic nose (Fig. 8) is significantly dependent on the application for which it is designed. In general, the electronic nose, is characterized by the presence of a vacuum system, a large number of gas sensors, a subsystem of acquisition and digitization and by a processing subsystem able to implement appropriate algorithms for classification or regression.

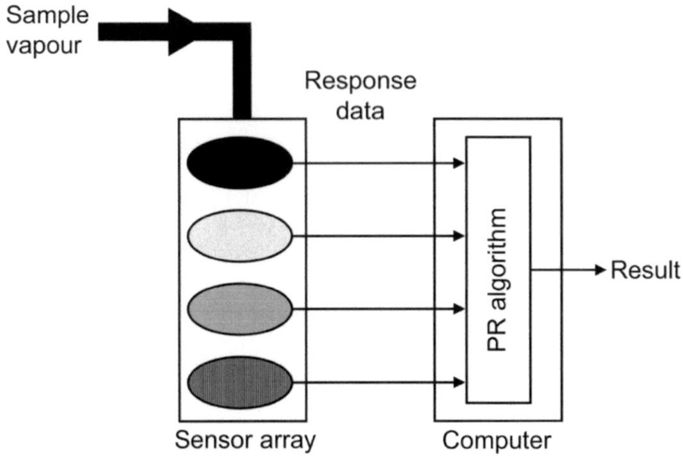

Fig. 8. A generalized structure of an electronic nose. (Deisingh et al. 2004)

The principle of working which operates the electronic nose is distinctly different from that of commonly used analytical instruments (e.g. gas chromatograph). The e-nose gives an overall assessment of the volatile fraction of the foodstuff that is, in large part responsible for the perception of the aroma of the investigated sample, without the need to separate and identify the various components. All the responses of the sensors resulted from the electronic nose creates a "map" of non-specific signals that constitute the profile of the food product, also called olfactory fingerprints.

The goal is to find a relationship between the set of independent variables, resulted from the sensor, and the set of dependent variables, characteristics of the sample. Chemometrics software required for data set processing, in environmental and food sectors, allows to process the data by methods of multivariate analysis such as PCA (Principal Component Analysis), LDA (Linear Discriminant Analysis), PLS (Partial Least Square Analysis), DFA (Discriminant Function Analysis). As example the Principal Component Analysis (PCA) is a method for detecting patterns in data sets and express them in order to highlight their similarities and/or differences. Example of electronic nose applications, in the food sector, could be: to monitor of foodstuff shelf life, to check certified quality or the trademark DOP, to make microbiological tests, to check controlled atmosphere in the packaging, to control fermentation stage or to identify the presence of components of the packaging transferred in the product, or to verify the state of cleaning of kegs (on-line measures).

2.4 Chemometrics in food sector

Chemometrics is an essential part of NIR and Vis/NIR spectroscopy in food sector. NIR and Vis/NIR instrumentation in fact must always be complemented with chemometric analysis to enable to extract useful information present in the spectra separating it both from not useful information to solve the problem and from spectral noise. Chemometric techniques most used are the principal component analysis (PCA) as a technique of qualitative analysis of the data and PLS regression analysis as a technique to obtain quantitative prediction of the parameters of interest (Naes et al., 2002; Wold et al., 2001; Nicolai et al., 2007; Cen & He, 2007).

The developed models should be tested using independent samples as validation sets to verify model accuracy and robustness. To evaluate model accuracy, the statistics used were the coefficient of correlation in calibration (r_{cal}), coefficient of correlation in prediction (r_{pred}), root mean square error of calibration (RMSEC), and root mean square error of prediction (RMSEP).

Correlation coefficients (r_{cal} and r_{pred}):

$$r_{cal} \text{ or } r_{pred} = \sqrt{1 - \frac{\sum_{i=1}^{n}(y_i - \hat{y}_i)^2}{\sum_{i=1}^{n}(y_i - \overline{y})^2}} \tag{2}$$

where y_i are the reference values, $\hat{y}_i$ are the values predicted by the PLS model, and $\bar{y}$ is the averaged reference value.

Standard errors of calibration and prediction (RMSEC and RMSEP):

$$\text{RMSEC or RMSEP} = \sqrt{\frac{\sum_{i=1}^{n}(y_i - \hat{y}_i)^2}{n}} \tag{3}$$

where n is the number of validated objects, and $\hat{y}_i$ and y_i are the predicted and measured values of the ith observation in the calibration or validation set, respectively. This value gives the average uncertainty that can be expected for predictions of future samples. The optimum calibrations should be selected based on minimizing the RMSEP. Percent errors (RMSEC% and RMSEP%) could be also calculated as: RMSEC (%) = RMSEC/averaged reference values of each parameter.

Prediction capacity of a model can be evaluated with the ratio performance deviation (RPD) (Williams & Sobering, 1996). The RPD is defined as the ratio of the standard deviation of the response variable to the RMSEP. RPD value > 2.5 means that the model has good prediction accuracy.

3. Applications

3.1 NIR and Vis/NIR spectroscopy

During the last 50 years, there has been a lot of emphasis on the quality and safety of the food products, of the production processes, and the relationship between the two (Burns and Ciurczak, 2001).

Near infrared (NIR) spectroscopy has proved to be one of the most efficient and advanced tools for monitoring and controlling of process and product quality in food industry. A lot of work has been done in this area. This review focuses on the use of NIR spectroscopy for the analysis of foods such as meat, fruit, grain, dairy products, oil, honey, wine and other areas, and looks at the literature published in the last 10 years.

3.1.1 Fruit and vegetables

Water is the most important chemical constituent of fruits and vegetables and water highly absorbs NIR radiation, so the NIR spectrum of such materials is dominated by water. Further, the NIR spectrum is essentially composed of a large set of overtones and combination bands. This, in combination with the complex chemical composition of a typical fruit or vegetable causes the NIR spectrum to be highly convoluted. Multivariate statistical techniques are required to extract the information about quality attributes which is buried in the NIR spectrum. Developments in multivariate statistical techniques such as partial least squares (PLS) regression and principal component analysis (PCA) are then applied to extract the required information from such convoluted spectra (Cozzolino et al., 2006b; McClure, 2003; Naes et al., 2004; Nicolai et al., 2007;).

The availability of low cost miniaturised spectrophotometers has opened up the possibility of portable devices which can be used directly on field for monitoring the maturity of fruit.

Guidetti et al. (2008) tested a portable Vis/NIR device (450-980 nm) for the prediction of ripening indexes (soluble solids content and firmness) and presence of compounds with functional properties (total anthocyanins, total flavonoids, total polyphenols and ascorbic acid) of of blueberries ('Brigitta' and 'Duke' varieties). Good predictive statistics were obtained with correlation coefficients (r) between 0.80 and 0.92 for the regression models built for fresh berries (Table 3). Similar results were obtained for the regression models for homogenized samples with r > 0.8 for all the indexes. Results showed that Vis/NIR spectroscopy is an interesting and rapid tool for assessing blueberry ripeness.

Dependent variable	LV	Calibration		Cross validation	
		r_{cal}	RMSEC	r_{cv}	RMSECV
TSS (°Brix)	4	0.86	0.78	0.85	0.79
Young's Module (MPa)	3	0.87	0.65	0.87	0.66
Total anthocyanins (mg/g f. w.)	4	0.87	0.31	0.87	0.31
Total flavonoids (mg cat/g)	4	0.87	0.37	0.86	0.37
Total polyphenols (mg cat/g f. w.)	11	0.82	0.20	0.81	0.20
Ascorbic acid (mg/100 g f. w.)	4	0.84	1.01	0.83	1.02

Table 3. Results of PLS models for fresh 'Duke' berry samples (r = coefficient of correlation; RMSEC = root mean square of the standard error in calibration; RMSECV = root mean square of the standard error in cross-validation; LV = latent variables). All data were preprocessed by second derivative of reduced and smoothed data.

3.1.2 Meat

In literature there are numerous applications of NIR spectroscopy for the analysis of meat quality. One of the most important aim is to monitor the freshness of meat products. Sinelli et al. in 2010 investigated the ability of Near Infrared spectroscopy to follow meat freshness decay. PCA was applied by authors to the data and was able to discriminate samples on the basis of storage time and temperature. The modelling of PC scores versus time allowed the setting of the time of initial freshness decay for the samples (6–7 days at 4.3 °C, 2–3 days at 8.1 °C and less than 1 day at 15.5 °C). Authors reported that results showed the feasibility of NIR for estimating quality decay of fresh minced beef during marketing.

Sierra et al. in 2007 conducted a study for the rapid prediction of the fatty acid (FA) profile of ground using near infrared transmittance spectroscopy (NIT). The samples were scanned in transmittance mode from 850 to 1050 nm. NIT spectra were able to accurately predict saturated R^2=0,837, branched R^2=0,701 and monounsaturated R^2=0,852 FAs. Results were considered interesting because intramuscular fat content and composition influence consumer selection of meat products.

Andrés et al. in 2007 implemented a study to evaluate the potential of visible and near infrared reflectance (NIR) spectroscopy to predict sensory characteristics related to the eating quality of lamb meat samples. A total of 232 muscle samples from Texel and Scottish Blackface lambs was analyzed by chemical procedures and scored by assessors in a taste panel and these parameters were predicted from Vis/NIR spectra. The results obtained by authors suggested that the more important regions of the spectra to estimate the sensory characteristics are related to the absorbance of intramuscular fat and water content in meat samples.

Even in the meat industry have been tried online applications of NIR spectroscopy. A study was conducted by Prieto et al. in 2009a to assess the on-line implementation of visible and near infrared reflectance (Vis/NIR) spectroscopy as an early predictor of beef quality traits, by direct application of a fibre-optic probe to the muscle immediately after exposing the meat surface in the abattoir. Authors reported good correlation results only for prediction of colour parameters while less good results were achieved for sensory parameters.

NIR spectroscopy could be used for the detection of beef contamination from harmful pathogens and the protection of consumer safety. Amamcharla et al. in 2010 investigated the potential of Fourier transform infrared spectroscopy (FTIR) to discriminate the Salmonella contaminated packed beef. Principal component analysis was performed on the entire spectrum (4000–500 cm^{-1}). Authors obtained encouraging classification results with different techniques and confirmed that NIR could be used for non-destructive discrimination of Salmonella contaminated packed beef samples from uncontaminated ones.

A review published by Prieto et al. in 2009b indicates that NIR showed high potential to predict chemical meat properties and to categorize meat into quality classes. But authors underlined also that NIR showed in different cases limited ability for estimating technological and sensory attributes, which may be mainly due to the heterogeneity of the meat samples and their preparation, the low precision of the reference methods and the subjectivity of assessors in taste panels.

3.1.3 Grains, bread and pasta

Grains including wheat, rice, and corn are main agricultural products in most countries. Grain quality is an important parameter not only for harvesting, but also for shipping (Burns and Ciurczak, 2001). In many countries, the price of grain is determined by its protein content, starch content, and/or hardness, often with substantial price increments between grades.

Measurement of carotenoid content of maize by Vis/NIR spectroscopy was investigated by Brenna and Berardo (2004). They generated calibrations for several individual carotenoids and the total carotenoid content with good results (R^2 about 0,9).

Several applications can be found in literature regarding the use of NIR for the prediction of the main physical and rheological parameters of pasta and bread. De Temmerman et al. in 2007 proposed near-infrared (NIR) reflectance spectroscopy for in-line determination of moisture concentrations in semolina pasta immediately after the extrusion process. Several pasta samples with different moisture concentrations were extruded while the reflectance spectra between 308 and 1704 nm were measured. An adequate prediction model was developed based on the Partial Least Squares (PLS) method using leave-one-out cross-validation. Good results were obtained with $R^2 = 0{,}956$ and very low level of RMSECV. This creates opportunities for measuring the moisture content with a low-cost sensor.

Zardetto & Dalla Rosa in 2006 studied the evaluation of the chemical and physical characteristics of fresh egg pasta samples obtained by using two different production methodologies: extrusion and lamination. Authors evaluated that it is possible to discriminate the two kinds of products by using FT-NIR spectroscopy. FT-NIR analysis results suggest the presence of a different matrix–water association, a diverse level of starch gelatinization and a distinct starch–gluten interaction in the two kinds of pasteurised samples.

The feasibility of using near infrared spectroscopy for prediction of nutrients in a wide range of bread varieties mainly produced from wheat and rye was investigated by Sørensen in 2009. Very good results were reported for the prediction of total contents of carbohydrates and energy from NIR data with R^2 values of 0.98 and 0.99 respectively.

Finally, a quick, non-destructive method, based on Fourier transform near-infrared (FT-NIR) spectroscopy for egg content determination of dry pasta was presented by Fodor et al. (2011) with good results.

3.1.4 Wine

Quantification of phenolic compounds in wine and during key stages in wine production is therefore an important quality control goal for the industry and several reports describing the application of NIR spectroscopy to this problem have been published.

Grape composition at harvest is one of the most important factors determining the future quality of wine. Measurement of grape characteristics that impact product quality is a requirement for vineyard improvement and for optimum production of wines (Carrara et al., 2008). Inspection of grapes upon arrival at the winery is a critical point in the wine production chain (Elbatawi & Ebaid, 2006).

An optical, portable, experimental system (Vis/NIR spectrophotometer) for nondestructive and quick prediction of ripening parameters of fresh berries and homogenized samples of grapes in the wavelength range 450-980 nm was built and tested by Guidetti et al. (2010) (Fig. 9). Calibrations for technological ripening and for anthocyanins had good correlation coefficients ($r_{cv} > 0.90$). These models were extensively validated using independent sample sets. Good statistical parameters were obtained for soluble solids content ($r > 0.8$, SEP < 1.24 °Brix) and for titratable acidity ($r > 0.8$, SEP < 2.00 g tartaric acid L^{-1}), showing the validity of the Vis/NIR spectrometer. Similarly, anthocyanins could be predicted accurately compared with the reference determination (Table 4). Finally, for qualitative analysis, spectral data on grapes were divided into two groups on the basis of grapes' soluble content and acidity in order to apply a classification analysis (PLS-DA). Good results were obtained with the Vis/NIR device, with 89% of samples correctly classified for soluble content and 83% of samples correctly classified for acidity. Results indicate that the Vis/NIR portable device could be an interesting and rapid tool for assessing grape ripeness directly in the field or upon receiving grapes in the wine industry.

Fig. 9. Images of spectral acquisition phases on fresh berries and on homogenized samples.

Parameter	Pretreat-ment[a]	LV	Calibration		Validation	
			r	RMSEC	r	RMSEP
TSS (°Brix)	MSC+d2	5	0.93	0.95	0.75	0.95
Titratable acidity (g tart. acid dm^{-3})	MSC+d2	6	0.95	1.16	0.85	1.12
pH	MSC+d2	5	0.85	0.08	0.80	0.13
PA (mg dm^{-3})	MSC+d2	5	0.95	80.90	0.78	129.00
EA (mg dm^{-3})	MSC+d2	3	0.93	57.70	0.84	77.70
TP (OD 280 nm)	MSC+d2	4	0.80	3.74	0.70	5.81

[a] MSC = multiplicative scatter correction, and d2 = second derivative.

Table 4. Results of PLS models for homogenized samples.

The application of some chemometric techniques directly to NIR spectral data with the aim of following the progress of conventional fermentation and maturation was investigated by Cozzolino et al. (2006b). The application of principal components analysis (PCA) allowed similar spectral changes in all samples to be followed over time. The PCA loading structure could be explained on the basis of absorptions from anthocyanins, tannins, phenolics, sugar and ethanol, the content of which changed according to different fermentation time points. This study demonstrated the possibility of applying NIR spectroscopy as a process analytical tool for the wine industry.

Urbano-Cuadrado et al. (2004) analysed by Vis/NIR spectroscopy different parameters commonly monitored in wineries. Coefficients of determination obtained for the fifteen parameters were higher than 0.80 and in most cases higher than 0.90 while SECV values were close to those of the reference method. Authors said that these prediction accuracies were sufficient for screening purposes.

Römisch et al. in 2009 presented a study on the characterization and determination of the geographical origin of wines. In this paper, three methods of discrimination and classification of multivariate data were considered and tested: the classification and regression trees (CART), the regularized discriminant analysis (RDA) and the partial least squares discriminant analysis (PLS-DA). PLS-DA analysis showed better classification results with percentage of correct classified samples from 88 to 100%.

Finally, PLS and artificial neural networks (ANN) techniques were compared by Janik et al. in 2007 for the prediction of total anthocyanin content in redgrape homogenates.

3.1.5 Other applications

The applications of Vis/NIR and NIR spectroscopy and their chemometric techniques are present in many other sectors of the food industry. In literature are reported works relating to the dairy, oil, coffee, honey and chocolate industry. In particular, interesting studies have been conducted by some authors for the application of NIR spectroscopy in detecting the geographical origin of raw materials and finished products, defending the protected designation of origin (PDO).

Olivieri et al. (2011) worked out the exploration of three different class-modelling techniques to evaluate classification abilities based on geographical origin of two PDO food products: olive oil from Liguria and honey from Corsica. Authors developed the best models for both Ligurian olive oil and Corsican honey by a potential function technique (POTFUN) with values of correctly classified around 83%.

González-Martín et al. in 2011 presented a work on the evaluation by near infrared reflectance (NIR) spectroscopy of different sensorial attributes of different type of cheeses, taking as reference data the evaluation of the sensorial properties obtained by a panel of eight trained experts. NIR spectra were collected with a remote reflectance fibre optic probe applying the probe directly to the cheese samples and the calibration equations were developed by using modified partial least-squares (MPLS) regression for 50 samples of cheese. Authors stated that obtained results can be considered good and acceptable for all the parameters analyzed (presence of holes, hardness, chewiness, creamy, salty, buttery flavour, rancid flavour, pungency and retronasal sensation).

The quality of coffee is related to the chemical constituents of the roasted beans, whose composition depends on the composition of green beans (i.e., un-roasted). Unroasted coffee beans contain different chemical compounds, which react amongst themselves during coffee roasting influencing the final product. For this reason, monitoring the row materials and the roasting process is very important. Ribeiro et al. in 2011 elaborated PLS models correlating coffee beverage sensory data and NIR spectra of 51 Arabica roasted coffee samples. Acidity, bitterness, flavour, cleanliness, body and overall quality of coffee beverage were considered. Results were good and authors confirmed that it is possible to estimate the quality of coffee using PLS regression models obtained by using NIR spectra of roasted Arabica coffees.

Da Costa Filho in 2009 elaborated a rapid method to determine sucrose in chocolate mass using near infrared spectroscopy. Data were modelled using partial least squares (PLS) and multiple linear regression (MLR), achieving good results (correlation coefficient of 0.998 and 0.997 respectively for the two chemometric techniques). Results showed that NIR can be used as rapid method to determine sucrose in chocolate mass in chocolate factories.

3.2 Image analysis

The chemical imaging spectroscopy is applied to various fields, from astronomy to agriculture (Baranowski et al., 2008, Monteiro et al., 2007, V. Smail, 2006), from the pharmaceutical industry (Lyon et al. 2002, Roggo et al., 2005) to medicine (Ferris et al. 2001, Zheng et al., 2004). But in recent years, has also found use for quality control and safety in food (Gowen et al., 2007b).

In general, classify or quantify the presence of compounds in a sample is the main purpose of the application in the food hyperspectral analysis. There already exist algorithms for classification and regression but, improved algorithms efficiency, could be a target, as well as create datasets, identify anomalies or objects with different spectral characteristics, compared hyperspectral image with those of data library. These goals can be achieved only if the experimental data are processed with chemometric methods. K-nearest neighbors and hierarchical clustering are examples of multivariate analysis that allow to get information from spectral and spatial data (Burger & Gowen, 2011). With the use of spectral image, which allows to obtain in a single determination, spectral and spatial information characterizing the sample, it is possible to identify which chemical species are present and how they are distributed in a matrix. Several chemometric techniques are available for the development of regression models (for example partial least squares regression, principal components regression, and linear regression) capable of estimating the concentrations of constituents in a sample, at the pixel level, allowing the spatial distribution or the mapping of a particular component in the sample analyzed. Moreover the hyperspectral image, combined with chemometric technique, is a powerful method to identify key wavelengths in order to develop of multispectral system, for on-line applications.

Karoui & De Baerdemaeker (2006) wrote a review about the analytical methods coupled with chemometric tools for the determination of the quality and identity of dairy products. Spectroscopic techniques (NIR, MIR, FFFS front face fluorescence spectroscopy, etc.), coupled with chemometric tools have many potential advantages for the evaluation of the identity dairy products (milk, ice cream, yogurt, butter, cheese, etc).

In another review Sankaran et al. (2010), compared the benefits and limitations of advanced techniques and multivariate methods to detect plant diseases in order to assist in monitoring health in plants under field conditions. These technologies include evaluation of volatile profiling (Electronic Nose), spectroscopy (fluorescence, visible and infrared) and imaging (fluorescence and hyperspectral) techniques for disease detection.

In literature it's possible find several examples of applications of spectroscopic image analysis. Hyperspectral imaging could be used as critical control points of food processing to inspect for potential contaminants, defects or lesions. Their absence is essential for ensuring food quality and safety. In some case the application on-line was achieved.

Ariana & Lu (2010), evaluated the internal defect and surface color of whole pickles, in a commercial pickle processing. They used a prototype of on-line hyperspectral imaging system, operating in the wavelength range of 400–1000 nm. Color of the pickles was modeled using tristimulus values: there were no differences in chroma and hue angle of good and defective pickles. PCA was applied to the hyperspectral images: transmittance images at 675–1000 nm were much more effective for internal defect detection compared to reflectance images for the visible region of 500–675 nm. A defect classification accuracy was of 86% compared with 70% by the human inspectors.

Mehl et al. (2002), used hyperspectral image analysis and PCA, like chemometrics technique, to reduce the information resulting from HIS and to identify three spectral bands capable of separating normal from contaminated apples. These spectral bands were implemented in a multispectral imaging system. On 153 samples, it's possible to get a good separation between normal and contaminated (scabs, fungal, soil contaminations, and bruises) apples was obtained for Gala (95%) and Golden Delicious (85%), separations were limited for Red Delicious (76%).

HSI application for damage detection on the caps of white mushrooms (Agaricus bisporus) was investigated from Gowen et al. (2007a). They employed a pushbroom line-scanning HSI instrument (wavelength range: 400–1000 nm). They investigated two data reduction methods. In the first method, PCA was applied to the hypercube of each sample, and the second PC (PC 2) scores image was used for identification of bruise-damaged regions on the mushroom surface. In the second method PCA was applied to a dataset comprising of average spectra from regions normal and bruise-damaged tissue. The second method performed better than the first when applied to a set of independent mushroom samples. Further, they (Gowen et al., 2009) identified mushrooms subjected to freeze damage using hyperspectral imaging. In this case they used Standard Normal Variate (SNV) transformation to pretreat the data, then they applied a procedure based on PCA and LDA to classify spectra of mushrooms into undamaged and freeze-damaged groups. The undamaged mushrooms and freeze-damaged mushrooms could be classified with high accuracy (>95% correct classification) after only 45 min thawing (at 23 ± 2 °C) at that time freeze–thaw damage was not visibly evident.

A study on fruits and vegetables (Cubero et al., 2010) used ultraviolet or near-infrared spectra to explore defects or features that the human eye is unable to see, with the aim of applying them for automatic inspection. This work present a summary of inspection systems for fruit and vegetables and the latest developments in the application of this technology to the inspection of internal and external quality of fruits and vegetables.

Li et al. (2011) detected common defects on oranges using hyperspectral (wavelength range: 400-1000) reflectance imaging. The disadvantage of studied algorithm is that it could not discriminate between different types of defects.

Bhuvaneswari et al. (2011) compared three methods (electronic speck counter, acid hydrolysis and flotation and near-infrared hyperspectral imaging) to investigate the presence of insect fragments (Tribolium castaneum_ Coleoptera: Tenebrionidae) in the semolina (ingredient for pasta and couscous). NIR hyperspectral imaging is a rapid, non-destructive method, as electronic speck counter, but they showed different correlation between insect fragments in the semolina and detection of specks in the samples: $R^2 = 0.99$ and 0.639-0.767 respectively. For NIR hyperspectral image technique, the prediction model were developed by PLS regression.

The most important features in meat are tenderness, juiciness and flavour. Jackmana et al., (2011) wrote a review about recent advances in the use of computer vision technology in the quality assessment of fresh meats. The researcher support that the best opportunities for improving computer vision solutions is the application of hyperspectral imaging in combination with statistical modelling. This synergy can provide some additional information on meat composition and structure. However, in parallel, new image processing algorithms, developed in other scientific disciplines, should be carefully considered for potential application to meat images.

Other applications concern the possibility of estimating a correlation between characteristics (physical or chemical) of the food and the spectra acquired with spectroscopic image. Moreover these techniques were able to locate and quantify the characteristic of interest within the image.

In most cases the range of wavelength used in applications of hyperspectral images is 400-1000 nm but Maftoonazad et al. (2010) used artificial neural network (ANN) modeling of hyperspectral radiometric (350-2500 nm) data for quality changes associated with avocados during storage. Respiration rate, total color difference, texture and weight loss of samples were measured as conventional quality parameters during storage. Hyperspectral imaging was used to evaluate spectral properties of avocados. Results indicated ANN models can predict the quality changes in avocado fruits better than the conventional regression models.

While Mahesh et al. (2011) used near-infrared hyperspectral images (wavelength range: 960–1700 nm), applied to a bulk samples, to classify the moisture levels (12, 14, 16, 18, and 20%) on the wheat. Principal components analysis (PCA) was used to identify the region (1260–1360 nm) with more information. The linear and quadratic discriminant analyses (LDA) and quadratic discriminant analysis (QDA) could classify the sample based on moisture contents than also identifying specific moisture levels with a god levels of accuracy (61- 100% in several case). Spectral features at key wavelengths of 1060, 1090, 1340, and 1450 nm were ranked at top in classifying wheat classes with different moisture contents.

Manley et al. (2011) used near infrared hyperspectral imaging combined with chemometrics techniques for tracking diffusion of conditioning water in single wheat kernels of different hardnesses. NIR analysers is a commonly, non-destructive, non-contact and fast solution for quality control, and a used tool to detect the moisture-content of carrot samples during storage but Firtha (2009) used hyperspectral system that is able to detect the spatial distribution of reflectance spectrum as well. Statistical analysis of the data has shown the optimal intensity function to describe moisture-content.

The intent of Junkwon et al. (2009), was to develop a technique for weight and ripeness estimation of palm oil (Elaeis guieensis Jacq. var. tenera) bunches by hyperspectral and RGB color images. In the hyperspectral images, the total number of pixels in the bunch was also counted from an image composed of three wavelengths (560 nm, 680 nm, and 740 nm), while the total number of pixels of space between fruits was obtained at a wavelength of 910 nm. Weight-estimation equations were determined by linear regression (LR) or multiple linear regression (MLR). As a result, the coefficient of determination (R^2) of actual weight and estimated weight were at a level of 0.989 and 0.992 for color and hyperspectral images, respectively. About the estimation of palm oil bunch ripeness the bunches was classified in 4 classes of ripeness (overripe, ripe, underripe, and unripe) (Fig. 10). Euclidean distances between the test sample and the standard 4 classes of ripeness were calculated, and the test sample was classified into the ripeness class. In the classification based on color image, (average RGB values of concealed and not-concealed areas), and by hyperspectral images (average intensity values of fruits pixels from the concealed area), the results of validation experiments with the developed estimation methods indicated acceptable estimation accuracy.

Fig. 10. Bunches of palm oil: a) unripe, b) underripe, c) ripe, and d) overripe (Junkwon et al., 2009)

Nguyen et al. (2011) illustrated the potential of combination of hyperspectral imaging chemometrics and image processing as a process monitoring tool for the potato processing industry. They predicted the optimal cooking time by hyperspectral imaging (wavelength range 400–1000 nm). By partial least squares discriminant analysis (PLS-DA), cooked and raw parts of boiled potatoes, were discriminated successfully. By modeling the evolution of the cooking front over time the optimal cooking time could be predicted with less than 10% relative error.

Yu H. & MacGregor J.F. (2003) applied multivariate image analysis and regression for prediction of coating content and distribution in the production of snack foods. Elaboration tools based on PCA and PLS was used for the extraction of features from RGB color images and for their use in predicting the average coating concentration and the coating distribution. On-line and off-line imaging were collected from several different snack food product lines and were used to develop and evaluate the methods. The better methods are now being used in the snack food industry for the on-line monitoring and control of product quality.

Siripatrawan et al. (2011) have developed a rapid method for the detection of *Escherichia coli* contamination in packaged fresh spinach using hyperspectral imaging (400–1000 nm) and chemometrics techniques. The PCA was implemented to remove redundant information of the hyperspectral data and artificial neural network (ANN) to correlate spectra with number of E. coli and to construct a prediction map of all pixel spectra of an image to display the number of E. coli in the sample.

In this study (Barbin et al. 2011) a hyperspectral imaging technique (range from 900 to 1700 nm) was developed to achieve fast, accurate, and objective determination of pork quality grades. The sample investigated were 75 pork cuts of *longissimus dorsi* muscle from three quality grades. Six significant wavelengths (960, 1074, 1124, 1147, 1207 and 1341 nm) that explain most of the variation among pork classes were identified from 2nd derivative spectra. PCA was carried out and the results indicated that pork classes could be precisely discriminated with overall accuracy of 96%. Algorithm was developed to produce classification maps of the investigated sample.

Valous et al. (2010) communicated perspectives and aspects, relating to imaging, spectroscopic and colorimetric techniques on the quality evaluation and control of hams. These no-contact and no-destructive techniques, can provide useful information regarding ham quality. Hams are considered a heterogenic solid system: varying colour, irregular shape and spatial distribution of pores. Fat-connective tissue, water, protein contribute to the microstructural complexity. This review paying attention on applications of imaging and spectroscopy techniques, for measuring properties and extracting features that correlate with ham quality.

In literature is present a review (Mathiassena et al., 2011) that focused the attention on application of imaging technologies (VIS/NIR imaging, VIS/NIR imaging spectroscopy, planar and computed tomography (CT) X-ray imaging, and magnetic resonance imaging) to inspection of fish and fish products.

Nicolai et al. (2007) wrote a review about the applications of non-destructive measurement of fruit and vegetable quality. Measurement principles are compared, and novel techniques (hyperspectral imaging) are reviewed. Special attention is paid to recent developments in portable systems. The problem of calibration transfer from one spectrophotometer to another is introduced, as well as techniques for calibration transfer. Chemometrics is an essential part of spectroscopy and the choice, of corrected techniques, is primary (linear or nonlinear regression, such as kernel-based methods are discussed). The principal objective of spectroscopy system applications in fruit and vegetables sector have focused on the nondestructive measurement of soluble solids content, texture, dry matter, acidity or disorders of fruit and vegetables. (root mean square error of prediction want to be achieved).

3.3 Electronic nose

The preservation of quality in post-harvest is the prerequisite for agri-food products in the final stages of commercialization. The fruit quality is related to the appearance (skin color, size, shape, integrity of the fruit), to the sensorial properties (hardness and crispness of the flesh, juicy, acid/sugars) and to safety (residues in fruit).

The agri-food products contain a variety of information, directly related to their quality, traditionally measured by means of tedious, time consuming and destructive analysis. For this reason, there is a growing interest in easy to use, rapid and non-destructive techniques useful for quality assessment.

Currently, electronic noses are mainly applied in the food industry to recognize the freshness of the products, the detection of fraud (source control, adulteration), the detection of contaminants.

An essential step in the analysis with an electronic nose, is the high performance of statistical elaboration. The electronic nose provides multivariated results that need to be processed using chemometric techniques. Even if the best performing programs are sophisticated and, consequently, require the operation of skilled personnel, most companies have implemented user-friendly software for data treatment in commercially available electronic noses (Ampuero & Bosset, 2003).

A commercial electronic nose, as a non-destructive tool, was used to characterise peach cultivars and to monitor their ripening stage during shelf-life (Benedetti et al. 2008). Principal component analysis (PCA) and linear discriminant analysis (LDA) were used to investigate whether the electronic nose was able to distinguish among four diverse cultivars. Classification and regression tree (CART) analysis was applied to characterise peach samples into the three classes of different ripening stages (unripe, ripe, over-ripe). Results classified samples in each respective group with a cross validation error rate of 4.87%.

Regarding the fruit and vegetable sector Torri et al. (2010) investigated the applicability of a commercial electronic nose in monitoring freshness of packaged pineapple slices during storage. The obtained results showed that the electronic nose was able to discriminate between several samples and to monitor the changes in volatile compounds correlated with quality decay. The second derivative of the transition function, used to interpolate the PC1 score trend versus the storage time at each temperature, was calculated to estimate the stability time.

Ampuero and Bosset (2003), presented a review about the application of electronic nose applied to dairy products. The present review deal with as examples the evaluation of the cheese ripening, the detection of mould in cheese, the classification of milk by trademark, by fat level and by preservation process, the classification and the quantification of off-flavours in milk, the evaluation of Maillard reactions during heating processes in block-milk, as well as the identification of single strains of disinfectant–resistant bacteria in mixed cultures in milk. For each application correspond the chemometric method to extrapolate the maximum information. PCA analysis was carried out in order to associate descriptors (chocolate, caramel, burnt and nutty), typical of volatiles generated by Maillard reactions during milk heating. In another case PCA showed a correctly classification of sample in function of the origin of off-flavours. In figure 11 is showed an example of result carried out by DFA (discriminant function analysis) statistical technique. In this case the aim of researcher was to classify samples of a given product by their place of production.

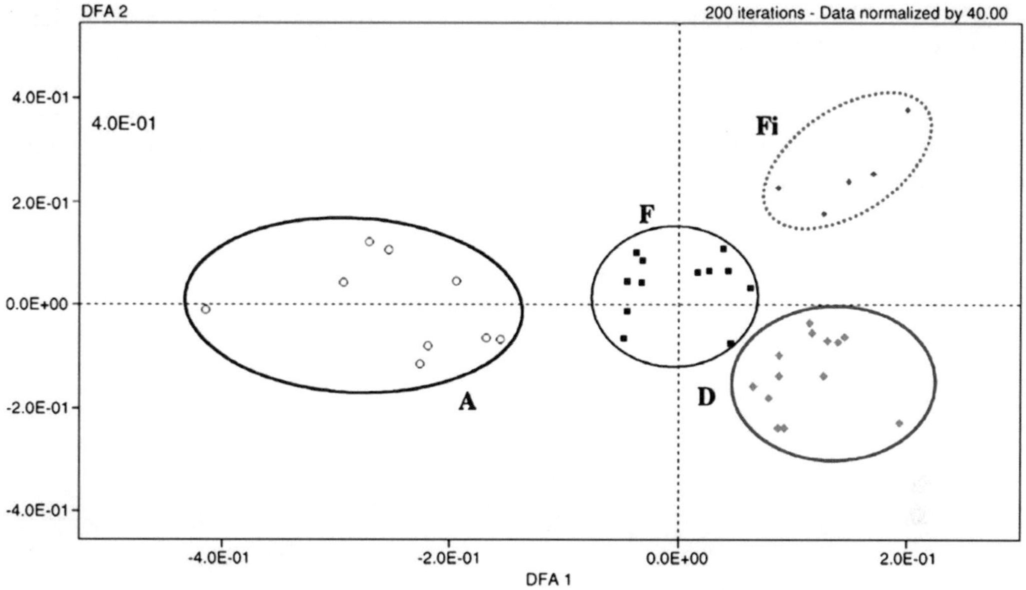

Fig. 11. Classification of Emmental cheese by the geographic origin performed with an electronic nose based on mass spectrometry. The graph shows DFA 1 vs. DFA 2 with 100% group classification based on five variables. No validation set was considered due to the limited number of samples. A: Austria, D: Germany, F: France, Fi: Finland (Pillonel et al, 2003)

The potential of electronic nose technique was investigated to monitoring storage time and the quality attribute of eggs by Yongwei et al. (2009). Using techniques of multivariate analysis was distinguished eggs under cool and room-temperature storage. Results showed that the E-nose could distinguish eggs of different storage time under cool and room-temperature storage by LDA, PCA, BPNN and GANN. Good distinction between eggs stored for different times were obtained by PCA and LDA (results by LDA were better than those obtained by PCA). By means BP neural network (BPNN) and the combination of a genetic algorithm and BP neural network (GANN) carried out good predictions for egg storage time (GANN demonstrated better correct classification rates than BPNN). The quadratic polynomial step regression (QPSR) algorithm established models that described the relationship between sensor signals and egg quality indices (Haugh unit and yolk factor). The QPST models showed an high predictive ability (R^2 = 0.91-0.93).

Guidetti et al. (2011) used electronic nose and infrared thermography to detect physiological disorders on apples (Golden Delicious and Stark Delicious). In particular the aim was to differentiate typical external apple diseases (in particular, physiological, pathological and entomological disorders). The applicability of the e-nose is based on the hypothesis that apples affected by physiopathology produce different volatile compounds from those produced by healthy fruits. The electronic nose data were elaborated by LDA in order to classify the apples into the four classes. Figure 12 shows how the first two LDA functions discriminate among classes. Considering Stark variety, function 1 seems to discriminate among the physiopathologies while function 2 discriminate the healthy apples from those with physiological disorders. The error rate and the cross validation error rate were of 2.6% and 26.3% respectively. In the case of Golden variety, along the first function there is the separation of Control samples from the apples affected by diseases, while in the vertical direction (function 2) there is an evident discrimination among the three physiopathologies. The error rate and the cross validation error rate were of 0.8% and 18% respectively.

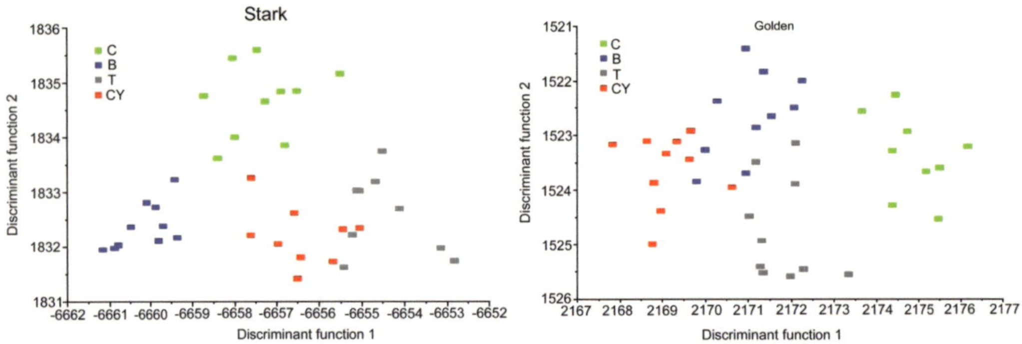

(a) (b)

Fig. 12. a) Canonical discriminant functions of LDA for Stark variety; b) Canonical discriminant functions of LDA for Golden variety. C=control, B=bitter pit, T=scab and CY=Cydia Pomonella.

Cerrato Oliveros et al. (2002) selected array of 12 metal oxide sensors to detected adulteration in virgin olive oils samples and to quantify the percentage of adulteration by electronic nose. Multivariate chemometric techniques such as PCA were applied to choose a set of optimally discriminant variables. Excellent results were obtained in the differentiation of adulterated and non-adulterated olive oils, by application of LDA, QDA. The models provide very satisfactory results, with prediction percentages >95%, and in some cases almost 100%. The results with ANN are slightly worse, although the classification criterion used here was very strict. To determine the percentage of adulteration in olive oil samples multivariate calibration techniques based on partial least squares and ANN were employed. Not so good results were carried out, even if there are exceptions. Finally, classification techniques can be used to determine the amount of adulterant oil added with excellent results.

4. Conclusion

This work shows the principal non-destructive applications for analysis in food sector. They are rapid techniques used in combination with chemometrics analysis for qualitative and quantitative analysis.

NIR spectroscopy is the technique that has been developed further in recent years. This success is because spectral measurement for one sample could be done in a few seconds. Numerous samples could be analyzed and multiindexes analysis can be carried out.

Compared with traditional methods, NIR and Vis/NIR are less expensive because of no demand of other materials such as chemical reagents except the electrical consumption. Many works are focused on the study of chemometrics. This is because an important challenge is to build robust calibration models, in fact it is important to apply chemometric methods able to select useful information from a great deal of spectral data. Moreover food researchers and analysts are looking for the sensitive wavelength in Vis/NIR region representing the characteristics of food products, with the aim of develop some simple and low-cost instruments (Cen & He, 2007).

HSI is the new frontier for optical analysis of foods. The performance of HSI instrumentation has developed such that a full hypercube can now be acquired in just a few seconds. In tandem with these developments, advances in component design have led to reductions in the size and cost of HSI systems. This has led to increased interest in their online implementation for quality monitoring in major industries such as food and pharmaceutical (Burger & Gowen, 2011). In future, with further improvement, the HSI system could meet the need of a commercial plant setting.

The equipment that the food industry has at its disposal is certainly complex and not easy to use. The chemometric approach has allowed, through different applicative researches, to arrive at algorithms that can support the analysis in the entire food chain from raw material producers to large retail organizations. Despite this, we are still faced with instrumentation with not easy usability and relatively high cost: the studies must move towards a more simplified instrumental approach through greater integration of hardware with software. The challenges are many: optimizing the information that you are able to extract from raw data and aimed at specific problems, simplify electronic components, increase the level of interaction tool/operator.

In conclusion the only way of an interdisciplinary approach can lead to the solution of a system that can provide at different level more immediate response and more food safety and quality.

5. References

Aleixos, N.; Blasco, J.; Navarrón, F. & Moltó, E. (2002). *Multispectral inspection of citrus in real-time using machine vision and digital signal processors*. Computers and Electronics in Agriculture. Vol.33, N.2, pp. 121-137

Amamcharla, J. K.; Panigrahi, S.; Logue, C. M.; Marchello, M. & Sherwood, J. S. (2010). *Fourier transform infrared spectroscopy (FTIR) as a tool for discriminating Salmonella typhimurium contaminated beef*. Sens. & Instrumen. Food Qual., Vol.4, pp. 1-12

Ampuero, S. & Bosset, J.O. (2003). *The electronic nose applied to dairy products: a review*. Sensors and Actuators B Vol.94, pp. 1-12

Andrés, S.; Murray, I.; Navajas, E.A.; Fisher, A.V.; Lambe, N.R. & Bünger, L. (2007). *Prediction of sensory characteristics of lamb meat samples by near infrared reflectance spectroscopy*. Meat Science Vol.76, pp. 509-516

Ariana, D.P. & Lu, R.A. (2010). *Evaluation of internal defect and surface color of whole pickles using hyperspectral imaging*. Journal of Food Engineering Vol.96, pp. 583-590

Baranowski, P.; Lipecki, J.; Mazurek, W. & Walczak, R.T. (2008). *Detenction of watercore in 'Gloster' apples using thermography*. Postharvest Biology and Technology Vol.47, pp. 358

Barbin, D.; Elmasry, G.; Sun, D.W.; Allen, P. (2011). *Near-infrared hyperspectral imaging for grading and classification of pork*. Meat Science. Article in press

Basilevsk, A. (1994) *Statistical factor analysis and related methods: theory and applications*. Wiley-Interscience Publication. ISBN 0-471-57082-6

Benedetti, S.; Buratti, S.; Spinardi, A.; Mannino, S. & Mignani, I. (2008). *Electronic nose as a non-destructive tool to characterize peach cultivars and to monitor their ripening stage during shelf-life*. Postharvest Biology and Technology Vol.47, pp. 181–188

Bhuvaneswari, K.; Fields, P. G.; White, N.D.G.; Sarkar, A. K.; Singh, C. B. & Jayas, D. S. (2011). *Image analysis for detecting insect fragments in semolina*. Journal of Stored Products Research Vol.47, pp. 20-24

Brenna, O.V. & Berardo, N. (2004). *Application of near-infrared reflectance spectroscopy (NIRS) to the evaluation of carotenoids content in maize*. J. Agric. Food Chem. Vol.52, 5577

Brosnan, T. & Sun, D.W. (2004). *Improving quality inspection of food products by computer vision: a review*. Journal of Food Engineering. Vol.61, pp. 3-16

Burger, J. & Gowen, A. (2011). *Data handling in hyperspectral image analysis*. Chemometrics and intelligent Laboratory Systems Vol.108, pp. 13-22

Burns, D.A. & Ciurczak, E.W. (2001). *Second ed. In: Handbook of Near-Infrared Analysis..*, Marcel Dekker, New York. Vol.27, N.28, pp. 729–782

Carrara, M.; Catania, P.; Vallone, M. & Piraino, S. (2008). *Mechanical harvest of grapes: Assessment of the physical-mechanical characteristics of the berry in order to improve the quality of wines*. In Proc. Intl. Conf. on Agricultural Engineering: Agricultural and Biosystems Engineering for a Sustainable World (AgEng 2008). CIGR

Cen, H. & He, Y. (2007). *Theory and application of near infrared reflectance spectroscopy in determination of food quality*. Trends in Food Science & Technology.Vol.18, pp. 72-83

Cerrato Oliveros, M.C.; Pérez Pavón, J. L.; Garcıa Pinto, C.; Fernández Laespada, M. E.; Moreno Cordero, B. & Forina, M. (2002). *Electronic nose based on metal oxide semiconductor sensors as a fast alternative for the detection of adulteration of virgin olive oils*. Analytica Chimica Acta Vol.459, pp. 219–228

Cozzolino, D.; Cynkar, W.; Janik, L.; Dambergs, R. G. & Gishen, M. (2006a). *Analysis of grape and wine by near infrared spectroscopy – A review*. J Near Infrared Spectros, Vol.14, pp. 279−289

Cozzolino, D.; Parker, M.; Dambergs, R.G.; Herderich, M. & Gishen, M. (2006b). *Chemometrics and visible-near infrared spectroscopic monitoring of red wine fermentation in a pilot scale*. Biotechnol. Bioeng. Vol. 95, pp. 1101

Cubero, S.; Aleixos, N.; Moltó, E.; Gómez-Sanchis, J. & Blasco, J. (2010). *Advances in Machine Vision Applications for Automatic Inspection and Quality Evaluation of Fruits and Vegetables*. Food Bioprocess Technol Vol.4, pp.487–504

Da Costa Filho, P. A. (2009). *Rapid determination of sucrose in chocolate mass using near infrared spectroscopy*. Analytica Chimica Acta Vol.631, pp. 206–211

De Temmerman, J.; Saeys, W.; Nicolai, B. & Ramon, H. (2007). *Near infrared reflectance spectroscopy as a tool for the in-line determination of the moisture concentration in extruded semolina pasta*. Biosystems Engineering Vol.97, pp. 313–321

Deisingh, A.K.; Stone, D.C. & Thompson, M. (2004). *Applications of electronic noses and tongues in food analysis*. International Journal of Food Science and Technology Vol.39, pp. 587–604

Diezak, J.D. (1988). *Microscopy and imagine analysis for R&D, Special report*. Food Technol. pp. 110-124

Du, C.J. & Sun, D.W. (2006). *Learning techniques used in computer vision for food quality evaluation: a review*. Journal of food engineering. Vol.72, N.1, pp. 39–55

Elbatawi, I. E. & Ebaid, M. T. (2006). *A new technique for grape inspection and sorting classification*. Arab Universities J. Agric. Sci. Vol.14, N.2, pp. 555-573

ElMarsy, G. & Sun, D.W. (2010). *Hyperspectral imaging for food quality, analysis and control.* Book, N.1, pp. 3-43 (http://elsevier.insidethecover.com/searchbook.jsp?isbn=9780123740854)

Ferris, D.; Lawhead, R.; Dickman, E.; Holtzapple, N.; Miller, J.; Grogan, S.; et al. 2001. *Multimodal hyperspectral imaging for the noninvasive diagnosis of cervical neoplasia.* Journal of Lower Genital Tract Disease Vol.5, N.2, pp. 65-72

Fessenden, R. J. & Fessenden, J. S. (1993). Chimica organica. Cap. 9: *Spettroscopia I: Spettri infrarossi, Risonanza Magnetica Nucleare*. Piccin Padova, Italy

Firtha, F. (2009). *Detecting moisture loss of carrot samples during storage by hyperspectral imaging system*. Acta Alimentaria Vol.38, N.1, pp. 55-66

Firtha, F.; Fekete, A.; Kaszab, T.; Gillay, B.; Nogula-Nagy, M.; Kovács, Z. & Kantor, D.B. (2008). *Methods for improving image quality and reducing data load of nir hyperspectral images*. Sensors 2008, 8, 3287-3298

Fodor, M.; Woller, A.; Turza, S. & Szigedi, T. (2011). *Development of a rapid, non-destructive method for egg content determination in dry pasta using FT-NIR technique.* Journal of Food Engineering 107, 195–199

Frank, I.E. & Todeschini, R. (1994). *The Data Analysis Handbook.* Elsevier. ISBN 0-444-81659-3, included in series: Data Handling in Science and Technology

Gardner, J.W. & Bartlett, P.N. (1994). *A brief history of electronic noses.* Sens. Actuat. B: Chem. Vol.18, pp. 211-220

González-Martín, M.I.; Severiano-Pérez, P.; Revilla, I.; Vivar-Quintana, A.M.; Hernández-Hierro, J.M.; González-Pérez, C. & Lobos-Ortega, I.A. (2011). *Prediction of sensory attributes of cheese by near-infrared spectroscopy*. Food Chemistry Vol.127, pp. 256–263

Gowen, A.A.; O'Donnell, C.P.; Cullen, P.J.; Downey, G. & Frias, J.M. (2007b). *Hyperspectral imaging - an emerging process analytical tool for food quality and safety control.* Trends in Food Science & Technology Vol.18, pp.590-598

Gowen, A.A.; O'Donnell, C.P.; Taghizadeh, M.; Cullen, P.J.; Frias, J.M. & Downey, G. (2007a). *Hyperspectral imaging combined with principal component analysis for bruise damage detection on white mushrooms (Agaricus bisporus)*. J. of chemometric DOI: 10.1002/cem.1127

Gowen, A.A.; Taghizadeh, M. & O'Donnell, C.P. (2009). *Identification of mushrooms subjected to freeze damage using hyperspectral imaging*. Journal of Food Engineering Vol.93, pp. 7-12

Guanasekaran, S. & Ding, K. (1994). *Using computer vision for food quality evaluation*. Food Technol. Vol.15, pp. 1-54;

Guidetti, R.; Beghi, R. & Bodria, L. (2010). *Evaluation of Grape Quality Parameters by a Simple Vis/NIR System*. Transaction of the ASABE, Vol.53 N.2, pp. 477-484, ISSN: 2151-0032

Guidetti, R.; Beghi, R.; Bodria, L.; Spinardi, A.; Mignani, I. & Folini, L. (2008). *Prediction of blueberry (Vaccinium corymbosum) ripeness by a portable Vis-NIR device.* Acta Horticulturae, n° 310, ISBN 978-90-66057-41-8, pp. 877-885

Guidetti, R.; Buratti, S. & Giovenzana, V. (2011). *Application of Electronic Nose and Infrared Thermography to detect physiological disorders on apples (Golden Delicious and Stark Delicious).* CIGR Section VI International Symposium on Towards a Sustainable Food Chain Food Process, Bioprocessing and Food Quality Management. Nantes, France - April 18-20, 2011

Jackmana, P.; Sun D.W. & Allen P. (2011). *Recent advances in the use of computer vision technology in the quality assessment of fresh meats.* Trends in Food Science & Technology Vol.22, pp. 185-197

Jackson, J.E. (1991). *A user's guide to principal components.* Wiley-Interscience Publication. ISBN 0-471-62267-2

Janik, L.J.; Cozzolino, D.; Dambergs, R.; Cynkar W. & Gishen, M. (2007). *The prediction of total anthocyanin concentration in red-grape homogenates using visiblenear- infrared spectroscopy and artificial neural networks.* Anal. Chim. Acta pp. 594-107

Junkwon, P.; Takigawa, T.; Okamoto, H.; Hasegawa, H.; Koike, M.; Sakai, K.; Siruntawineti, J.; Chaeychomsri, W.; Sanevas, N.; Tittinuchanon, P. & Bahalayodhin, B. (2009). *Potential application of color and hyperspectral images for estimation of weight and ripeness of oil palm (elaeis guineensis jacq. var. tenera).* Agricultural Information Research Vol.18, N.2, pp. 72–81

Karoui, R. & De Baerdemaeker, J. (2006). *A review of the analytical methods coupled with chemometric tools for the determination of the quality and identity of dairy products.* Food Chemistry Vol.102, pp. 621–640;

Li, J.; Rao, X. & Ying, Y. (2011). *Detection of common defects on oranges using hyperspectral reflectance imaging.* Computers and Electronics in Agriculture Vol.78, pp. 38–48

Lu, R. & Ariana, D. (2002). *A Near-Infrared Sensing Technique for Measuring Internal Quality of Apple Fruit.* Applied Engineering in Agricolture, Vol.18, N.5, pp. 585-590

Lunadei, L. (2008). *Image analysis as a methodology to improve the selection of foodstuffs.* PhD thesis in "Technological innovation for agro-food and environmental sciences", Department of Agricultural Engineering, Università degli Studi di Milano

Lyon, R. C.; Lester, D. S.; Lewis, E. N.; Lee, E.; Yu, L. X.; Jefferson, E. H.; et al. (2002). *Near-infrared spectral imaging for quality assurance of pharmaceutical products: analysis of tablets to assess powder blend homogeneity.* AAPS PharmSciTech Vol.3, N.3, pp. 17

Maftoonazad, N.; Karimi, Y.; Ramaswamy, H.S. & Prasher, S.O. (2010). *Artificial neural network modeling of hyperspectral radiometric data for quality changes associated with avocados during storage.* Journal of Food Processing and Preservation ISSN 1745-4549

Mahesh, S.; Jayas, D. S.; Paliwal, J. & White, N. D. G. (2011). *Identification of wheat classes at different moisture levels using near-infrared hyperspectral images of bulk samples.* Sens. & Instrumen. Food Qual. Vol.5, pp. 1–9

Manley, M.; Du Toit, G. & Geladi, P., (2011). *Tracking diffusion of conditioning water in single wheat kernels of different hardnesses by near infrared hyperspectral imaging.* Analytica Chimica Acta Vol.686, pp. 64–75

Massart, D.L.; Buydens, L.M.C.; De Jong, S.; Lewi P.J. & Smeyers-Verbek, J. (1998). *Handbook of Chemometrics and Qualimetrics: Part B.* Edited by B.G.M. ISBN: 978-0-444-82853-8, included in series: Data Handling in Science and Technology

Massart, D.L.; Vandeginste, B.G.M.; Buydens, L.M.C.; De Jong, S.; Lewi P.J. & Smeyers-Verbek, J. (1997). *Handbook of Chemometrics and Qualimetrics: Part A.* Elsevier, ISBN: 0-444-89724-0, included in series: Data Handling in Science and Technology

Mathiassena, J. R.; Misimib, E.; Bondøb, M.; Veliyulinb, E. & Ove Østvik, S. (2011). *Trends in application of imaging technologies to inspection of fish and fish products.* Trends in Food Science & Technology Vol.22, pp. 257-275

McClure, W. F. (2003). *204 years of near infrared technology: 1800 - 2003.* Journal of Near Infrared Spectroscopy, Vol.11, pp. 487−518

Mehl, P. M.; Chao, K.; Kim, M.; Chen, Y. R. (2002). *Detection of defects on selected apple cultivars using hyperspectral and multispectral image analysis.* Applied Engineering in Agriculture Vol.18, N.2, pp. 219-226

Monteiro, S.; Minekawa, Y.; Kosugi, Y.; Akazawa, T. & Oda, K. (2007). *Prediction of sweetness and amino acid content in soybean crops from hyperspectral imagery*. ISPRS Journal of Photogrammetry and Remote Sensing Vol.62, N.1, pp. 2–12

Naes, T.; Isaksson, T.; Fearn, T. & Davies, T. (2002). *A user-friendly guide to multivariate calibration and classification*. Chichester, UK: NIR Publications ISBN 0-9528666-2-5

Nguyen, D.; Trong, N.; Tsuta, M.; Nicolaï, B.M.; De Baerdemaeker, J. & Saeys, W. (2011). *Prediction of optimal cooking time for boiled potatoes by hyperspectral imaging*. Journal of Food Engineering Vol.105, pp. 617–624

Nicolai, B. M.; Beullens, K.; Bobelyn, E.; Peirs, A.; Saeys, W.; Theron & K. I., Lammertyna J. (2007). *Non-destructive measurement of fruit and vegetable quality by means of NIR spectroscopy: A review*. Postharvest Biology and Technology, Vol.46, pp. 99–118

Oliveri, P.; Di Egidio, V.; Woodcock, T. & Downey, G. (2011). *Application of class-modelling techniques to near infrared data for food authentication purposes*. Food Chemistry Vol.125, pp. 1450–1456

Osborne, B.G.; Fearn, T. & Hindle, P.H. (1993). *Practical NIR Spectroscopy with Applications in Food and Beverage Analylis*. Cap. 4: Fundamentals of near infrared instrumentation, pp. 73-76. Longman Scientific & Technical

Pillonel, L.; Ampuero, S.; Tabacchi, R. & Bosset, J.O. (2003). *Analytical methods for the determination of the geographic origin of Emmental cheese, volatile compounds by GC–MS–FID and electronic nose, Eur.* J. Food Res. Technol. Vol.216, pp. 179–183

Prieto, N.; Roehe, R.; Lavín, P.; Batten, G. & Andrés, S. (2009b). *Application of near infrared reflectance spectroscopy to predict meat and meat products quality: A review*. Meat Science Vol.83, pp. 175–186

Prieto, N.; Ross, D.W.; Navajas, E.A.; Nute, G.R.; Richardson, R.I.; Hyslop, J.J.; Simm, G. & Roehe, R. (2009a). *On-line application of visible and near infrared reflectance spectroscopy to predict chemical–physical and sensory characteristics of beef quality*. Meat Science Vol.83, pp. 96–103

Ribeiro, J.S.; Ferreira, M.M.C. & Salva, T.J.G. (2011). *Chemometric models for the quantitative descriptive sensory analysis of Arabica coffee beverages using near infrared spectroscopy*. Talanta Vol.83, pp.1352–1358

Riva, M. (1999). Introduzione alle tecniche di Image Analysis. http://www.distam.unimi.it/image_analysis/image0.htm

Roggo, Y.; Edmond, A.; Chalus, P. & Ulmschneider, M. (2005). *Infrared hyperspectral imaging for qualitative analysis of pharmaceutical solid forms*. Analytica Chimica Acta, Vol.535 N.1-2, pp. 79-87

Römisch, U.; Jäger, H.; Capron, X.; Lanteri, S.; Forina, M. & Smeyers-Verbeke, J. (2009). *Characterization and determination of the geographical origin of wines. Part III: multivariate discrimination and classification methods*. Eur. Food Res. Technol. Vol.230, pp. 31–45

Russ, J. C.; Stewart, W. & Russ, J. C., J. C. (1988). *The measurement of macroscopic image*. Food Technol. Vol.42, pp. 94-102

Sankaran, S.; Mishra, A.; Ehsani, R. & Davis, C. (2010). *A review of advanced techniques for detecting plant diseases*. Computers and Electronics in Agriculture Vol.72, pp. 1–13

Sierra, V.; Aldai, N.; Castro, P.; Osoro, K.; Coto-Montes, A. & Oliva, M. (2007). *Prediction of the fatty acid composition of beef by near infrared transmittance spectroscopy*. Meat Science Vol.78, pp. 248–255

Sinelli, N.; Limbo, S.; Torri, L.; Di Egidio, V. & Casiraghi, E. (2010). *Evaluation of freshness decay of minced beef stored in high-oxygen modified atmosphere packaged at different temperatures using NIR and MIR spectroscopy*. Meat Science Vol.86, pp. 748–752

Siripatrawan, U.; Makino, Y.; Kawagoe, Y. & Oshita, S. (2011). *Rapid detection of Escherichia coli contamination in packaged fresh spinach using hyperspectral imaging*. Talanta Vol.85, pp. 276–281

Smail, V.; Fritz, A. & Wetzel, D. (2006). *Chemical imaging of intact seeds with NIR focal plane array assists plant breeding*. Vibrational Spectroscopy Vol.42, N.2, pp. 215-221

Sørensen, L.K. (2009). Application of reflectance near infrared spectroscopy for bread analyses. Food Chemistry Vol.113, pp. 1318–1322

Stark, E.K. & Luchter, K. (2003). *Diversity in NIR Instrumentation, in Near Infrared Spectroscopy.*: Proceeding oh the 11th International Conference. NIR Publication, Chichester, UK, pp. 55-66

Torri, L.; Sinelli, N. & Limbo S. (2010). *Shelf life evaluation of fresh-cut pineapple by using an electronic nose.* Postharvest Biology and Technology Vol.56, pp. 239–245

Urbano-Cuadrado, M.; de Castro, M.D.L.; Perez-Juan, P.M.; Garcia-Olmo, J. & Gomez-Nieto, M.A. (2004). *Near infrared reflectance, spectroscopy and multivariate analysis in enology – Determination or screening of fifteen parameters in different types of wines.* Anal. Chim. Acta Vol.527, pp. 81-88

Valous, N. A.; Mendoza, F. & Sun, D.W. (2010). *Emerging non-contact imaging, spectroscopic and colorimetric technologies for quality evaluation and control of hams: a review.* Trends in Food Science & Technology Vol.21, pp. 26-43

Williams, P. C. & Sobering, D. (1996). *How do we do it: A brief summary of the methods we use in developing near infrared calibrations.* In A. M. C. Davies & P. C. Williams (Eds.), Near infrared spectroscopy: the future waves (pp. 185–188). Chichester: NIR Publications

Wilson, A. D. & Baietto, M. (2009). *Applications and advances in electronic-nose technologies.* Sensors Vol.9, pp. 5099-5148

Wold, S.; Sjöström, M. & Eriksson, L. (2001). *PLS-regression: a basic tool of chemometrics.* Chemom. Intell. Lab. Syst. Vol.58, pp. 109–130

Yongwei, W.; Wang, J.; Zhou, B. & Lu, Q. (2009). *Monitoring storage time and quality attribute of egg based on electronic nose.* Analytica Chimica Acta Vol.650, pp. 183–188

Yu, H. & MacGregor, J.F., (2003). *Multivariate image analysis and regression for prediction of coating content and distribution in the production of snack foods.* Chemometrics and Intelligent Laboratory Systems Vol.67, pp. 125–144

Zardetto, S. & Dalla Rosa, M. (2006). *Study of the effect of lamination process on pasta by physical chemical determination and near infrared spectroscopy analysis.* Journal of Food Engineering Vol.74, pp. 402–409

Zheng, C.; Sun, D.W. & Zheng, L. (2006). *Recent developments and applications of image features for food quality evaluation and inspection: a review.* Trends in food Science &Technology. Vol.17, pp. 642-655

Zheng, G.; Chen, Y.; Intes, X.; Chance, B. & Glickson, J. D. (2004). *Contrast-enhanced near-infrared (NIR) optical imaging for subsurface cancer detection.* Journal of Porphyrins and Phthalocyanines Vol.8, N.9, pp. 1106-1117

11

Metabolomics and Chemometrics as Tools for Chemo(bio)diversity Analysis - Maize Landraces and Propolis

Marcelo Maraschin et al.*
Plant Morphogenesis and Biochemistry Laboratory, Federal University of Santa Catarina, Florianopolis, SC, Brazil

1. Introduction

Developments in analytical techniques (GC-MS, LC-MS, ^{1}H-, ^{13}C-NMR, FT-MS, e.g.) are progressing rapidly and have been driven mostly by the requirements in the healthcare and food sectors. Simultaneous high-throughput measurements of several analytes at the level of the transcript (transcriptomics), proteins, (proteomics), and metabolites (metabolomics) are currently performed, producing a prodigious amount of data. Thus, the advent of *omic* studies has created an information explosion, resulting in a paradigm shift in the emphasis of analytical research of biological systems. The traditional approaches of biochemistry and molecular cell biology, where the cellular processes have been investigated individually and often independent of each other, are giving way to a wider approach of analyzing the cellular composition in its entirety, allowing achieving a *quasi*-complete metabolic picture.

The exponential growth of data, largely from genomics and genomic technologies, has changed the way biologists think about and handle data. In order to derive meaning from these large data sets, tools are required to analyze and identify patterns in the data, and allow data to be placed into a biological context. In this scenario, biologists have a continuous need for tools to manage and analyze the ever-increasing data supply. Optimal use of the data set, primarily of chemical nature, requires effective methods to analyze and manage them. It is obvious that all *omic* approaches will rely heavily upon bioinformatics for the storage, retrieval, and analysis of large data sets. Thus, and taking into account the multivariate nature of analysis in *omic* technologies, there is an increase emphasis in research on the application of chemometric techniques for extracting relevant information.

* Shirley Kuhnen[1], Priscilla M. M. Lemos[1], Simone Kobe de Oliveira[1], Diego A. da Silva[1], Maíra M. Tomazzoli[1], Ana Carolina V. Souza[1], Rúbia Mara Pinto[2], Virgílio G. Uarrota[1], Ivanir Cella[2], Antônio G. Ferreira[3], Amélia R. S. Zeggio[1], Maria B.R. Veleirinho[4], Ivone Delgadillo[4] and Flavia A. Vieira[4]
[1] *Plant Morphogenesis and Biochemistry Laboratory, Federal University of Santa Catarina, Florianopolis, SC, Brazil;*
[2] *EPAGRI – Florianópolis, SC, Brazil;*
[3] *NMR Laboratory, Federal University of São Carlos, São Carlos-SP;*
[4] *Chemistry Department, University of Aveiro – Campus Santiago, Aveiro - Portugal*

Metabolomics* and chemometrics† have been used in a number of areas to provide biological information beyond the simple identification of cell constituents. These areas include:

a. Fingerprinting of species, genotypes or ecotypes for taxonomic or biochemical (gene discovery) purposes;
b. Monitoring the behavior of specific classes of metabolites in relation to applied exogenous chemical and/or physical stimuli;
c. Studying developmental processes such as establishment of symbiotic associations or fruit ripening;
d. Comparing and contrasting the metabolite content of mutant or transgenic plants with that of their wild-type counterparts.

In general sense, strategies to obtain biological information in the above mentioned areas have focused on the analysis of metabolic differences that evidence responses to a range of extrinsic (ambient) and intrinsic (genetic) stimuli. Since no single analytical method has been found to obtain a complete picture of the metabolome of an organism, an association of advanced analytical techniques (GC-MS, LC-MS, FTIR, ^{1}H-, ^{13}C-NMR, FT-MS, e.g.) coupled to chemometrics, e.g., univariate (ANOVA, correlation analysis, regression analysis) or multivariate (PCA, HA, PLS) statistical techniques, has been performed in order to rapidly identify up- or down-regulated endogenous metabolites in complex matrices such as plant extracts, flours, starches, and biofluids, for instance. Plant extracts are recognized to be a complex matrix containing a wide range of primary and secondary metabolites that vary according to the environmental condition, genotype, developmental stage, and agronomic traits, for example. Such a complex matrix has long been used to characterize plant genotypes growing in a given geographic region and/or subjected to external stimuli, giving rise to additional information of interest, e.g., plant genetic breeding programs, local biodiversity conservation, food industry, and quality control in drug development/production processes.

In the former case, programs for genetic breeding of plants have often focused on the analysis of landraces‡ genotypes (i.e., creole and local varieties), aiming at to identify individuals well adapted to specific local environmental conditions (soil and climate) and with superior agronomic performance and biomass yield. Indeed, the analysis and exploitation of the local genotypes' diversity has long been used as a strategy to improve agronomic traits by conventional breeding methods in plant crops of economical interest, as well as for stimulating the preservation of plant genetic resources. Taking into consideration that a series of primary (e.g., proteins and starch) and secondary metabolites (alkaloids, phenolic acids, and carotenoids, for instance) are well recognized compounds associated to the plants' adaptation mechanisms to their surroundings ecological factors, metabolomics and chemometrics have emerged as an interesting approach for helping the selection of

* *Metabolomics*: constitutes a quantitative and qualitative survey of the whole metabolites of an organism as well as a tissue, thus it reflects the genome and proteome of a sample.

† *Chemometrics*: according to the definition of the Chemometrics Society, it is the chemical discipline that uses mathematical and statistical methods to design or select optimal procedures and experiments, and to provide maximum chemical information by analyzing chemical data.

‡ Landraces are genotypes with a high capacity to tolerate biotic and abiotic stress, resulting in high yield stability and an intermediate yield level under a low input agricultural system.

superior genotypes, as further described in the first part of this chapter for maize landraces developed and cultured in southern regions of Brazil.

In a second part of this chapter is described the adoption of a typical metabolomic platform, i.e., FTIR and UV-visible spectroscopies coupled to chemometrics, for discriminating propolis samples produced in southern Brazil, a region of huge plant biodiversity. Propolis is typically a complex matrix and has been recognized for its broad pharmacological activities (anti-inflammatory, antibacterial, antifungal, anticancer, and antioxidant, e.g.) since ancient times. Propolis (registration number chemical abstracts service - CAS 9009-62-5) is a beekeeping resinous and complex product, with a variable physical appearance, collected and transformed by honey bees, *Apis mellifera*, from the vegetation they visit. It may be ochre, red, brown, light brown or green; some are friable and steady, while the others are gummy and elastic.

Phenolics such as flavonoids and phenol-carboxylic acids are strategic components in propolis to render it bioactive against several pathogenic microorganisms, for instance as bacteriostatic and/or bactericidal agents. The flora (buds, twigs, bark, and less importantly, flowers) surrounding the hive is the basic source for the phenolics stuff and thus exerts an outstanding importance on the propolis final composition and on its physical, chemical, and biological properties. Although the wax component is an unquestionable supplement provided by the bee secretory apparatus by far less is known about the degree of intensity that these laborious insects play changing all the other chemical constituents collected in the Nature including minor ingredients like essential oils (10%), most of them responsible for the delicate and pleasant odor. All this flora contribution to propolis and the exact wax content may then explain physical properties such as color, taste, texture, melting point, and more importantly, from the health standpoint, a lot of pharmaceutical applications. However, for purpose of industrial applications, the propolis hydroalcoholic extract needs to meet specific composition in order to guarantee any claimed pharmacological activity. One common method used by the industry for quality control is analyzing the propolis sample for the presence of chemical markers known to be present in the specific propolis product they market. Even though this has been the acceptable method for quality control, the presence of the chemical markers do not always guarantee an individual is getting the actual propolis stated by the product label, especially if the product has been spiked with the chemical markers. The quantitation method for the chemical markers will confirm the compounds presence, but it may not confirm the presence of the propolis known to contain the chemical markers. Authentication of the propolis material may be possible by a chemical fingerprint of it and, if possible, of its botanical sources. Thus, chemical fingerprinting, i.e., metabolomics and chemometrics, is an additional method that has been claimed to be included in the quality control process in order to confirm or deny the propolis sample quality being used for manufacturing of a derived product of that resinous and complex matrix. The second part of this chapter aims to demonstrate the possibility of a FTIR and UV-vis metabolomic-chemometrics approach to identify and classify propolis samples originating from nineteen geographic regions (Santa Catarina State, southern Brazil) in different classes, on the basis of the concerted variation in metabolite levels detected by those spectroscopic techniques. Exploratory data analysis and patterns of chemical composition based on, for instance, principal component analysis, as well as discriminating models will be described in order to unravel propolis chemotypes produced in southern Brazil.

2. Maize: metabolomic and chemometric analyses for the study of landraces

Maize (*Zea mays* L.) was chosen as a model for metabolomic analysis because although most of this cereal produced worldwide is used for animal feeding, an important amount is also used in human diet and for industrial purposes, providing raw material for food, pharmaceuticals, and cosmetics production. The maize grain is composed of several chemicals of commercial value and the diversity of its applications depends on the differences in relative chemical composition, e.g. protein, oil, and starch contents, traits that show prominent genetic components (Baye et al., 2006; White, 2001). Over the last centuries, farmers have created thousands of maize varieties suitable for cultivation in numerous environments. Accordingly, it seems consensual that the maize landraces' phenotypes, e.g., morphological and agronomic traits and special chemical characteristics of grains are resultant of the domestication process. Thus, high throughput metabolomic analysis of maize genotypes could improve metabolic singularities knowledge about landraces, helping their characterization and evaluation, and indicating new alternatives for their use. In this context, to distinguish metabolic profiles it is necessary to consider the use of diverse analytical tools, such as spectroscopic and chromatographic techniques for instance. Techniques that are reproducible, stable with time, and do not require complex sample preparation such as infrared vibrational spectroscopy and nuclear magnetic resonance spectroscopy are desirable for metabolic profiling.

2.1 Metabolic profiling of maize landraces through FTIR-PCA – integral and degermed flours

Vibrational spectroscopy, and particularly Fourier transform infrared spectroscopy (FTIR) is thought to be interesting as one aims at discriminating and classifying maize landraces according to their chemical traits. FTIR is a physicochemical method that measures the vibrations of bonds within functional groups and generates a spectrum that can be regarded as a metabolic fingerprint. It is a flexible method that can quickly provide qualitative and quantitative information with minimal or no sample preparation of complex biological matrices (Ferreira et al., 2001). By other hand, a FTIR spectrum is complex, containing many variables per sample and making visual analysis very difficult. Hence, to extract useful information from the whole spectra, multivariate data analysis is needed, particularly through the determination of the principal components (PCA - Fukusaki & Kobayashi, 2005). Such a multivariate analysis technique could allow the characterization of the sample relationships (scores plans or axis) and the recovery of their subspectral profiles (loadings). This approach was applied to classify flour samples from whole (integral) and degermed maize grains of twenty-six landraces developed and cultivated by small farmers in the far-west region of Santa Catarina State, southern Brazil (Anchieta County - 26°31′11″S, 53°20′26″W).

Previously to multivariate analysis, FTIR spectra were normalized, baseline-corrected in the region of interest by drawing a straight line before resolution enhancement (k factor of 1.7) was applied using Fourier self deconvolution (Opus v. 5.0, Bruker Biospin, GmbH, Rheinstetten, Germany). Chemometric analysis used normalized, baseline-corrected (3000–600 cm^{-1}. 1700 data points) and deconvoluted spectra, which were transferred via a JCAMP format (OPUS v. 5.0, Bruker Biospin GmbH, Rheinstetten, Germany) into the data analysis software for PCA (The Unscramble v. 9.1, CAMO Software Inc., Woodbridge, USA).

Previously to PCA analysis each spectrum within the (3000–600 cm^{-1}) region was standard normal deviates corrected.

Figure 1 shows a PCA scores scatter plot for flour samples from whole and degermed grains using the whole FTIR spectral window data set (3000–600 cm^{-1}). The scores scatter plot (PC1 vs. PC2) that contains 93% of the data set variability shows a clear discrimination among flour samples of whole and degermed grains.

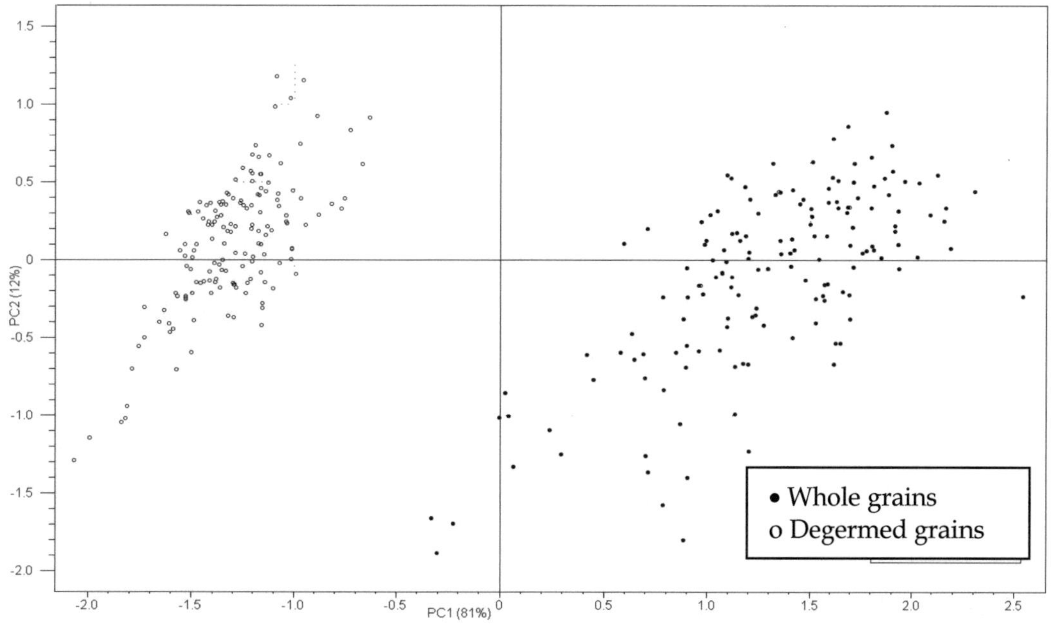

Fig. 1. Principal component analysis scores scatter plot of the FTIR data set in the spectral window of 3000–600 cm^{-1} wavenumber of landrace maize flours of whole and degermed grain cultivated in the southern Brazil.

The samples of whole grains grouped in PC1+ axis seemed to be more discrepant in their chemical composition, appearing more scattered through the quadrants of the PCA representation. Figure 2 shows the loadings plot of PC1, revealing the most important wavenumbers which explain the distinction of the samples previously found (scores scatter plot). The loadings indicated a prominent effect of the lipid components (2924, 2850, and 1743 cm^{-1}) for the segregation observed. The two major structures of the grains are the endosperm and the germ (embryo) that constitute approximately 80 and 10% of the mature kernel dry weight, respectively. The endosperm is largely starch (approaching 90%) and the germ contains high levels of oil (30%) and protein (18% - Boyer & Hannah, 2001).

The greatest chemical diversity observed in whole grains can be explained by genetic variation of embryos resulting from sexual reproduction. Some authors suggest that the high level of genetic and epigenetic diversity observed in maize could be responsible for its great adaptation capacity to a wide range of ecological factors. Lemos (2010) analyzing the metabolic profile of maize landraces' leaf tissues from Anchieta County (southern Brazil) found a prominent chemical variability among individuals of same variety, although inter-variety variability has also been observed.

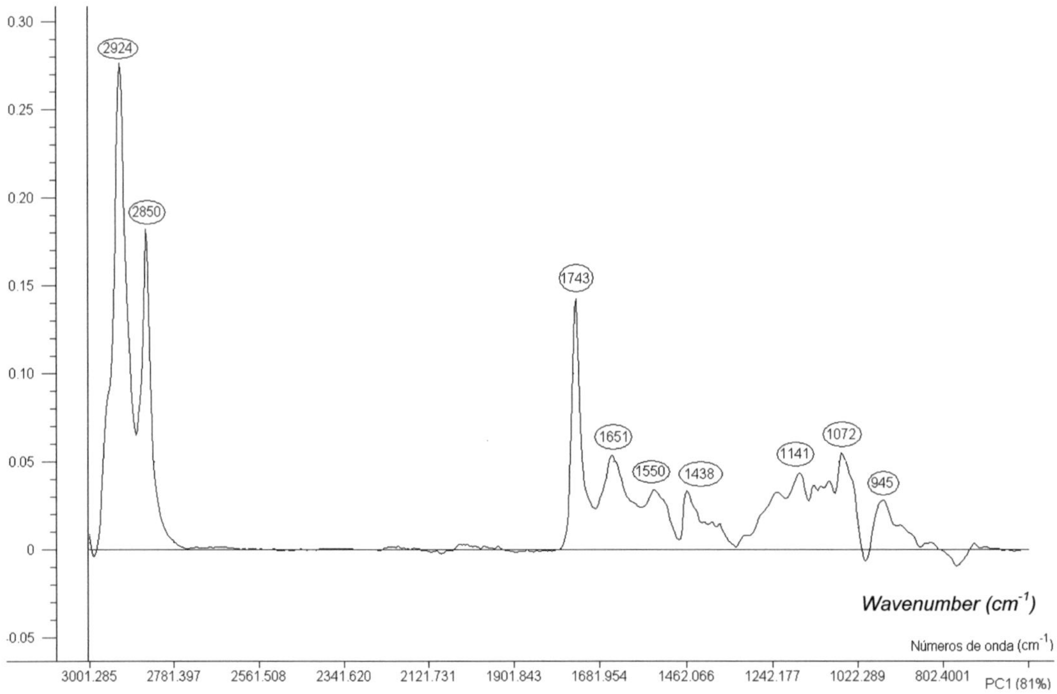

Fig. 2. PC1 loadings plot of the FTIR spectra of maize flours of whole and degermed grains in the 3000–600 cm^{-1} wavenumber region.

2.2 Starch recognition pattern of maize landraces by NMR spectroscopy and PCA

The composition of maize grains can be heterogeneous for both the quantity and quality of compounds from endosperm as starch, protein, and oil. In this context, metabolomics coupled to chemometrics approach was successfully applied to the discrimination of starches from the studied twenty-six maize landraces. The starches were extracted from flours with distilled water (1: 70, w/v) under reflux (80ºC, 1 h), precipitated with ethyl alcohol (12 h, 4ºC), and oven-dried (55ºC until constant weight). Samples (50 mg) were dissolved in DMSO-δ_6 (0.6 mL) and ^{1}H-NMR spectra obtained under standard conditions. Sodium-3-trimethylsilylpropionate (TMSP-2, 2, 3, 3-d_4) was used as internal reference (δ_{ppm} 0.0). Spectra were processed using 32768 data points by applying an exponential line broadening of 0.3 Hz for sensitivity enhancement before Fourier transformation and were accurately phased, baseline adjusted, and converted into JCAMP format to build the data matrix. All calculations were carried out using the Pirouette software (v. 3.11, InfoMetrix, Woodinville, Washington, USA). The PCA analysis of the whole ^{1}H-NMR data set (32.000

data points) was performed including the spectra of amylose and amylopectin standards. The chemical structures and the purity of the standards of amylose and amylopectin were confirmed by ^{13}C-NMR spectroscopy. The PC1 (32%) vs. PC2 (28%) scores scatter plot allowed a clear segregation of the amylopectin standard and a discrimination for the maize flours samples in two groups (Fig. 3) by PC1 axis.

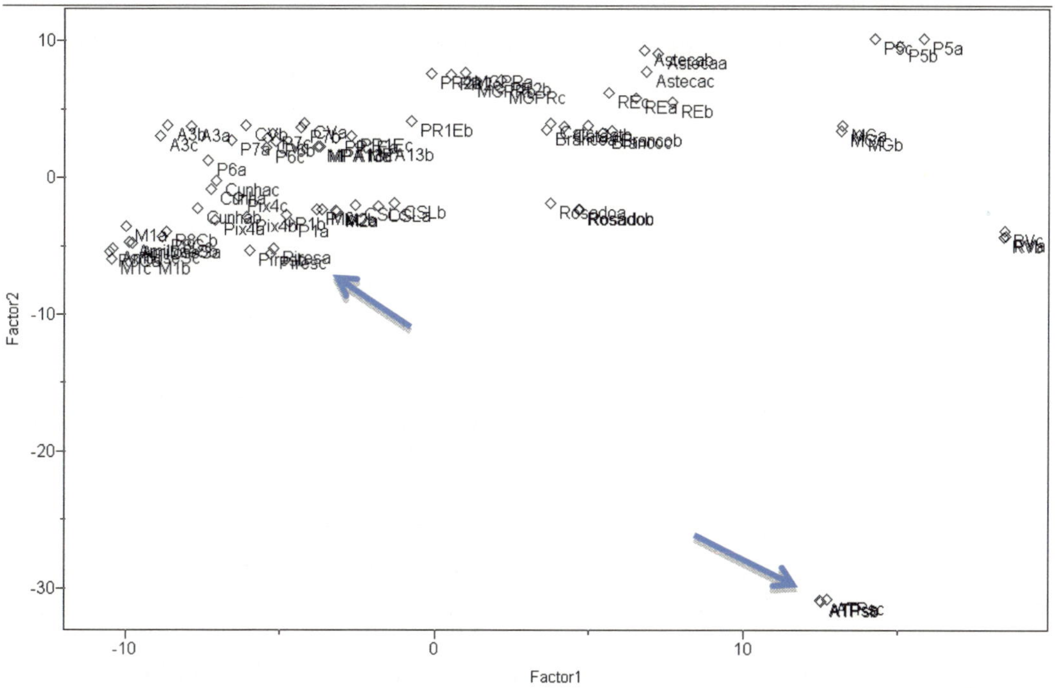

Fig. 3. Classification plot of starch fractions of maize landraces for ^{1}H-NMR after PCA analysis. The arrows shows the amylose and amylopectin stardands.

The Roxo 41 variety located closer to the amylopectin standard suggesting the predominance of that polysaccharide in relation to amylose in its starchy fraction. This result is in accordance to PCA analysis of the IR data set of the fingerprint region of carbohydrates that diagnosed the starch granules from Roxo 41 with superior amylopectin amount in respect to their amylose content (data not shown). By other hand, the amylose standard was grouped with MPA1 and Rajado 8 Carreiras varieties, suggesting that the starch granules contain superior amount of that polysaccharide in its starchy fraction.

3. Propolis: ATR-FTIR and UV-visible spectrophotometry coupled to chemometrics as an analytical approach for pattern recognition

3.1 Attenuated total reflectance-Fourier transform infrared spectroscopy

Propolis (registration number chemical abstracts service - CAS 9009-62-5) is a sticky colored material, which honeybees collect from different plants exudates and modify in its hypopharyngeal glands, being used in the hive to fill gaps and to protect against invaders as insects and microorganisms. Raw propolis usually contains 50% resin and balsam, 30% wax,

10% aromatic oils, 5% pollen, and 5% other substances as inorganic salts and amino acids. This resin has been used by humanity since ancient civilizations like Egyptian, Assyrian, Greek, Roman, and Inca. In these days, a number of studies have confirmed important biological activities such as antibacterial, antifungal, antiviral, antioxidant, anti-inflammatory, hepatoprotective, and antitumoral (for review see Bankova, 2009; Banksota et al., 2001; Castaldo & Capasso, 2002).

The aspect, texture and the chemical composition of propolis is highly variable and depends on the climate, season, bee species and mainly the local flora which is visited by bees to collect resin (Markham et al., 1996). For this reason, comparing propolis samples from distinct regions might be the same as to compare extracts of two plants that belong to different taxonomical families (Bankova, 2005).

Propolis from Europe is the best known type of propolis. In European regions with temperate climate bees obtain resin mainly from the buds of *Poplus* species and the main bioactive components are flavonoids (Greenaway et al., 1990). In tropical countries, the botanical resources are much more variable in respect to temperate zones, so bees find much more possibilities of collecting resins and hence the chemical composition of tropical propolis are more variable and distinct from European ones (Sawaya et al., 2011). Different compounds have been reported in tropical propolis such as terpenoids and prenylated derivatives of *p*-coumaric acids in Brazilian propolis (Marcucci, 1995), lignans in Chilean samples (Valcic et al., 1998), and polyisoprenylated benzophenones in Venezuelan, Brazilian, and Cuban propolis (Cuesta-Rubio et al., 1999; Marcucci, 1995).

In order to be accepted officially into the main stream of the healthcare system and for industrial applications, propolis needs chemical standardization that guarantees its quality, safety, efficacy, and provenance. The chemical diversity mainly caused by the botanical origin makes the standardization difficult. Since the chemistry and biological activity of propolis depends on its geographical origin, a proper method to discriminate its origin is needed (Bankova, 2005).

Chromatographic methods (HPLC, TLC, GC, e.g.) are largely used to identification and quantification of propolis compounds, but it its becoming clear that to separate and evaluate all constituents of propolis is an almost impossible task (Sarbu & Mot, 2011). Even thought the presence of the chemical markers are considered an acceptable method for quality control, not always is guarantee about what is stated by the product label, especially if the product has been spiked with the chemical markers. Besides, literature has demonstrated that is not possible to ascribe the pharmacological activity solely to a unique compound and until now no single propolis component has shown to possess anti-bacterial activity higher than total extract (Kujumgiev et al., 1999; Popova et al., 2004). Thus, a possibility is offered by the fingerprinting methods that can analyze in a non-selective way the propolis samples as a whole.

Poplar propolis, for example, can be distinguished by UV-visible spectrophotometric determination of all three important components (flavones and flavonols, flavonones and dihydroflavonols, and total phenolics - Popova et al., 2004), but some constraints regarding such an analytical approach has been claimed for propolis from tropical regions (Bankova & Marcucci, 2000).

The search for faster screening methods capable of characterizing propolis samples of different geographic origins and composition has lead to the use of direct insertion mass spectrometric fingerprinting techniques (ESI-MS and EASI-MS), which has proven to be a fast and robust method for propolis characterization (Sawaya et al., 2011), although this analytical approach can only detect compounds that ionize under the experimental conditions. Similarly, Fourier transform infrared vibrational spectroscopy (FTIR) has also demonstrated to be valuable to chemically characterize complex matrices such as propolis (Wu et al., 2008).

In order to achieve the goal of treat propolis sample as a whole than just be focused only in marker compounds, chemometric methods are being considered an important tool to analyze the huge data sets generated by non-selective analytical techniques such as UV-vis, MS, NMR, and FT-IR, generating information not only about chemical composition of propolis but also discriminating its geographical origin.

Authentication of propolis material may be possible by a chemical fingerprint of it and, if possible, of its botanical sources. Thus, chemical fingerprinting, i.e., metabolomics and chemometrics, is an additional method that has been claimed to be included as a quality control method in order to confirm or deny the propolis sample being used for the manufacturing of a derived product of that resinous and complex matrix.

Over the last decades, infrared (IR) vibrational spectroscopy has been well established as a useful tool for structure elucidation and quality control in several industrial applications. Indeed, the development of Fourier transform (FT) IR and attenuated total reflectance (ATR) techniques have also evolved allowing rapid IR measurements of organosolvent extracts of plant tissues, edible oils, and essential oils, for example (Damm et al., 2005; Lai et al., 1994; Schulz & Baranska, 2007). In consequence of the strong dipole moment of water, IR spectroscopy applications have mostly focused on the analysis of dried or non-aqueous plant matrices and currently IR methods are widely used as a fast analytical technique for the authentication and detection of adulteration of vegetable oils.

ATR-FTIR spectroscopy was applied to propolis samples collected in the autumn-2010 and originated from nineteen geographic regions of Santa Catarina State (southern Brazil) in order to gain insights as to the chemical profile of those complex matrices. FTIR spectroscopy measures the vibrations of bonds within functional groups and generates a spectrum that can be regarded as a metabolic fingerprint. Similar IR spectral profiles (3000 - 600 cm^{-1}, figure 4) were found by a preliminary visual analysis for purpose of an exploratory overview of data, revealing typical signals of e.g., lipids (2910 – 2845 cm^{-1}), monoterpenes (1732, 1592, 1114, 1022, 972 cm^{-1}), sesquiterpenes (1472 cm^{-1}), and sucrose (1122 cm^{-1} - Schulz & Baranska, 2007) for all the studied samples. However, we were not able to identify by visual inspection of the spectra a clear picture regarding a discriminating effect of any primary or secondary metabolites among the propolis samples.

A FTIR spectrum is complex, containing many variables per sample and making visual analysis very difficult. Hence, to extract extra useful information, i.e., latent variables, from the whole spectra chemometric analysis was performed considering the whole FTIR data set using principal components analysis (PCA) for an exploratory overview of data. This method could reveal similarity/dissimilarity patterns among propolis samples, simplifying

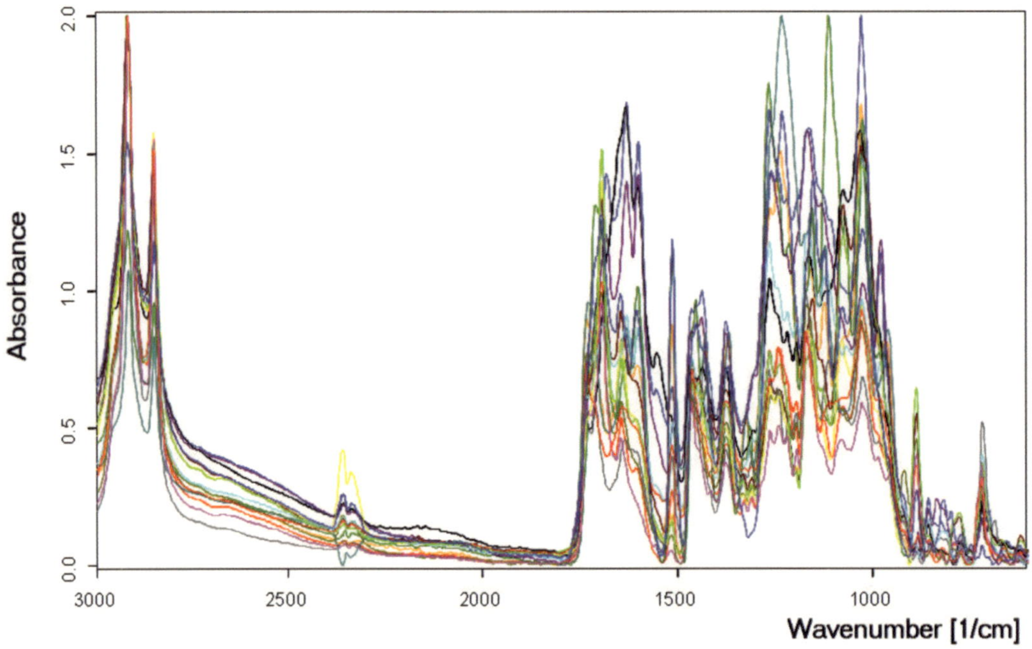

Fig. 4. IR spectral profile of propolis samples (autumn, 2010) produced in southern Brazil, according to the geographic regions of origin in Santa Catarina State. IR spectra are shown from top to bottom following the geographic precedence, i.e. 19 counties, of the propolis samples: Angelina *(**ANG**)*, Balneário Gaivotas *(**BG**)*, Bom Retiro *(**BR_1 and BR_2**)*, Caçador *(**Cç**)*, Campo-Erê *(**CE**)*, Canoinhas *(**CA**)*, Campos Novos *(**CN**)*, Descanso *(**DS**)*, José Boiteux *(**JB**)*, Porto União *(**PU**)*, Serra Alta *(**SA**)*, São Joaquim *(**SJ_1** and **SJ_2**)*, São José do Cerrito *(**SJC**)*, Urupema *(**URU**)*, Vidal Ramos *(**VR**)*, Florianópolis *(**FLN**)*, and Xaxim *(**XX**)*.

data dimensions and results interpretation, without missing the more relevant information associated to them (Fukusaki & Kobayashi, 2005; Leardi, 2003). The covariance was choose for matrix construction in PCA calculation, since all variables considered were expressed in the same unit. By doing so, the magnitude differences were maintained, i.e., data were not standardized and variables contribution to samples distribution along axes was direct proportional to their magnitude. For the purpose of the propolis chemical profile analysis this kind of information is thought to be very useful, because wavenumber with higher absorbances (higher metabolites concentration) contribute more significantly with objects distribution into PCA, introducing quantitative information beside the compositional information of the sample data.

The principal component analysis (PCA) of the whole spectral data (3000 - 600 cm^{-1}, 1700 data points) revealed that PC1 and PC2 defined 88% of the variability from the original IR data and a peculiar pattern of lipids (2914 cm^{-1} and 2848 cm^{-1} - C-H stretching vibrations) for the samples from the northern region (*NR*) of Santa Catarina State. The climate in that region is typically mesothermic, humid subtropical with a mild summer and an annual temperature average of 17.2ºC - 19.4ºC. On the other hand, the propolis produced in the highlands (1360m altitude, annual maximum and minimum temperatures average of 18.9ºC and 9.3ºC, respectively) were discrepant regarding their monoterpene (1114 cm^{-1} and 972 cm^{-1} - ϖ-CH2) and sesquiterpene (1472 cm^{-1} - δ CH_2) compounds (Schulz & Baranska, 2007) - Figure 5. In despite of NR_1 *and* NR_2 propolis samples have grouped in PC1- and PC2+, they differ somewhat in respect to their chemical composition, an effect attributed to the flora composition found in those regions, e.g., mostly Atlantic Rainforest in NR_1 as NR_2 shows extensive areas covered by artificial reforestations i.e., *Eucalyptus* spp and *Pinus* spp, furnishing distinct raw materials for propolis production.

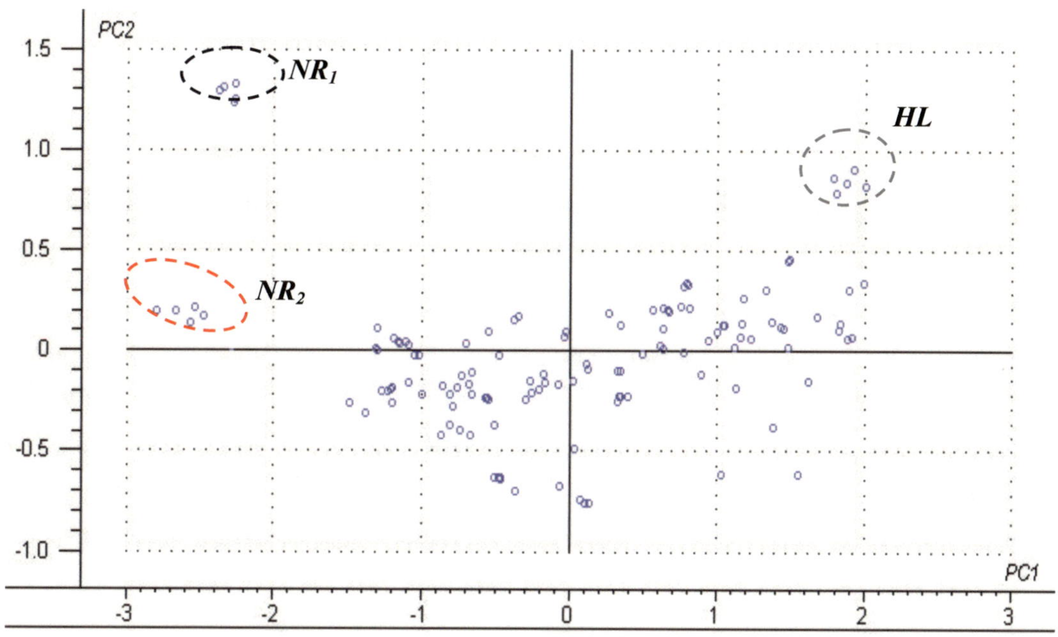

Fig. 5. Principal component analysis scores scatter plot of the FTIR data set in the spectral window of 3000–600 cm^{-1} wavenumber (1700 data points) of propolis samples produced in the southern Brazil (Santa Catarina State). NR_1, NR_2 and *HL* refer to propolis samples originated from northern and highland regions of Santa Catarina State. The calculations were carried out using The Unscrambler software (v. 9.1, Oslo - Norway). PC1 and PC2 accounts for 88% of the variance preserved.

Further chemometric analysis took into consideration the fact that propolis is a very well known source of phenolic compounds, e.g., phenolics acids and flavonoids. Indeed, phenolic compounds occur ubiquitously in most plant species and take part of the chemical constitution of propolis worldwide. IR spectroscopy allows to identify phenolic compounds since they demonstrate strong IR bands due to C-H wagging vibrations between 1260 - 1180 cm^{-1} and 900 - 820 cm^{-1} (Schulz & Baranska, 2007). The principal component calculations were performed for both 1260 - 1180 cm^{-1} and 900 - 820 cm^{-1} spectral windows and PC1 and PC2 resolved about 96% of the spectral data variability. An interesting discrimination profile was detected where samples from the far-west (*FW*) region grouped distinctly in respect to northern (*NR_1*) ones, which also differed from the highlands (*HL*) propolis samples. Such findings can be explained in any extension based on the flora composition of the studied geographic regions. In the northern and far-west regions of Santa Catarina State the Atlantic Rainforest is typically found, but the floristic composition varies according to the altitude, e.g., 240 m altitude - *NR_1* and 830 m - *FW*. Besides, as a mesothermic humid subtropical climate is found in *NR_1*, the *FW* region is characterized by a temperate climate that determines a discrepant composition of plant species. Finally, the *HL* region (1360m altitude, temperate climate) is covered by the Araucaria Forest, where parana pine (*Araucaria angustifolia, Gymnospermae, Araucariaceae*) occurs as a dominant plant species. *A. angustifolia* produces a resinous exudate rich in guaiacyl type lignans, fatty acids, sterols (Anderegg & Rowe, 2009), phenolics, and terpenic acids that is thought to be used by honey bee (*Apis mellifera*) for the propolis production. Since the plant species populations influence the propolis chemical composition, the discrimination profile detected by ATR-FTIR coupled to chemometrics seems to be an interesting analytical approach to gain insights as to the effect of the climatic factors and floristic composition on the chemical traits of that complex matrix.

3.2 Ultraviolet-visible scanning spectrophotometry

Combination of UV-visible spectrophotometric wavelength scans and chemometric (PCA) analysis seems to be a simple and fast way to prospect plant extracts. This analytical strategy revealed to be fruitful for discrimination of *habanero* peppers according to their content of capsaicinoids, substances responsible for the pungency of their fruits (Davis et al., 2007).

Chemometric analysis was performed considering the absorbance values of the total UV-visible data set (200 ηm to 700 ηm, 450 data points) for the propolis samples in study, by using principal components analysis (PCA) for an exploratory overview of data.

In a first approach, principal components analysis (PCA) was tested by both correlation and covariance matrices of calculations. If correlation is used, the data set is standardized (mean-centered and columns scaled to the unit of variance), decreasing the effect of differences in magnitude between variables and leading to a distribution of objects (*eigenvalues*) with equal influence from all variables. On the other hand, if covariance is used, data is only mean-centered; retaining its original scale. The resulting distribution is then determined either by composition and magnitude of variables, leading to a PCA representation more influenced by larger observed values (Manetti et al., 2004). A similar distribution of objects was found for both correlation and covariance matrices in PC calculations, as PC1 and PC2 resolved 91.2% and 96.3%, respectively of the variability of the spectrophotometric data set. Thus, the covariance matrix was chosen for PCA calculations, since all variables considered were expressed in the same unit. By doing so, the magnitude differences were maintained, i.e., data were not standardized and variables contribution to samples distribution along axes was direct proportional to their magnitude. For the purpose of the chemical profile analysis of the propolis samples this kind of information is thought to be very useful, because wavelengths with higher absorbances (higher metabolites concentration) contribute more significantly with objects distribution into PCA, introducing quantitative information beside the compositional information of the sample data.

Principal component analysis was performed using The Unscrambler software (v. 9.1, Oslo - Norway) and revealed mostly a distribution of the propolis samples along the PC1 axis (91% sample total variability), as PC2 (5% sample total variability) seemed to be lesser discriminator of the objects. A clear separation of the samples according to the east-west axis of Santa Catarina State could be found, where propolis produced near coastal regions (*CR_1 and CR_2*) grouped in PC1+/PC2-, as the sample from the far-west region *(FW)* was detected in the opposite side of PC1 axis, along with the samples from the northern region (*PU, Cç,* and *CA* - Figure 6). Interestingly, propolis samples from the counties *SJ, URU, BR* (highlands counties), and *ANG,* which shown geographic proximities and a certain common floral composition, seemed to be similar in their chemical profiles as determined by UV-visible scanning spectrophotometry, grouping in PC1+/PC2+.

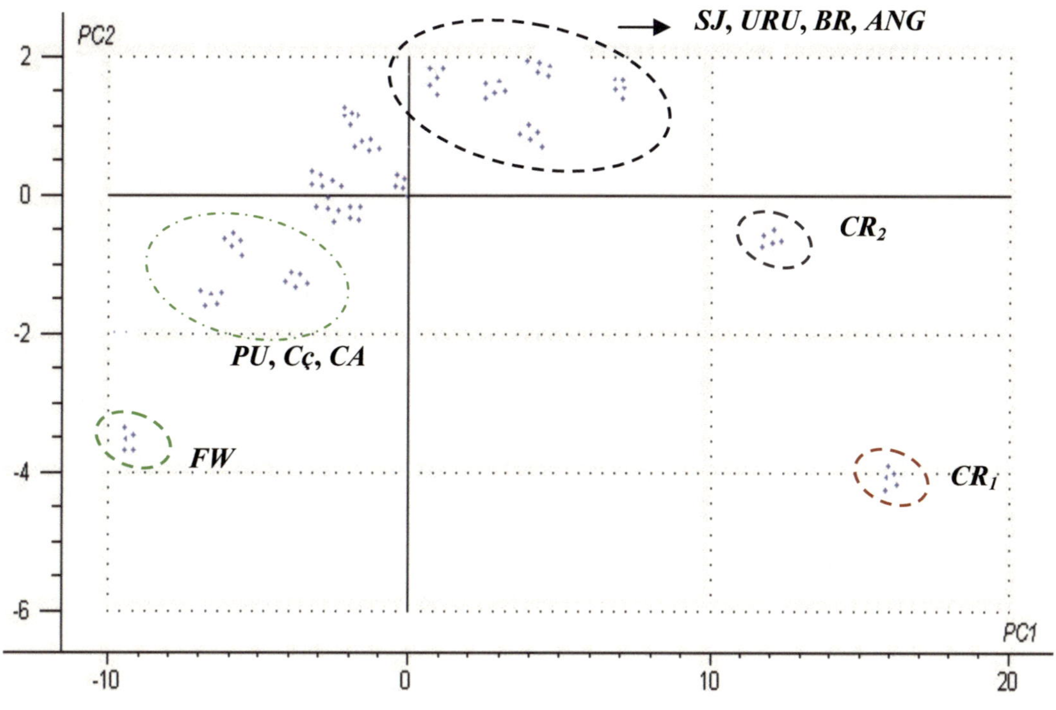

Fig. 6. Principal component analysis scores scatter plot of the UV-visible data set in the spectral window of 200 ηm to 700 ηm (450 data points) of propolis samples produced in the southern Brazil (Santa Catarina State). *CR_1, CR_2,* and *FW* refer to propolis samples originated from coastal (*BG* and *FLN* Counties) and far-west (*CE* County) regions, respectively, of Santa Catarina State. The sample grouping of propolis with similar UV-visivel scanning profiles regarding their (poly)phenolic composition is detached in the PC1+/PC2+ quadrant. PC1 and PC2 resolved 96% of the total variability of the spectral data set.

High loadings associated to the wavelengths 394 ηm, 360 ηm, 440 ηm, and 310 ηm seemed to influence the observed distribution of the propolis samples and could be associated to the presence of (poly)phenolic compounds. In fact, the λ_{max} for the cinnamic acid and its derivatives is near 310-320 ηm as for the flavonols is usually around 360 ηm (Tsao & Deng, 2004). Further chemical analysis of the total content of phenolics ad flavonoids in the propolis originated from the counties *SJ*, *URU*, *BR*, and *ANG* revealed similar contents, with average concentrations of 1411.52 μg/ml and 4.61 μg/ml of those secondary metabolites, respectively, in the hydroalcoholic (70: 30, v/v) extract. Such findings differed ($P<0.05$ - *Tukey* test) in respect to the concentrations detected for the propolis samples produced in the coastal (793.67 μg/ml - total phenolics and 2.82 μg/ml - flavonoids) and far-west (952.97 μg/mL - total phenolics and 0.59 μg/ml flavonoids) regions of Santa Catarina State, corroborating the PCA results herein shown.

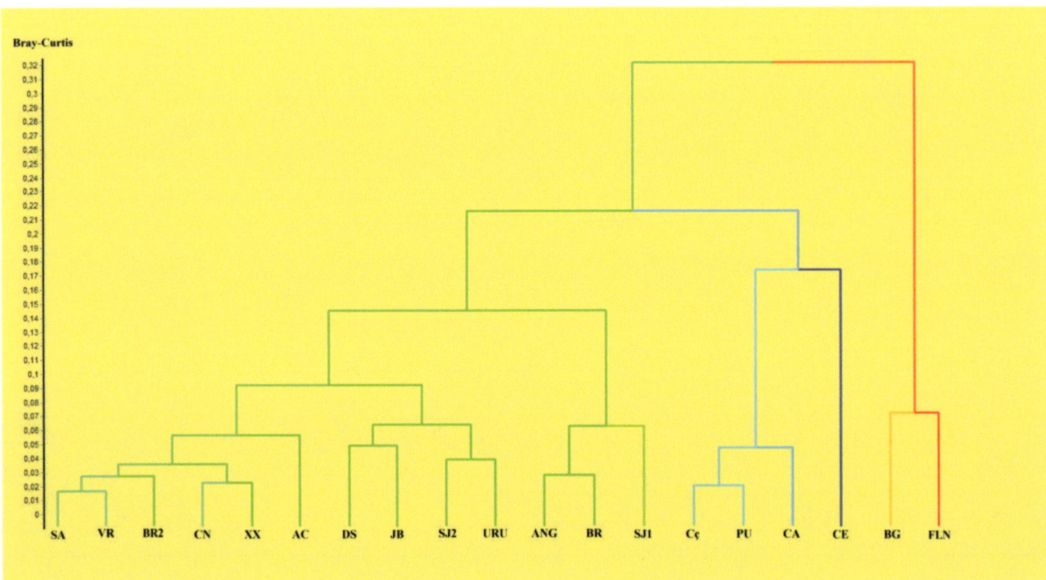

Fig. 7. Dendrogram of propolis samples using average linkage with Bray-Curtis dissimilarity measure. Data calculations were based on the absorbance values for the UV-visible spectral window of 200 ηm to 700 ηm of propolis samples produced in Santa Catarina State - southern Brazil, autumn-2010.

In order to check the chemical similarity pattern of propolis samples detected by PCA, further cluster analysis of the whole absorbance UV-vis data set, i.e., absorbance values of 200 ηm to 700 ηm (450 data points), was performed by using the Unweighted Pair Group Method with Arithmatic Mean (UPGMA) based on Bray-Curtis dissimilarity coefficient. UPGMA is a simple agglomerative or hierachical clustering method used in for the creation of phonetic trees, i.e., phenograms, hierarchical trees or dendrograms that indicate the similarity degree among samples/objects of interest, so that observations in the same cluster are similar in some sense. In UPGMA method after two objects with the least dissimilarity fuse together an arithmetic average of the dissimilarity of this new cluster and the rest of the objects is calculated. This leads to a reduction in the size of the original dissimilarity matrix. The procedure continues with the dissimilarity matrix being correspondingly reduced. When the average between an object and a cluster is calculated, the method gives equal weights to the members of the clusters when averaging, i.e., unweighted. Thus, in the progressive reduction of the dissimilarity matrix, only relationships between groups are considered, which are given equal weighting and this leads to loss of information about the relationships between pairs of objects (Legendre, 1998; Singh, 2008).

The hierarchical tree of the similarity of chemical profiles of the propolis samples is shown in figure 7. The findings suggest a resemblance of grouping as found by PCA calculations in respect to the *SJ*, *URU*, *BR*, and *ANG* samples, as well as for the propolis originated from the coastal (*BG* and *FLN*) and northern regions (*CA*, *PU*, and *Cç*). Additionally, UPGMA analysis also discriminate the propolis produced in the western (*AC*, *XX*, and *CN*) and far-west regions.

4. Conclusions

The chemo(bio)diversity analysis of maize landraces and propolis produced in southern regions of Brazil was successfully assessed by using a typical metabolomic platform involving spectroscopic techniques (FTIR, ^{1}H- and ^{13}C-NMR, and UV-visible) and chemometrics. The huge amount of data afforded by those spectroscopic techniques was analyzed using multivariate statistical methods such as principal component analysis and cluster analysis allowing obtaining extra information on the metabolic profile of the complex matrices in study.

The analytical approach described showed to be suitable when ones aim to discriminate maize flour samples from whole and degermed maize, an issue thought to be important for the food, cosmetic, and pharmaceutical industries regarding the usage and quality control process of that raw material. Similarly, the classification of maize landraces according to their starch traits is considered technologically relevant in order to optimize the usage of non-chemically modified starches in industrial process, for instance.

The classification of Brazilian propolis as to their chemical profiles and geographic regions seems to be relevant because that biomass is typically quite complex, making difficult and expensive to perform a complete characterization in that sense. By doing so, the propolis produced in southern Brazil might be better evaluated as to their potential usage in cosmetic and pharmaceutical industry, taking into consideration their secondary metabolite constituents, e.g., mono/sesquiterpenes and phenolics. The coupling of chemometrics-spectroscopic techniques used is thought to be essential to allow detecting peculiar chemical traits of the propolis samples according to their geographic regions in a simple and fast way.

5. Acknowledgment

Authors are indebt to FAPESC, CNPq, and CAPES for financial support and fellowships.

6. References

Anderegg, RJ & Rowe, JW. (2009). Lignans, the major component of resin from Araucaria angustifolia knots. *International Journal of the Biology, Chemistry, Physics and Technology of Wood,* Vol. 28, pp.171–175. ISSN 0018-3830

Bankova, V. (2005). Chemical diversity of propolis and the problem of standardization. *Journal of Ethnopharmacology,* Vol. 100, pp.114-117. ISSN: 0378-8741

Bankova, V & Marcucci, MC. (2000). Standardization of propolis: present status and perspectives. *Bee World,* Vol. 81, pp.182-188. ISSN: 0005-772X

Banksota, AH., Tezuka, Y & Kadota, S. (2001). Recent progress in pharmacological research of propolis. *Phytotherapy Research,* Vol. 15, pp.561-571. ISSN: 1099-1573

Baye, TM., Pearson, TC & Settles, AM. (2006). Development of a calibration to predict maize seed composition using single kernel near infrared spectroscopy. *Journal of Cereal Science,* Vol. 43, pp.236–243. ISSN: 0733-5210

Boyer, CD & Hannah, C. (2001). Kernel mutants of corn. In: Specialty corns. HALLAUER, AR. (Ed.). 2nd ed. pp. 153, CRC Press, London.

Castaldo, S & Capasso, F. (2002). Propolis, an old remedy used in modern medicine. *Fitoterapia,* Vol. 73, pp.S1-S6. ISSN: 0367-326X

Cuesta-Rubio, O., Cuellar, AC., Rojas, N., Velez, HC., Rastrelli, L & Aquino, R. (1999). A polyisoprenylated benzophenone from Cuban propolis. *Journal of Natural Products,* Vol. 62, pp.1013-1015. ISSN: 0974-5211

Damm, U., Lampen, P., Heise, HM., Davies, AN & Mcintyre, PS. (2005). Spectral variable selection for partial least squares calibration applied to authentication and quantification of extra virgin olive oils using Fourier transform Raman spectroscopy. *Applied Spectroscopy,* Vol. 59, pp.1286-1294. ISSN: 0003-7028

Davis, CB., Markey, CE., Busch, MA & Busch, KW. (2007). Determination of capsaicinoids in habanero peppers by chemometric analysis of UV spectral data. *Journal of Agricultural and Food Chemistry,* Vol. 55, pp. 5925-5933. ISSN: 0021-8561

Ferreira, D., Barros, A., Coimbra, MA & Delgadillo, I. (2001). Use of FTIR spectroscopy to follow the effect of processing in cell wall polysaccharide extracts of a sun-dried pear. *Carbohydrate Polymers,* Vol. 45, pp.175–182. ISSN: 0144-8617

Fukusaki, E & Kobayashi, A. (2005). Plant metabolomics: potential for practical operation. *Journal of Bioscience and Bioengineering,* Vol. 100, pp.347–354. ISSN: 1389-1723

Greenaway, W., Scaysbrook, T & Whately, FR. (1990). The composition and plant origin of propolis: a report of work at Oxford. *Bee World,* Vol. 71, pp. 107-118. ISSN: 0005-772X

Kujumgiev, A., Tsvetkova, I., Serkedjieva, YU., Bankova, V., Christov, R & Popov, S. (1999). Antibacterial, antifungal and antiviral activity of propolis of different geographical origins. *Journal of Ethnopharmacology,* Vol. 64, pp. 235-240. ISSN: 0378-8741

Lai, YW., Kemsley, EK & Wilson, J. (1994). Potential of Fourier transform infrared spectroscopy for the authentication of vegetable oils. *Journal of Agricultural and Food Chemistry,* Vol. 42, pp.1154-1159. ISSN: 0021-8561

Leardi, R. (2003). Chemometrics in data analysis. In: A user-friendly guide to multivariate calibration and classification. Naes, T., Isaksson, T., Fearn, T & Davies, T (eds). NIR Publications, West Sussex.

Legendre, P. (1998). *Numerical Ecology*. Elsevier Science, New York..

Lemos PMM (2010). Análise do metaboloma foliar parcial de variedades locais de milho (*Zea mays* L.) e dos efeitos anti-tumoral *in vitro* e na morfogênese embrionária de Gallus domesticus. PhD thesis, Federal University of Santa Catarina, Brazil.

Manetti, C., Bianchetti, C., Bizarri, M., Casciani, L., Castro, C., D´Ascenzo, G., Delfini, M., DI Cocco, ME., Laganà, A., Miccheli, A., Motto, M & Conti, F. (2004). NMR-based metabonomic study of transgenic maize. *Phytochemistry,* Vol. 65, pp.3187-3198. ISSN: 0031-9422

Marcucci, MC. (1995). Propolis: chemical composition, biological properties and therapeutic activity. *Apidologie,* Vol. 26, pp.83-99. ISSN: 1297-9678

Markham, KR., Mitchell, KA., Wilkins, AL., Daldy, JA & Lu, Y. (1996). HPLC and CG-MS identification of the major organic constituents in New Zealand propolis. *Phytochemistry* Vol. 42, pp.205-211. ISSN: 0031-9422

Popova, M., Bankova, V., Butovska, D., Petkov, V., Damynova, BN., Sabatini, AG., Marcazzan, GL & Bogdanov, S (2004). Validated methods for the quantifications of biologically active constituents of poplar-type propolis. *Phytochemical Analysis,* Vol. 15, pp.235-240. ISSN: 1099-1565

Sarbu, C & Mot, AC. (2011). Ecosystem discrimination and fingerprinting of Romain propolis by hierarchical fuzzy clustering and image analysis of TLC patterns. *Talanta,* Vol. 85, pp.1112-1117. ISSN: 0039-9140

Sawaya, ACHF., Silva, IB & Marcucci, MC. (2011). Analytical methods applied to diverse types of Brazilian propolis. *Chemistry Central Journal,* Vol. 5, pp.1-10. ISSN: 1752-153X

Schulz, H & Baranska, M. (2007). Identification and quantification of valuable plant substances by IR and Raman spectroscopy. *Vibrational Spectroscopy,* Vol. 43, pp.13-25. ISSN: 0924-2031

Singh, W. (2008). Robustness of three hierarchical agglomerative clustering techniques for ecological data. M.Sc. thesis, University of Iceland, Iceland.

Tsao, R & Deng, Z. (2004). Separation procedures for naturally occurring antioxidant phytochemicals. *Journal of Chromatography B,* Vol. 812, pp.85-99. ISSN: 1570-0232

Valcic, S., Montenegro, G & Timmermann, BN. (1998). Lignans from Chilean propolis. *Journal of Natural Products,* Vol. 61, pp.771-775. ISSN: 0974-5211

White, PJ. (2001). Properties of corn starch. In: Specialty corns. HALLAUER, AR. (Ed.). 2nd ed. pp. 189, CRC Press, London.

Wu, YW., Sun, SQ., Zhao, Y., Li, Q & Zhou, J. (2008). Rapid discrimination of extracts of Chinese propolis and poplar buds by FT-IR and 2D IR correlation spectroscopy. *Journal of Molecular Structure,* Vol. 884, pp.48-54. ISSN: 0022-2860

12

Using Principal Component Scores and Artificial Neural Networks in Predicting Water Quality Index

Rashid Atta Khan[2], Sharifuddin M. Zain[2], Hafizan Juahir[1], Mohd Kamil Yusoff[1] and Tg Hanidza T.I.[1]
[1]*Department of Environmental Science, Faculty of Environmental Study, University Putra Malaysia, Serdang*
[2]*Chemistry Department, Faculty of Science, University of Malaya, Kuala Lumpur Malaysia*

1. Introduction

The management of river water quality is a major environmental challenge. One of the major challenges is in determining point and non-point sources of pollutants. Industrial and municipal wastewater discharges can be considered as constant polluting sources, unlike surface water runoff which is seasonal and highly affected by climate. According to Aiken et al. (1982), 42 tributaries in Peninsular Malaysia are categorized as very polluted including the Langat River. Until 1999, there were about 13 polluted tributaries and 36 polluted rivers due to human activities such as, industry, construction and agriculture (Department of Environment, Malaysia (DOE), 1999). In 1990, there were 48 clean rivers classified as clean but the number is reduced to 32 rivers in 1999 (Rosnani Ibrahim, 2001).

Surface water pollution is identified as the major problem affecting the Langat River Basin in Malaysia. Increase in developing areas within the river basin has in turn increased pollution loading into the Langat River. To avoid further degradation, the DOE have installed telemetric stations along the river basin to continuously monitor the water quality. As a result, abundant data were collected since 1988. There are 927 monitoring stations located within 120 river basins throughout Malaysia. Water quality data were used to determine the water quality status and to classify the rivers based on water quality index (WQI) and Interim National Water Quality Standards for Malaysia (INWQS). WQI provides a useful way to predict changes and trends in the water quality by considering multiple parameters. WQI is calculated from six selected water quality variables, namely dissolved oxygen (DO), biochemical oxygen demand (BOD), chemical oxygen demand (COD), suspended solid (SS), ammonical nitrogen (AN) and pH (DOE, 1997). It is a well-known phenomenon that the contribution of pollution loading into river systems from the environment involves a complex interaction of many factors (e.g. chemical, physical and meteorological interaction). These primary pollutants are emitted from land use activities surrounding the river basin (e.g. agriculture, forest, urban, industrial and others) Rapid urbanization along the Langat River plays an important role in the increase of point source

(PS) and non-point source (NPS). In view of this complex interaction, use of modelling techniques to solve this problem, is needed. However, the problem of obtaining models that adequately represent the dynamic behaviour of field data is not easy. Lack of good understanding and description of the phenomena involved, the availability of reliable and complete field data set and the estimation of the numerous parameters involved are the major factors contributing to this problem. Beck (1986) noted that, increase in model complexity will undoubtedly increase the number of parameters, leading to the problems of identification.

Applications of ANN (Artificial Neural Networks) to environmental problems are becoming more common (Silverman and Dracup, 2000; Scardi, 2001; Recknagel et al., 2002; Bowden et al., 2005; Muttil and Chau, 2007). The applications of ANN, which are computing systems that were originally designed to simulate the structure and function of the brain (Rumelhart et al, 1986) is a relatively new concept in environmental modeling. If trained properly, a neural network model is capable of 'learning' linear as well as the nonlinear features in the data (Elsner and Tronis, 1992).

ANN consists of a set of simple processing units (neurons) arranged in a defined architecture and connected by weighted channels which act to transform remotely-sensed data into a classification. The classification techniques of ANN are unlike the conventional ones. It is distribution-free, may sometimes use small training sets (Hepner et al., 1990) and, once trained; it is rapid computationally, which will be of value in processing large data sets (Gershon and Miller, 1993). Furthermore, ANNs have been shown to be able to map land cover more accurately compared to many widely used statistical classification techniques (Benediktsson et al., 1990; Foody et al., 1995) and alternatives such as evidential reasoning (Peddle et al., 1994).

It has been proposed that the best tool to model non-linear environmental relationship is ANN (Zhang and Stanley, 1997; Jain and Indurthy, 2003). Research have been undertaken at Imperial College, London which attempts to investigate the capability of ANN approach in modelling spatial and temporal variations in river water quality (Clarici, 1995). ANNs were used as a predictive model to predict cyanobacteria Anabaena spp. in the River Murray, South Australia (Maier et al., 1998). DeSilets et al. (1992), have also used ANN to predict salinity. Ha and Stenstrom (2003), proposed a neural network approach to examine the relationship between storm water quality and various types of land use.

ANN has been successfully applied on the study of river water quality in Malaysia (Zarita Zainudin, 2001; Mohd Ekhwan Toriman and Hafizan Juahir, 2003; Hafizan Juahir et al., 2003a,b; Hafizan et al, 2004a,b; 2005; Ruslan Rainis et al., 2004). An approach for identifying possibilities of water quality improvement could be developed by using this concept. Such information could provide opportunities for better river basin management to control river water pollution in Malaysia. In the Malaysian context, Hafizan Juahir et al. (2003a) showed that the ANN model gives a better performance compared to the autoregressive integrated moving average (ARIMA) model in forecasting DO. The use of ANN for river regulation (Mohd. Ekhwan Toriman and Hafizan Juahir, 2003) and the application of the second order back propagation method (Hafizan Juahir et al., 2004a) on water quality of the Langat River have also been demonstrated.

In natural environment, water quality is a multivariate phenomenon, at least as reflected in the multitude of constituents which are used to characterize the quality of water body. Water quality is very difficult to model because of the different interactions between pollutants and meteorological variables. The principal component analysis (PCA) is one of the approaches to avoid this problem and has received increasing attention as an accepted method in environmental pattern recognition (Simeonov et al., 2003; Wunderline et al., 2001; Helena et al., 2000; Loska and Wiechula, 2003)

The objective of this study is to use the PCA method to classify predictor variables according to their interrelation, and to obtain parsimonious prediction model (i.e., model that depend on as few variables as necessary) for WQI with other physico-chemical and biological data as predictor variables to model the water quality of the Langat river. For this purpose, principal component scores of 23 physico-chemical and biological water quality parameters were generated and selected appropriately as input variables in ANN models for predicting WQI.

2. Methodology

2.1 The data and monitoring sites

The water quality data in this study were obtained from seven stations along the main Langat River (Fig. 1).

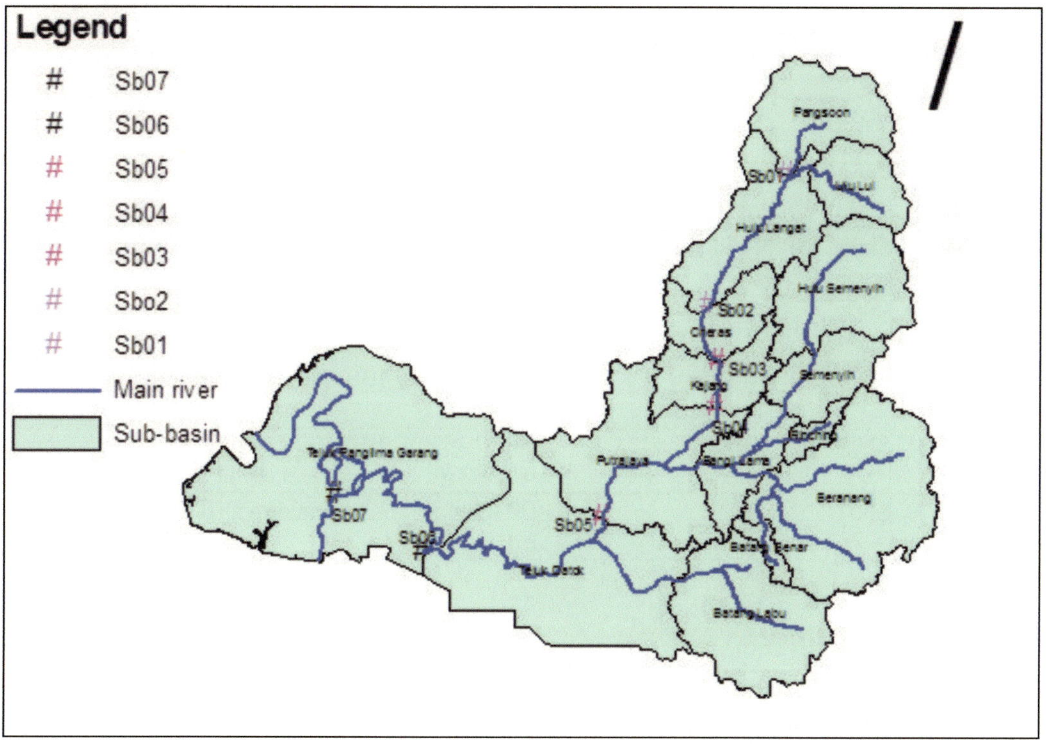

Fig. 1. Data from seven water quality stations (Sb) were selected in this study along the main river.

The water quality monitoring stations are manned by the DOE and Ministry of Natural Resource and Environment of Malaysia. The selected stations are illustrated in Table 1. The data used in the study is from September 1995 to May 2002. Seven sites were chosen, namely, Teluk Panglima Garang (site 7), Teluk Datok (site 6), Putrajaya (site 5), Kajang (site 4), Cheras (site 3), Hulu Langat (site 2), Pangsoon and Ulu Lui (site 1). Sites 3 to site 7 are located in the region of high pollution load as there are a several wastewater drains situated in the middle and downstream of the Langat River basin. Site 2 is partly situated in the middle stream region, designated as moderately polluted. Site 1 and a part of site 2 are located upstream of the Langat River, in an area of relatively low river pollution. It is worth mentioning here that some stations have missing data and not all stations were consistently sampled.

Although there are 30 water quality parameters available, only 23 completely monitored parameters were selected. A total of 254 samples were used for the analysis. The 23 water quality parameters were dissolved oxygen (DO), biological oxygen demand (BOD), electrical conductivity (EC), chemical oxygen demand (COD), ammoniacal nitrogen (AN), pH, suspended solids (SS), temperature (T), salinity (Sal), turbidity (Tur), dissolved solid (DS), total solid (TS), nitrate (NO), chlorine (Cl), phosphate (PO), zinc (Zn), calcium (Ca), iron (Fe), potassium (K), magnesium (Mg), sodium (Na), E.coli and coliform.

DOE Station No.	**Study Code**	**Distance From Estuary (km)**	**Grid Reference**	**Location**
2814602	Sb07	4.19	2°. 52.027′N 101° 26.241′E	Kampung Air Tawar (penghujung jalan)
2815603	Sb06	33.49	2° 48.952′N 101° 30.780′E	Telok Datuk, near Banting Town
2817641	Sb05	63.43	2° 51.311′N 101° 40.882′E	Bridge at Kampung Dengkil
2918606	Sb04	81.14	2° 57.835′N 101° 47.030′E	Near West Country Estate
2917642	Sb03	86.94	2° 59.533′N 101° 47.219′E	Kajang bridge
3017612	Sb02	93.38	3° 02.459′N 101° 46.387′E	Junction to Serdang, Cheras at Batu 11
3118647	Sb01	113.99	3° 09.953′N 101° 50.926′E	Bridge at Batu 18

Table 1. DOE sampling station at study area.

2.2 Principal component analysis

In this work, PCA was performed on the above mentioned water quality parameters to rank their relative significance and to describe their interrelation patterns. Chosen PC scores of the 23 water quality parameters were used as input variables in ANN model to predict the WQI. The principal components (PCs) can be expressed as

$$z_{ij} = a_{i1}x_{1j} + a_{i2}x_{2j} + ... + a_{im}x_{mj} \quad (1)$$

Where z is the component score, a is the component loading, x the measured value of variable, i is the component number, j is the sample number and m is the total number of variables.

The PCs generated by PCA are sometimes not readily interpreted; therefore, it is advisable to rotate the PCs by varimax rotation. Varimax rotation ensures that each variable is maximally correlated with only one PC and a near zero association with the other components (Abdul-Wahab et al., 2005; Sousa *et al.*, 2007). Varimax rotations applied on the PCs with eigenvalues more than 1 are considered significant (Kim and Mueller, 1987) where the typical criteria are 75-95% of total variance (Chen and Mynett, 2003). The rotations were carried out, in order to obtain new groups of variables. Variables with communality greater than 0.7 are considered, having significant factor loadings (Stevens, 1986).

2.3 Artificial neural networks for WQI prediction

In this work, the back propagation (BP) ANN was used in the development of all the prediction models. The Activation Transfer Function of a back-propagation network is usually a differentiable Sigmoid (S-shape) function, which helps to apply the non-linear mapping from inputs to outputs. A three layer back-propagation ANN is used in this study. The number of input and output neurons is determined by the nature of the problem under study. In this study, the networks were trained, tested and validated with one hidden layer and 1 to 10 hidden neurons. This choice was based on the work of Jiang et al. (2004), who found that the results with one hidden layer was better than that of two hidden layers, and the best performance was obtained using a structure with 3 to 6 neurons in the hidden layer. The output neuron (layer) gives the predicted WQI value.

Two different types of ANN models were developed. In the first type, prediction was performed based on the original PCs. In the second type of ANNs developed, scores of rotated (varimax rotation) PCs (ANN-RPCs) with eigenvalues greater than 1 were selected as input. For this model, prediction of WQI was performed using two to six rotated principal components separately.

The original PCs and rotated PCs (RPCs) data sets consist of 305 observations (305 rows) and are divided into training, testing and validating phases for WQI prediction. The ANN predicted WQI values are compared to the WQI values calculated using the DOE-WQI formula which is based on 6 water quality parameters, namely the DO, COD, BOD, AN, SS and pH (DOE, 1997). The input data matrix consists of 23 water quality variables (column) and 305 observations (rows) [23×305]. The observed data for each station is arranged according to time of observation from September 13, 1995 to June 7, 2002. Table 2 describes the data structure. The validation data is at least 10% of the whole data set, with 75% training set and 25% testing set data (Kuo et al., 2007).

No. of Observations	Input parameters								Output
	$Input_1$	$Input_2$	$Input_3$	.	.	.	.	$Input_{23}$	$Output_1$
1	$Obs_{1,1}$	$Obs_{1,2}$	$Obs_{1,3}$	.	.	.	.	$Obs_{1,23}$	$O_{1,1}$
2	$Obs_{2,1}$	$Obs_{2,2}$	$Obs_{2,3}$	.	.	.	.	$Obs_{2,23}$	$O_{2,1}$
.	.	.	.	.	.	.	.	.	.
.	.	.	.	.	.	.	.	.	.
.	.	.	.	.	.	.	.	.	.
...	....	....	...	....	....	....	....	...	.
									...
305	$Obs_{305,1}$	$Obs_{305,2}$	1					$Obs_{305,23}$	$O_{305,1}$

Table 2. The data structure for ANN prediction model.

2.4 Determination of model performance

The model's behaviour in both learning (training and testing) and validating phase, is evaluated using the following statistical methods; the correlation coefficient (R) at 95% confidence limit, given by equations;

$$\text{Coefficient of correlation (R), } r = \frac{\left[\sum_{i=1}^{n} x_i\hat{x}_i - \frac{1}{n}\left(\sum x_i\right)\left(\sum \hat{x}_i\right)\right]^2}{\sqrt{\left[\sum x_i^2 - \frac{1}{n}\left(\sum x_i\right)^2\right]\left[\sum \hat{x}_i - \frac{1}{n}\left(\sum \hat{x}_i\right)^2\right]}} \tag{2}$$

and the mean bias error or residual error given by;

$$\text{Mean bias error (MBE), } MBE = \frac{1}{n}\sum_{i=1}^{n}(\hat{x}_i - x_i) \tag{3}$$

Where $\hat{x}_i$ and x_i represent observed values and the corresponding forecast values for i =1,2,…..,n.

The prediction performance evaluated using these two methods are used to evaluate the accuracy of the forecast and for comparing the forecasting ability of each approach.

The 95% confidence limit is used to determine that the predicted output lie within the confidence range. It is assumed that a predicted value fall into an interval within which there is an associated uncertainty. According to Wackerly et al. (1996), this uncertainty is derived from the residual errors that have already been calculated within that range of values. If the residual errors are randomly distributed, there is a general rule of thumb which states that they will lie within two standard deviations of their mean with a probability of 0.95. This method was used in the measurements of the ANN prediction performance conducted by some researchers (Bishop, 1995; Tibshirani, 1996; Shao et al., 1997; Zhang et al., 1998; Lowe and Zapart, 1999; Townsend and Tarassenko, 1999)

ANN models and statistical analyses were carried out using MATLAB 7.0 and XLSTAT2008 (Excel2003 add-in) for Windows.

3. Results and discussion

Post PCA, out of the 23 principal components generated, only six PCs with eigenvalues higher than 1 (Table 3) were selected for the ANN input parameters. Selected PCs explained 75.1% of the total variation. Furthermore, communality values were high for the selected PCs, for example, the values are 93% for Cond., 95% for Sal, 98% for DS and TS (Table 4). These results further confirm the choice of the selected number of PCs (Stevens, 1986).

For the first six rotated PCs (RPCs), the loadings from PCA are given in Table 4. The highest correlations between variables are noted in bold. For instance, Cond., Sal, DS, TS, Cl, Ca, K, Mg and Na, have high correlations with RPC1. Eighteen variables with strong loadings were included in the six selected RPCs. Significant variables in RPC1 are Cond., Sal., DS, TS, Cl, Ca, K, Mg, and Na; in RPC2 they are DO, BOD and AN; in RPC3 they are SS and Tur and in RPC4, NO_3^- and PO_4^{3-}. The only meaningful loads in RPC5 and RPC6 are pH and Zn.

	PC1	PC2	PC3	PC4	PC5	PC6
Eigenvalue	9.074	2.387	2.067	1.492	1.225	1.026
Variability (%)	39.451	10.380	8.987	6.488	5.326	4.459
Cumulative %	39.451	49.830	58.817	65.305	70.631	75.091

Table 3. Descriptive statistics of selected original PCs with eigenvalues more than 1.

Variables	RPC1	RPC2	RPC3	RPC4	RPC5	RPC6	Communalities
DO	-0.205	**-0.722**	-0.121	0.046	0.485	-0.066	0.82
BOD	0.035	**0.740**	0.071	0.110	0.110	0.022	0.58
COD	0.340	0.103	0.081	-0.166	0.268	0.326	0.34
SS	-0.042	-0.009	**0.920**	0.010	-0.025	0.017	0.85
pH	0.189	-0.109	-0.204	0.020	**0.792**	-0.083	0.72
AN	-0.092	**0.797**	-0.151	0.161	0.023	-0.032	0.69
T	0.337	0.368	-0.242	-0.298	-0.317	0.208	0.54
Cond.	**0.963**	0.022	-0.043	0.035	0.013	-0.022	0.93
Sal.	**0.974**	0.023	-0.038	0.030	0.008	-0.004	0.95
Tur.	-0.031	-0.007	**0.863**	0.011	-0.140	-0.035	0.77
DS	**0.988**	0.017	-0.034	0.013	0.009	-0.005	0.98
TS	**0.985**	0.017	0.069	0.014	0.007	-0.003	0.98
NO_3^-	0.018	0.033	0.107	**0.688**	-0.126	0.300	0.59
Cl	**0.986**	0.010	-0.029	-0.004	0.020	0.005	0.97
PO_4^{3-}	0.023	0.312	-0.106	**0.700**	0.112	-0.073	0.62
Zn	-0.019	0.044	-0.011	0.186	-0.128	**0.767**	0.64
Ca	**0.980**	0.028	-0.026	-0.043	-0.024	0.039	0.97
Fe	-0.080	0.043	0.475	0.540	0.066	0.192	0.57
K	**0.984**	0.004	-0.031	-0.004	0.004	0.010	0.97
Mg	**0.974**	0.000	-0.022	-0.028	-0.002	0.037	0.95
Na	**0.986**	0.002	-0.025	-0.020	0.005	0.017	0.97
COLI	-0.254	0.361	0.097	-0.424	0.457	0.056	0.60
COLIFORM	-0.032	0.049	-0.025	0.042	-0.077	-0.517	0.28

Table 4. Rotated factor loadings using six PCs.

Using the original principal component scores as inputs, the best architecture consist of a three layer network with 23 input neurons, 10 neurons in the hidden layer and one neuron in the output layer. Considering RPC scores as inputs, the best architectures were achieved with almost the same number of hidden neurons. The hidden neurons consist of 9 and 10 neurons respectively. Training was carried out for a maximum 10000 iterations. Selection of the network was performed at maximum correlation coefficient (R) and 95% confidence limit.

Table 5 and Figure 2 illustrate the prediction performances of ANN models using different combinations of PC scores as input variables. ANN using the first 2 PCs (PC1 and PC2) does not perform very well as far as accuracy is concerned for all the training, testing and validation phases. It is observed that the prediction performance of the validation phase is slightly worse compared to the training and testing phases. It is important to point out that for this model, the cumulative percentage in explaining the variance given by these two RPCs is only 49.8%. None of the strong loading variables contains the variables forming the WQI equation. DO, BOD and pH loadings in PC2 explain only 10.4% of the total variance.

Based on the results, it is apparent that the WQI prediction performance increases with the increase in number of input variables. The highest accuracy in predicting WQI is given by model ANN-RPC6, which contains six RPCs with 75.1% variation explained, giving an R^2 value of 0.64 (training), 0.87 (testing), and 0.72 (validation) respectively.

Model	No.of PC	R squared			MBE		
		Training	Testing	Validation	Training	Testing	Validation
ANN-RPC2 (2 inputs)	2	0.43	0.70	0.32	28.01	-167.90	-40.71
ANN-RPC3 (3 inputs)	3	0.60	0.78	0.61	64.95	-109.68	6.60
ANN-RPC4 (4 inputs)	4	0.53	0.79	0.47	0	-165.04	-89.78
ANN-RPC5 (5 inputs)	5	0.53	0.79	0.47	140.12	-143.75	-44.77
ANN-RPC6 (6 inputs)	6	0.64	0.87	0.72	67.93	-58.57	-44.61
ANN-PC23 (23 original PC inputs)	23	0.60	0.85	0.66	-18	-81.59	-49.83

Table 5. The prediction performances of the different ANN models.

From table 5, it can be observed that the prediction performance of the ANN model using original PCs (23 input PC scores) is not significantly different from the RPC models. However, as RPC models use fewer variables and is far less complex, the advantage over the ANN-PC23 model is obvious. Comparing the MBE values, it is generally observed that the signs for the validation phases are negative for both the un-rotated and rotated PC models. This is an indication that the predicted WQI values are consistently underestimated in both approaches.

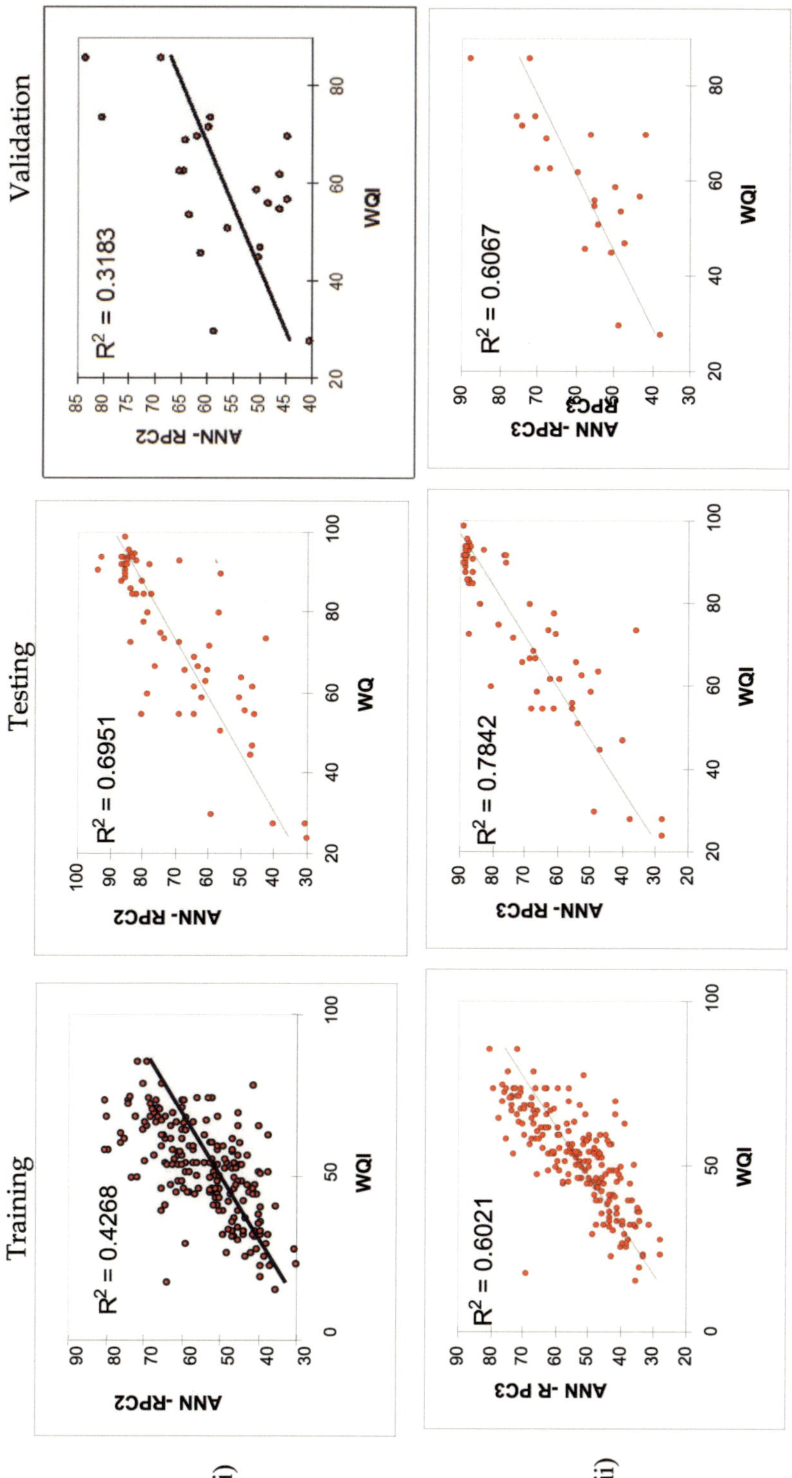
Training
Testing
Validation
i)
ii)
$R^2 = 0.4268$
$R^2 = 0.6951$
$R^2 = 0.3183$
$R^2 = 0.6021$
$R^2 = 0.7842$
$R^2 = 0.6067$
ANN -RPC2
ANN- RPC2
ANN -R PC3
ANN- RPC3
ANN -RPC3
WQI
WQ

Part I

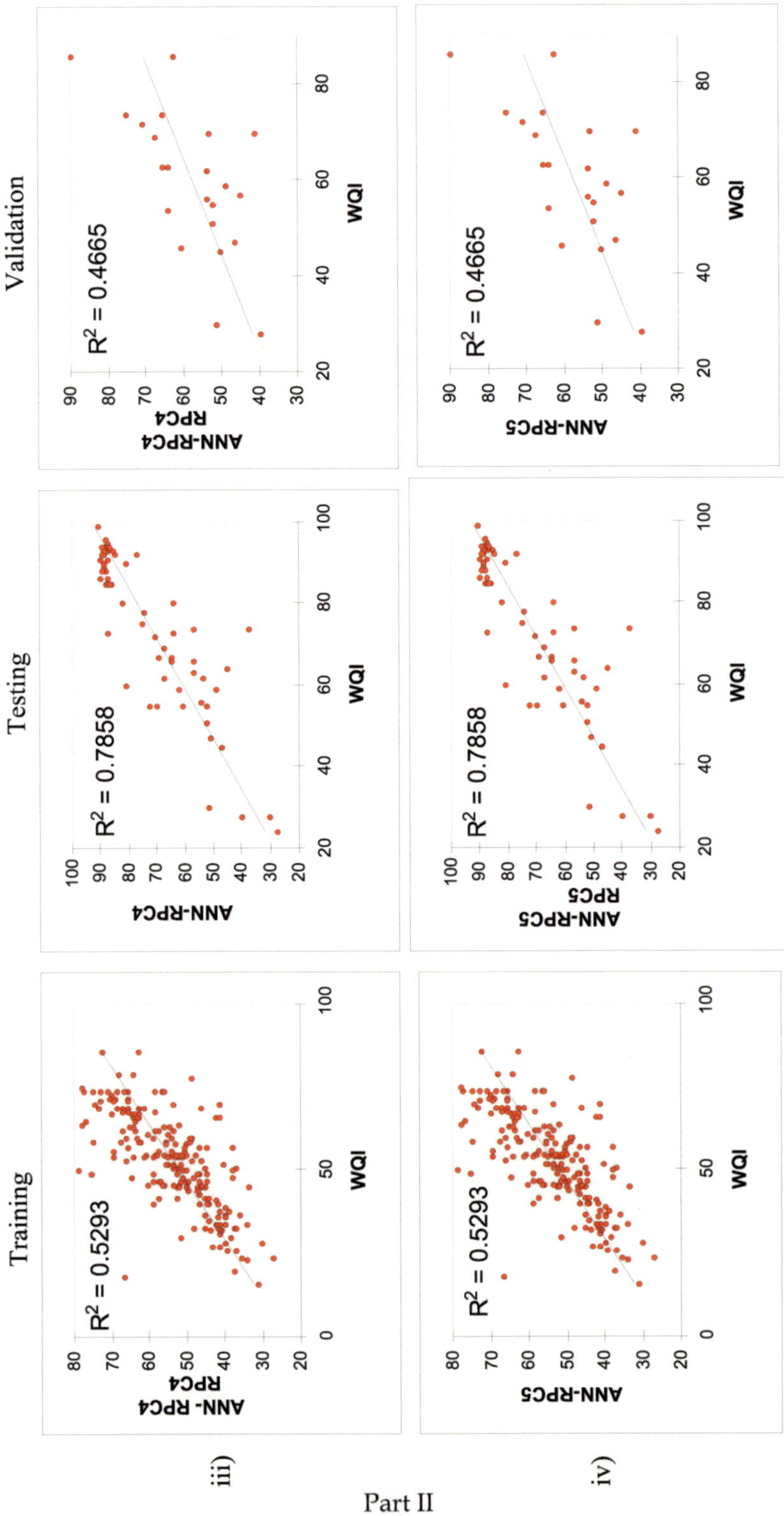
Training
Testing
Validation
iii)
iv)
Part II
$R^2 = 0.5293$
$R^2 = 0.7858$
$R^2 = 0.4665$
ANN- RPC4
ANN-RPC4
ANN-RPC5
WQI

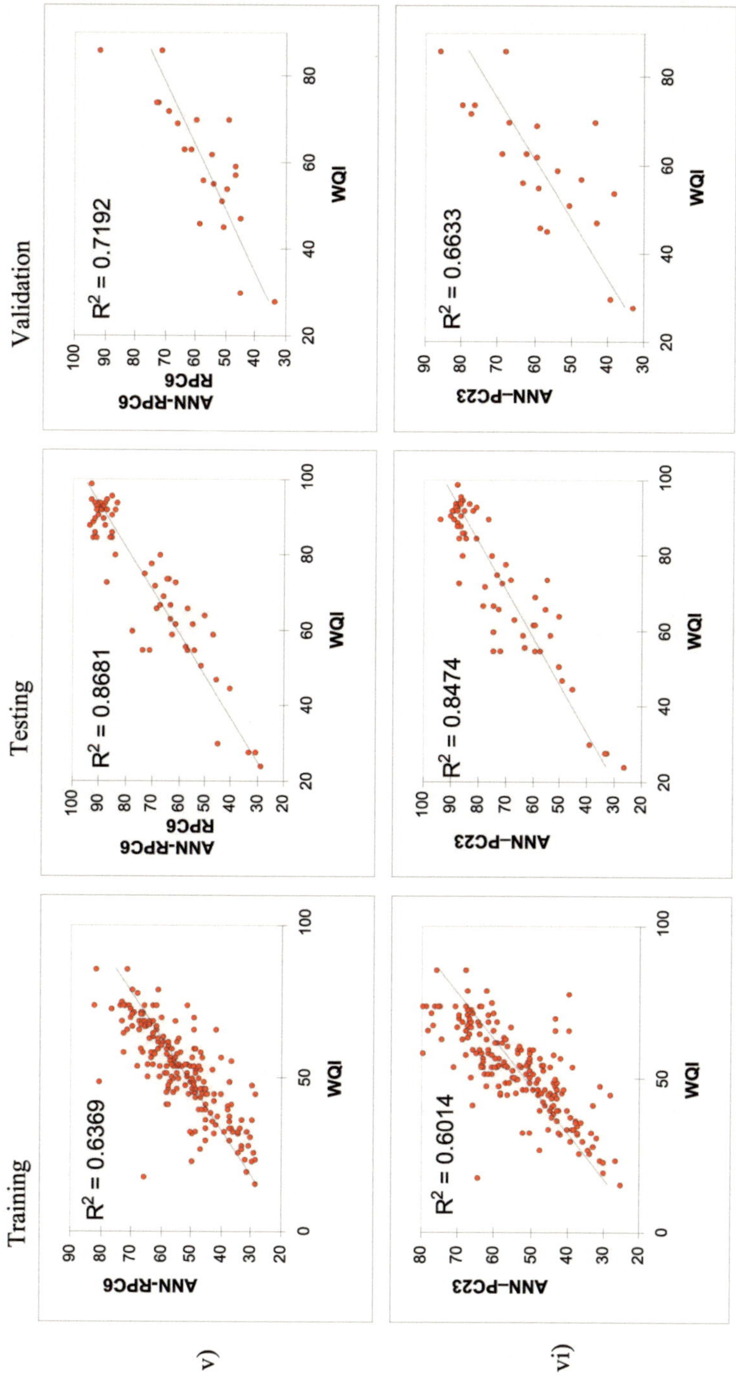

Part III

Fig. 2. The prediction performances for different combination of PC scores during training, testing and validation phases : (i) 2 RPCs, (ii) 3 RPCs, (iii) 4 RPCs, (iv) 5 RPCs, (v) 6 RPCs and, (vi) 23 original PCs.

This study also attempts to allocate 95% confidence interval on the WQI prediction produced by the best ANN model. Figure 3, 4 and 5 show the comparison between predicted values and the upper (UL) and lower limits (LL) lying within 95% confidence interval. This was carried out for ANN-RPC6 and ANN-PC23 models. It can be seen that only 4.3% out of the 305 predicted values were identified beyond the 95% confidence limit (1% fall below the LL and 3.3% fall beyond the UL) for ANN-RPC6. For ANN-PC23, 25% of the 305 observations fall beyond the upper and lower of 95% confidence interval limit (14% fall below the LL and 11.8% fall beyond the UL). This basically shows that by using reduced rotated PC scores as input, better results can be obtained without losing information. It is thus apparent that ANN prediction using scores of varimax rotated PCs result in a more accurate WQI prediction.

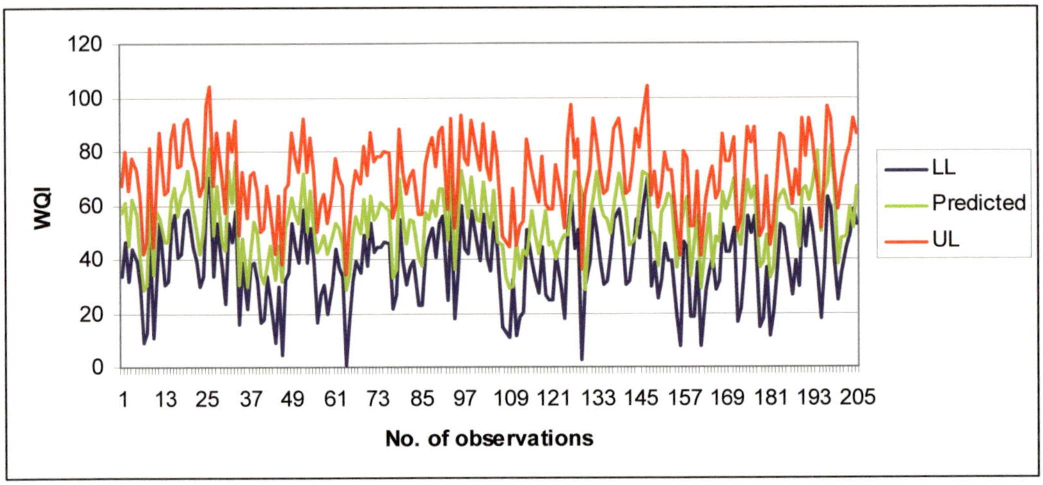

(a)

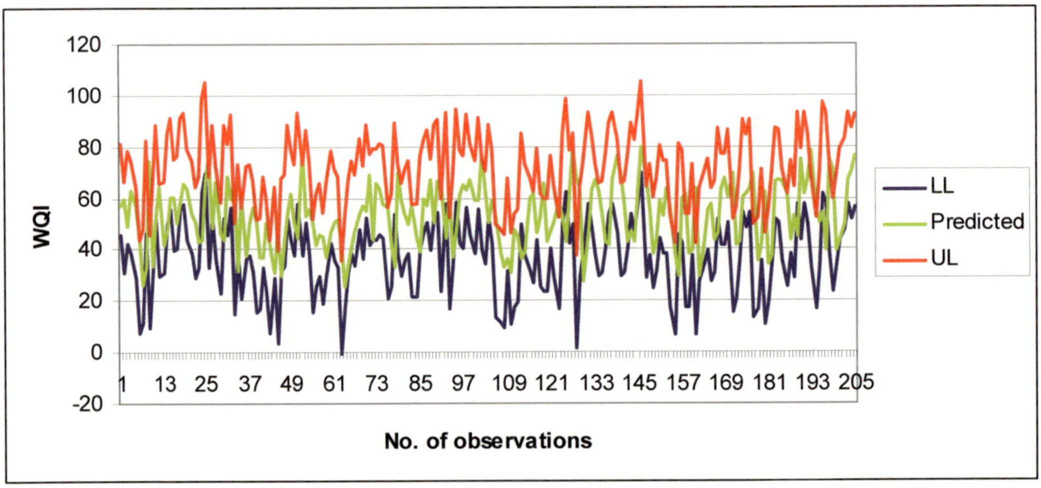

(b)

Fig. 3. Predicted WQI within the 95% confidence interval during training phase using (a) six rotated PCs, and (b) 23 original PCs.

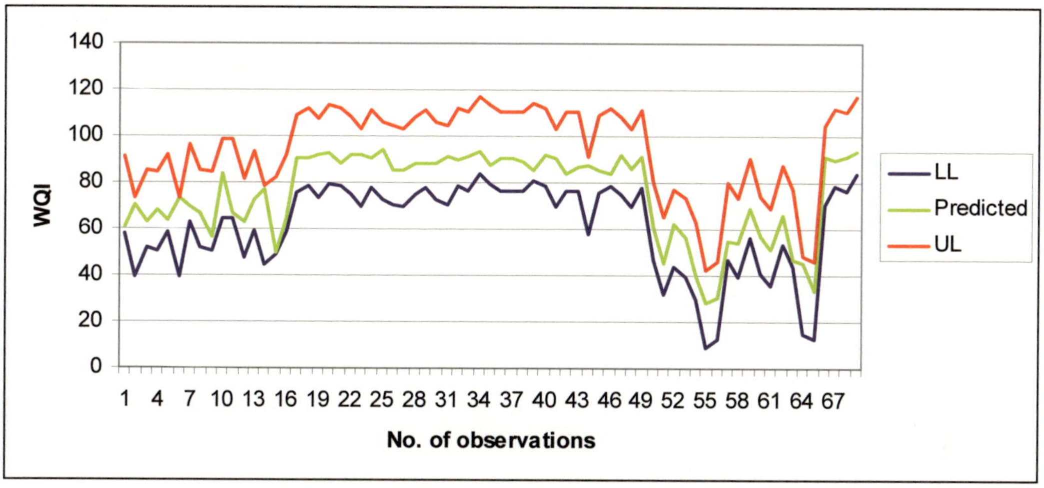

(a)

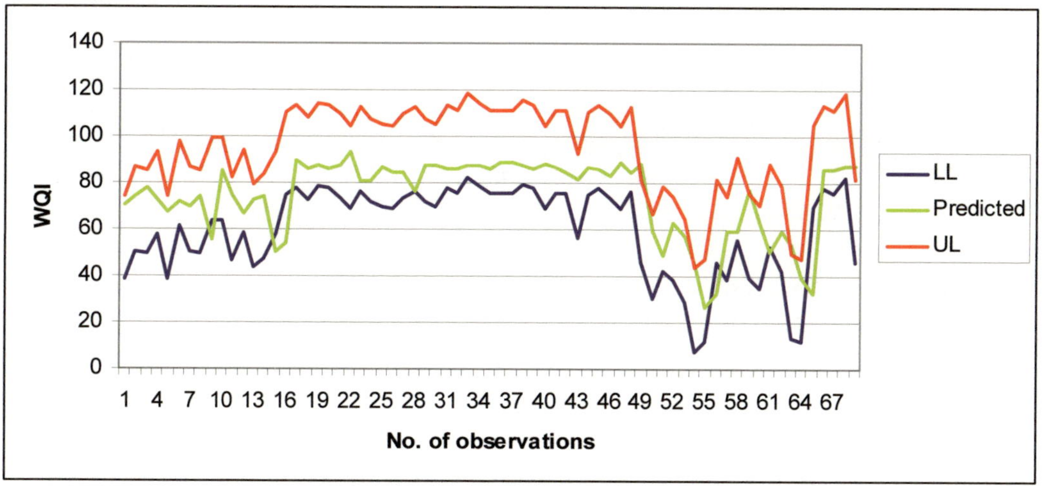

(b)

Fig. 4. Predicted WQI within the 95% confidence interval during testing phase using (a) six rotated PCs, and (b) 23 original PCs.

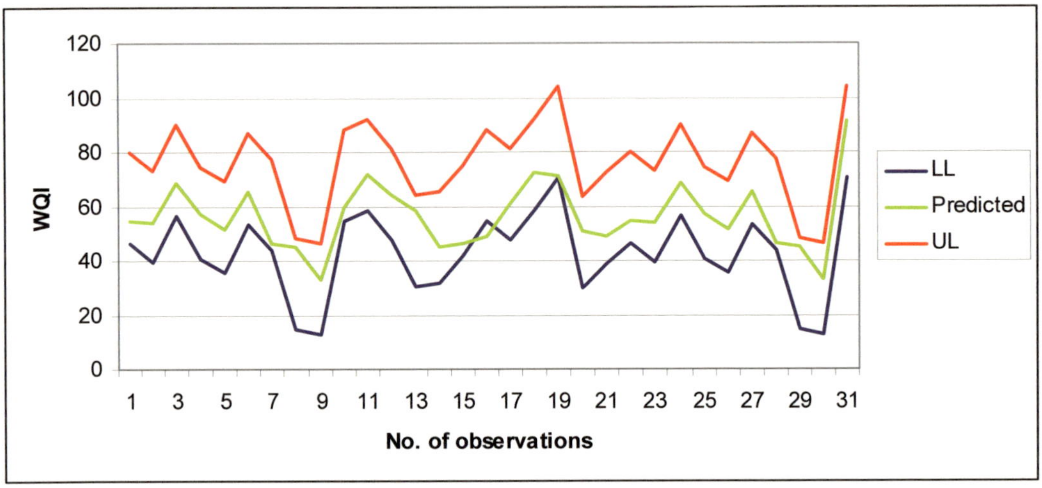

(a)

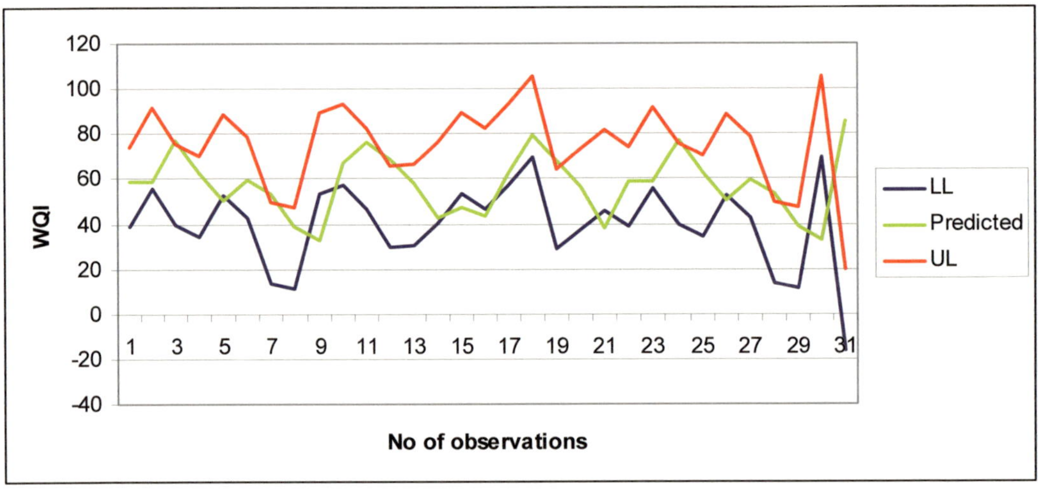

(b)

Fig. 5. Predicted WQI within the 95% confidence interval during validation phase using (a) six rotated PCs, and (b) 23 original PCs.

4. Conclusion

In this work, a combination of PCA and ANN is used to predict WQI based on 23 historical water quality parameters. The original predictors were selected based on the available Malaysian DOE data. To obtain the latent variables as inputs into the ANN, two different approaches were used; one based on un-rotated original PCs and the other based on varimax rotated PCs.

Using six PCs, significant loadings are observed for Cond, Sal, DS, TS, Cl, Ca, K, Mg and Na in PC1, DO, BOD and AN in PC2, SS and Tur in PC3, NO3- and PO43- in PC4, pH in PC5 and Zn in PC6. ANN models based on these 6 PC scores can predict WQI with acceptable accuracy (within 95% confidence limit). Moreover, the ANN model using the 23 original PCs as input, do not render the prediction more accurate, even with a complex network structure. The use of rotated PC scores based models is clearly more effective and efficient due to the elimination of collinearity and reduction of predictor variables without losing important information.

5. Acknowledgment

The authors acknowledge the financial and technical support for this project provided by the Ministry of Science, Technology and Innovation and Universiti Putra Malaysia under the Science Fund Project no. 01-01-04-SF0733. The authors wish to thank,the Department of Environment, and Department of Irrigation and Drainage, Ministry of Natural Resources and Environment of Malaysia, Institute for Development and Environment (LESTARI), Universiti Kebangsaan Malaysia, Universiti Malaya Consultancy Unit (UPUM) and Chemistry Department of Universiti Malaya, who have provided us with secondary data and valuable advice.

6. References

Abdul-Wahab, S.A., Bakheit, C.S. and Al-Alawi, S.M., 2005. Principal component and multiple regression analysis in modeling of ground-level ozone and factors affecting its concentrations. *Environmental Modelling & Software* 20, p.1263-1271.

Aiken, R.S., Leigh, C.H., Leinbach, T.R., and Moss, M.R., 1982. Development and Environment in Peninsular Malaysia. McGraw-Hill International Book Company: Singapore.

Beck, M.B., 1986. Identification, estimation and control of biological waste-water treatment processes. *IEE Proceeding* 133, p.254-264

Benediktsson, J.A., Swain, P.H., and Ersoy, O.K., 1990. Neural network approaches versus statistical methods in classification of multisource remote sensing data. *I.E.E.E. Transactions on Geoscience and Remote Sensing,* 28, 540-551

Bishop, C. M., 1995. Neural Networks for Pattern Recognition. Clarendon Press, Oxford Bowden, G.J., Dandy, G.C. and Maier, H.R., 2005. Input determination for neural network models in water resources applications. Part 1-background and methodology. *Journal of Hydrology* 301, p.75-92.

Chen, Q. and Mynett, A.E., 2003. Integration of data mining techniques and heuristic knowledge in fuzzy logic modeling of eutrophication in Taihu Lake. *Ecological Modelling* 162, p.55-67.

Clarici, E., 1995. Environmental Modelling Using Neural Networks, PhD Thesis, Imperial College.

Department of Environment Malaysia, DOE 1997. Malaysia environmental quality reports 1999. Kuala Lumpur: Ministry of Science, Technology and Environment.

Department of Environment Malaysia, DOE 1999. Malaysia environmental quality reports 1999. Kuala Lumpur: Ministry of Science, Technology and Environment.

DeSilets, L., Golden, B., Wang, Q., and Kumar, R., 1992. Predicting salinity in the Chesapeake Bay using backpropagation. *Computer and Operations Research,* 19, p.227-285

Elsner, J.B., and Tronis, A.A., 1992. Nonlinear prediction, chaos, and noise. Bull. *Am. Meterol. Soc.* 73(1), p.303-314.

Foody, G.M., McCulloch, M.B., and Yates, W.B., 1995. Classification of remotely sensed data by an artificial neural network: issues related to training data characteristics. *Photogrammetric Engineering and Remote Sensing.*

Ha, H. and Stenstrom, M. K., 2003. Identification of land use with water quality data in stormwater using a neural network. *Water Research*, 37, p.4222-4230

Hafizan Juahir, Sharifuddin M. Zain, Zainol Mustafa, and Azme Khamis, 2001. Dissolved oxygen forecasting due to landuse activities using time series analysis at Sungai Lang at, Hulu Lang at, Selangor. *Ecological Environmental Modelling, Proceeding of the National Workshop*, Universiti Sains Malaysia, 3-4 September, p.157-164.

Hafizan Juahir, Sharifuddin M. Zain, Mohd. Ekhwan Toriman, M. Nazari Jaafar and W. Klaewtanong, 2003a. Performance of autoregressive integrated moving average and neural network approaches for forecasting dissolved oxygen at Langat River Malaysia. *Urban Ecosystem Studies In Malaysia: A study of change.* Universal Publishers, p. 145-165.

Hafizan Juahir, Sharifuddin M. Zain, M. Nazari Jaafar, Zainal Majid and M. Ekhwan Toriman, 2003b. Land use temporal changes: application of GIS and statistical analysis on the impact of water quality at Langat River Basin, Malaysia. presented *in 2nd Annual Asian Conference of Map Asia 2003*, 17-19, Oct., PWTC Kuala Lumpur.

Hafizan Juahir, Sharifuddin M. Zain, M. Nazari Jaafar and Zainal Ahmad, 2004a. An Application of Second order backpropagation method in Modeling River Discharge at Sungai Langat, Malaysia. *Water Environmental Planning: Towards integrated planning and management of water resources for environmental risks*, IIUM, p.300-307.

Hafizan Juahir, Sharifuddin M. Zain, M. Ekhwan Toriman and Mazlin Mokhtar, 2004b. Application of Artificial Neural Network Model In the Predicting Water Quality Index. *Jurnal Kejuruteraan Awam*, 16 (2), p.42-55

Helena, B., Pardo, R., Vega, M., Barrado, E., Fernandez, J.M., Fernandez, L., 2000. Temporal evaluation of grounwater composition in an alluvial aquifer (Pisuerga river, Spain) by principal component analysis. *Water Research* 34, 807-816.

Hepner, G.F., Logan, T., Ritter, N., and Bryant, N., 1990. Artificial Neural Network classification using a minimal training set: comparison to conventional supervised classification. *Photogrammetric Engineering and Remote Sensing*, 56, 469-473

Jain, A. & Prasad Indurthy, S. K. V. (2003) Comparative Analysis of Event-based Rainfall-runoff Modeling Techniques-Deterministic, Statistical, and Artificial Neural Networks, *Journal of Hydrologic Engineering*, 8, p. 93-98.

Jiang, D., Zhang, Y., Hu, X., Zeng, Y., Tan, J. And Shao, D., 2004. Progress in developing an ANN model for air pollution index forecast. *Atmospheric Environment* 38, p.7055-7064.

Kim, J.,-O., and Mueller, C.W., 1987. Introduction to factor analysis: what it is and how to do it. *Quantitative Applications in the Social Sciences Series.* Sage University Press, Newbury Park.

Kuo, J.-T., Hsieh, M.-H., Lung, W.-S. And She, N., 2007. Using artificial neural network for reservoir eutrophication prediction. *Ecological Modelling*, 200, p.171-177

Loska, K. and Wiechula, D., 2003. Application of principal component analysis for the estimation of source of heavy metal contamination in surface sediments from the Rybnik Reservoir. *Chemosphere* 51, p.723-733.

Lowe, D. and Zapart, C., 1999. Point-Wise Confidence Interval Estimation by Neural Networks: A Comparative Study based on Automotive Engine Calibration. *Neural Computing & Applications*, Vol. 8, p.77-85.

Mohd. Ekhwan Toriman and Hafizan Juahir, 2003. Artificial Neural Network Modelling For Langat River Discharge: Implication For River Restoration. Pertandingan Minggu Penyelidikan dan Inovasi UKM, Pusat Pengurusan Penyelidikan, 3-5 Julai.

Muttil, N. and Chau, K.,-W., 2007. Machine-learning paradigms for selecting ecologically significant input variables. *Engineering Application of Artificial Intelligence* 20, p.735-744.

Peddle, D.R., Foody, G.M., Zhang, A., Franklin, S.E., and Ledrew, E.F., 1994.

Multisource image classification II: An empirical comparison of evidential reasoning and neural network approaches. *Canadian Journal of Remote Sensing*, 12, 277-302.

Recknagel, F., Bobbin, J., Whigham, P., and Wilson, H., 2002. Comparative application of artificial neural networks and genetic algorithms for multivariate time-series modeling of algal blooms in freshwater lakes. *Journal of Hydroinformatics* 4(2), p.125-134.

Rosnani Ibrahim, 2001. River Water quality Status In Malaysia. *Proceedings National Conference On Sustainable River Basin Management* In Malaysia, 13 & 14 November 2000, Kuala Lumpur, Malaysia.

Rumelhart, D. E., Hinton, E. and Williams, J., 1986. Learning internal representation by error propagation. *Parallel Distributed Processing*, 1, p. 318-362.

Ruslan Rainis, Kamarul Ismail and Hafizan Juahir, 2004. Modeling The Relationship Between River Water Quality Index (WQI) and Land Uses Using Artificial Neural Networks (ANN). Presented in JSPS Seminar, December 15-17, Kyoto, Japan.

Scardi, M., 2001. Advances in neural network modeling of phytoplankton primary production. *Ecological Modelling* 146, p.33-45.

Schalkoff, R., 1992. Pattern Recognition: Statistical, Structural and Neural Approaches. New York, Wiley

Shao, R., Martin, E.B., Zhang, J. and Morris, A.J., 1997. Confidence bounds for neural network representations. *Computers and Chemical Engineering*, 21, p.1173-1178.

Silverman, D., and Dracup, J.A., 2000. Artificial neural networks and long-range precipitation in California. *Journal of Applied Meteorology* 31(1), p.57-66.

Simeonov, V., Stratis, J.A., Samara, C., Zachariadis, G., Voutsa, D., Anthemidis, A., Sofoniou, M. and Kouimtzis, Th., 2003. Assessment of the surface water quality in Northern Greece. *Water Research* 37, p.4119-4124.

Sousa, S.I.V., Martins, F.G., Alvim-Ferraz, M.C.M. and Pereira, M.C., 2007. Multiple linear regression and artificial neural networks based on principal components to predict ozone concentrations. *Environmental Modelling & Software* 22, p.97-103.

Stevens, J., 1986. Applied Multivariate Statistics for the Social Science. Hill Sdale: New Jersey, USA, p.515.

Tibshirani, R., 1999. A comparison of some error estimates for neural network models. *Neural Computation*, 8, p.152-163.

Townsend, N.W. and Tarassenko, L., 1999. Estimations of error bounds for neural-network function approximators. *IEEE Trans Neural Netwoks*, 10(2), p.217.

Wackerly, D.D., Mendenhall, W and Scheaffer, R.L., 1996. Mathematical Statistics with Applications. 5th. Ed., Duxbury Press: Belmont, USA.

Wunderlin, D.A., Diaz, M.P., Ame, M.V., Pesce, S.F., Hued, A.C., Bistoni, M.A., 2001. Pattern recognition techniques for the evaluation of spatial and temporal variations in water quality. A case study: Suquia river basin (Cordoba, Argentina). *Water Research* 35, 2881-2894.

Zarita Zainuddin, 2004. Modelling Nonlinear Relationship in Ecology and Biology using Neural Networks. In Koh Hock Lye and Yahya Abu Hassan, Ecological Environmental Modelling (ECOMOD 2001): *Proceedings of the National Workshop*, 3-4 September, USM, p.88-95

Zhang, J., Morris, A.J., Martin, A.J. and Kiparissides, C., 1998. Prediction of polymer quality in batch polymerization using robust neural networks. *Chemical Engineering Journal*. 69, p.135-143.

Zhang, Q. & Stanley, S. J.,1997. Forecasting raw-water quality parameters for North Saskatchewan River by neural network modeling. *Water Resource*, 31, p. 2340-2350.

13

PARAFAC Analysis for Temperature-Dependent NMR Spectra of Poly(Lactic Acid) Nanocomposite

Hideyuki Shinzawa[1], Masakazu Nishida[1], Toshiyuki Tanaka[2], Kenzi Suzuki[3] and Wataru Kanematsu[1]
[1]*Research Institute of Instrumentation Frontier, Advanced Industrial Science and Technology (AIST)*
[2]*Mikawa Textile Research Center, Aichi Industrial Technology Institute (AITEC)*
[3]*Department of Chemical Engineering, Graduate School of Engineering, Nagoya University*
Japan

1. Introduction

This chapter provides a tutorial on the fundamental concept of Parallel factor (PARAFAC) analysis and a practical example of its application. PARAFAC, which attains clarity and simplicity in sorting out convoluted information of highly complex chemical systems, is a powerful and versatile tool for the detailed analysis of multi-way data, which is a dataset represented as a multidimensional array. Its intriguing idea to condense the essence of the information present in the multi-way data into a very compact matrix representation referred to as scores and loadings has gained considerable popularity among scientists in many different areas of research activities.

The basic idea of PARAFAC is so flexible and general that its application is not limited to a particular field of spectroscopy confined to a specific electromagnetic probe. Examples of the application include fluorescence (Christensen *et al.*, 2005; Rinnan *et al.*, 2005), IR (Wu *et al.*, 2003), NMR (Bro *et al.*, 2010), UV (Ebrahimi et al., 2008; Van Benthem et al., 2011) and mass spectroscopy (Amigo *et al.*, 2008). The first part of this chapter covers the theoretical background of trilinear decomposition of three-way data by PARAFAC with comparison to bilinear decomposition of two-way data by Principal component analysis (PCA).

In the second part of this chapter, an illustrative example of PARAFAC analysis for three-way data obtained in an actual laboratory experiment is presented to show how PARAFAC trilinear model can be constructed and analyzed to derive in-depth understanding of the system from the data. Thermal deformation of several types of poly lactic acid (PLA) nanocomposites undergoing grass-to-rubber transition is probed by cross-polarization magic-angle (CP-MAS) NMR spectroscopy. Namely, sets of temperature-dependent NMR spectra are measured under varying clay content in the PLA nanocomposite samples. While temperature strongly affects molecular dynamics of PLA, the clay content in the samples also influences the molecular mobility. Thus, NMR spectra in this study become a three-way dataset described as a function of both temperature and clay content. Details of the effects of the temperature and clay content on the physical state of nanocomposite are elucidated by using PARAFAC trilinear model.

2. PARAFAC

2.1 Multi-way data

So, what does a multi-way data look like? It is insightful first to note the data structures of two-way and three-way data. Schematic descriptions of two-way and three-way data based on external perturbation(s) are shown in Fig. 1. In a general spectroscopic measurement, external perturbations are applied to the system of interest to induce the response to the stimuli. Characteristic response of the system is presented in the form of spectrum. For example, when the thermal behaviour of a sample is studied by a spectroscopic method, such as IR, Raman and NMR, the sample is heated up to undergo thermal deformation and its molecular level variation induced by the stimulus is captured at each spectral variable, e.g. wavenumber. The spectral dataset thus obtained will be represented as a two-way array with the *(i,j)*th element denoting the spectral intensity value at the *i*th temperature and the *j*th wavenumber.

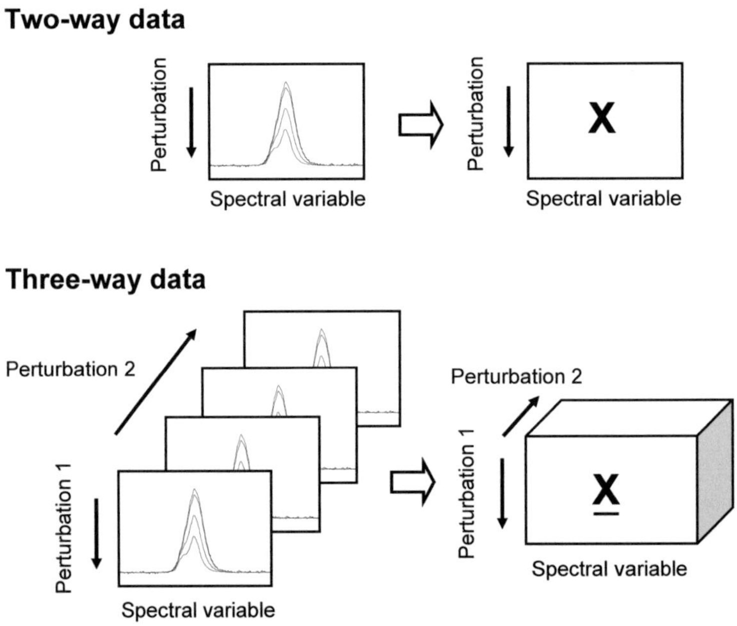

Fig. 1. Schematic illustration of two-way and three-way data.

Now, let us consider another experiment with one more perturbation. As described above, stimulation of a single sample ends up with two-way data array. But what if we still have some more samples, whose properties (e.g. concentration) are different? We will repeat a similar experiment for every single sample. This generates multiple two-way data. Thus, the entire dataset eventually becomes a stuck of the multiple two-way data like a cube, which contains two dimensions concerning applied two perturbations. Such spectral dataset is described as a three-way array with the *(i,j,k)*th element denoting the spectral intensity value at the *i*th concentration, the *j*th temperature, and the *k*th wavenumber. For example, the samples will show the variation of their molecular structure depending on the temperature. This may be also influenced by the change in the concentration. Thus the spectral intensities of the samples are potentially influenced by the temperature as well as concentration.

2.2 PARAFAC model

It is possible to condense the essence of the information present in multi-way data into a very compact matrix representation referred to as scores and loadings. The basic hypothesis of factor analysis techniques is that the improved proxy of the original data matrix can be reconstructed from only a limited number of significant factors. Thus, while the score and loading matrices contain only a small number of factors, it effectively carries all the necessary information about spectral features and, eventually, it becomes possible to sorting out the convoluted information content of highly complex chemical systems. The detailed analysis of such matrices potentially brings useful insight into building a mechanistic model for understanding complex phenomena studied by spectroscopic method.

Principal component analysis (PCA) is mathematical decomposition of two-way data in terms of the orthogonal set of dominant factors, i.e., eigenvectors (Smilde *et al.*, 2004; Shinzawa *et al.*, 2010). Two-way data decomposition by PCA results in yielding two matrices called scores and loadings which complementarily represent the entire features broadly distributed in the two-way data as follows,

$$\mathbf{X} = \mathbf{T}\mathbf{P}^{t} + \mathbf{E}_{\mathbf{PCA}} \tag{1}$$

where **T** and **P** are PCA score and loading matrices consisting of *r* vectors, respectively. The rank *r* corresponds to the number of principal components representing the significant portion of the information contained within the data matrix **X**. The selection of *r* is somewhat arbitrary. It is usually set to be a number, as small as possible but sufficiently large enough such that there are no obvious spectral features found in the residual matrix $\mathbf{E}_{\mathbf{PCA}}$. The residual matrix $\mathbf{E}_{\mathbf{PCA}}$ is the portion of the original data, which is not accounted for by the first *r* principal components used for the data representation. The two matrices **T** and **P** complementally represent the entire features broadly distributed in **X**. Namely, **T** holds abstract information concerning the relationship among the samples and **P** contains summary of variable, e.g. wavenumber which provides chemically or physically meaningful interpretation to the pattern observed in **T**. For example, PCA of the two-way data based on temperature-dependent spectra provides **T** describing similar or dissimilar thermal behaviour of the sample during the perturbation period and corresponding **P** represent information on key molecular structure associated with such similar or dissimilar thermal behaviour of the sample.

For even more data, PARAFAC is used to decompose a multi-way data and Fig. 2 illustrates graphical representation of PARAFAC operation to decompose a three-way data into score and loading vectors. PARAFAC is utilized to decompose the multi-way data into a linear combination of score and loading matrices (Smilde *et al.*, 2004; Bro, 2004). The information on behavior induced by the perturbations is effectively described by score vectors and corresponding loading vectors provide chemically or physically meaningful interpretation

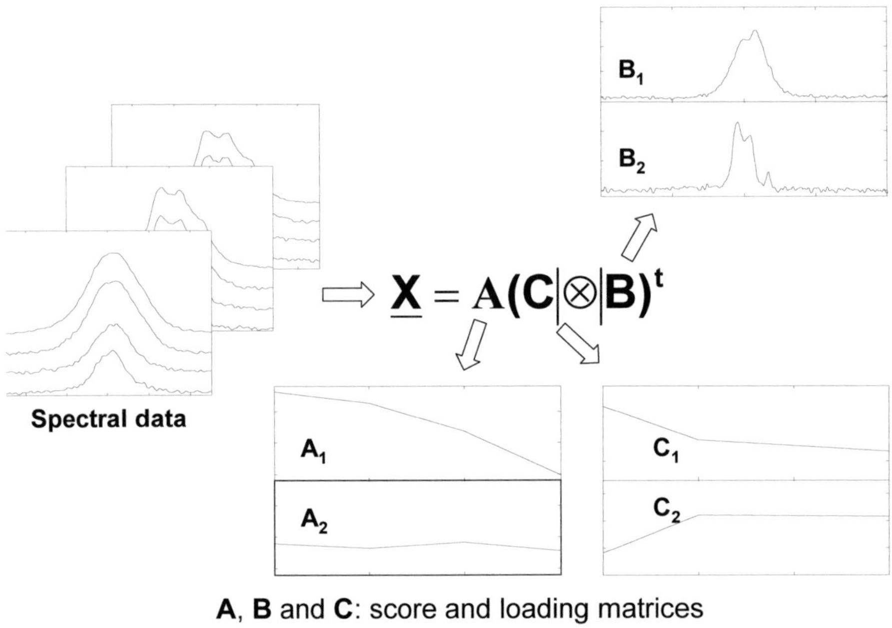

Fig. 2. Schematic illustration of PARAFAC trilinear model.

to the patterns observed in the scores of the PARAFAC trilinear model. Namely, by using PARAFAC operation, $I \times J \times K$ array matrix $\underline{\mathbf{X}}$ can be expressed in terms of a product of score and loading matrices, **A**, **B**, and **C**, and a residual matrix **E** as follows

$$\mathbf{X}^{(I \times JK)} = \mathbf{A}(\mathbf{C}|\otimes|\mathbf{B})^{\mathrm{t}} + \mathbf{E}^{(I \times JK)} \tag{2}$$

where $(I \times JK)$ refers to the way that the multi-way array is unfolded. The notation $|\otimes|$ means Khatri-Rao product which operate Kronecker product $|\otimes|$ on partitioned matrices defined as

$$\mathbf{C}|\otimes|\mathbf{B} = [\mathbf{c}_1 \otimes \mathbf{b}_1 \quad \mathbf{c}_2 \otimes \mathbf{b}_2 \quad \cdots \quad \mathbf{c}_F \otimes \mathbf{b}_F] \tag{3}$$

In PARAFAC analysis, the set of matrices **A**, **B** and **C** are usually obtained by iteratively solving alternating least-squares (ALS) problems $\min_{\mathbf{A,B,C}} \left\| \mathbf{X}^{(I \times JK)} - \mathbf{A}(\mathbf{C}|\otimes|\mathbf{B})^{\mathrm{t}} \right\|$ over **A** for fixed **B** and **C**, as well as the minimization over **B** or **C** in the similar matrix operation manner under appropriate model constraints, such as the non-negativity of concentration and spectral intensity (Bro & de Jong, 1997; Bro & Sidiropoulos, 1998). General procedure of PARAFAC becomes as follows,

Initialize **B** and **C** to obtain **Z** as

$$\mathbf{Z} = \mathbf{C} \otimes \mathbf{B} \tag{4}$$

A is given by

$$\mathbf{A} = \mathbf{X}^{(I \times JK)}\mathbf{Z}(\mathbf{Z}^{t}\mathbf{Z})^{+} \tag{5}$$

where the superscript + means the Moore-Penrose inverse. Then update **Z** as

$$\mathbf{Z} = \mathbf{C} \otimes \mathbf{A} \tag{6}$$

B is obtained as

$$\mathbf{B} = \mathbf{X}^{(J \times IK)}\mathbf{Z}(\mathbf{Z}^{t}\mathbf{Z})^{+} \tag{7}$$

Update **Z** as

$$\mathbf{Z} = \mathbf{B} \otimes \mathbf{A} \tag{8}$$

C is given by

$$\mathbf{C} = \mathbf{X}^{(K \times IJ)}\mathbf{Z}(\mathbf{Z}^{t}\mathbf{Z})^{+} \tag{9}$$

If the residual between the original $\underline{\mathbf{X}}$ and reconstructed $\underline{\mathbf{X}}$ by Eq. 2 is greater than error criteria, one repeats Eqs. (4)-(9) until convergence.

The initial estimates for **B** and **C** is important to obtain sufficient minimization of the error criteria (Shinzawa *et al.*, 2007, 2008a & 2008b). Although ALS algorithm usually offers an eventual convergence to the optimal solution with a sufficiently large number of iterations, it sometimes reaches the suboptimal local minimum (Jiang *et al.*, 2003 & 2004). Unfortunately, such local convergence does not usually offer a global minimum, but it may just be stuck in a local minimum, producing insufficient solution. The major cause of the suboptimal local convergence may be a poor initial estimation. One possible solution for this problem is to select proper initial estimate which is less sensitive to the presence of a local minimum, e.g. via signaler value decomposition (Bro & de Jong, 1997; Bro & Sidiropoulos, 1998; Wang *et al.*, 2006; Awa *et al.*, 2008).

3. Example

3.1 PLA nanocomposite

A pertinent example for PARAFAC analysis based on NMR spectra of PLA nanocomposites is provided here to show how certain useful information can be effectively extracted from an actual laboratory experiment.

Fig. 3 shows the molecular structure of PLA. PLA polymer is made up of many long chains consisting of the repeat unit shown in the figure. PLA is derived from renewable resources, such as corn starch via fermentation and it is biodegradable under the right conditions, such as the presence of oxygen (Tsuji *et al.*, 2010). Thus, PLA is a possible candidate of a new class of renewable polymers as a substitute for the petrochemical polymers. However, the physical properties of PLA are inadequate for the replacement of conventional commodity plastics in many applications.

Nanocomposite is a technique to improve the physical strength, thermal resistance and gas barrier by the dispersion of nanoclay into the polymer (Katti *et al.*, 2006). The improvement of such polymer properties by using nanocomposite is one of the primary areas of interest due to its potential applications. The polymer nanocomposites are generally formed by the addition of a small amount of nanoclay dispersion.

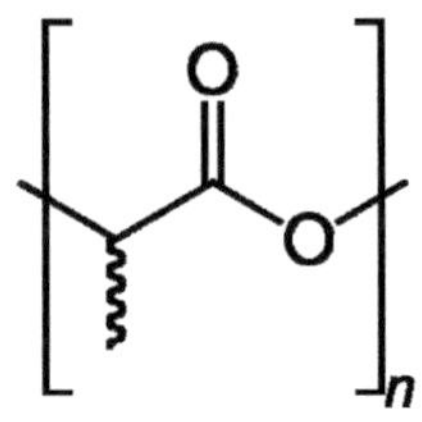

Fig. 3. Molecular structure of PLA.

Fig. 4 shows a schematic illustration of polymer nanocomposite. A typical form of the nanocomposite is intercalated nanocomposite, in which the unit cells of clay structure are expanded by the insertion of polymer into the interlayer spacing, while the periodicity of clay crystal structure is maintained. Most commonly, montmorillonite (MMT) is used as clay due to its highly expansive characteristic (Suguna Lakshmi *et al.*, 2008; Cervantes-Uc *et al.*, 2009). The MMT unit cell is composed of aluminum octahedra sandwiched between two silica tetrahedra with the unit cell dimension of about 1 nm in thickness. For facilitating better miscibility of hydrophobic polymer with the clay and increasing the spacing of the interlayer clay gallery, it is often treated with organic modifiers which are generally long carbon chain compounds with alkylammonium or alkylphosphonium cations.

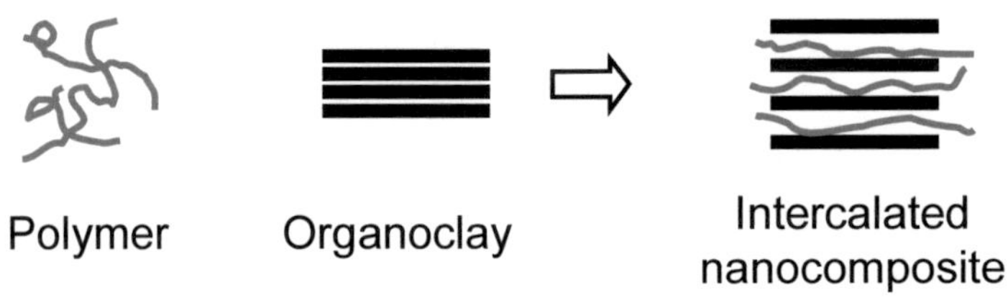

Fig. 4. Schematic illustration of polymer nanocomposite.

PLA nanocomposite samples used in this study were prepared with PLA (Teramac®, Unitika) and organically modified clay (S-BEN W®, Hojun). The samples were put into a Labo-plastomill consisting of a 30C150 kneader and an R100 mixer (Toyo Seiki Seisaku-sho, Ltd., Tokyo) to melt-blend at 190 °C and 50 rpm for about 10 minutes. Pellets thus obtained were pressed into 0.2 mm thick sheet sandwiched between two thick Teflon® films by a hot press at 190 °C.

Fig. 5 represents the effect of nanocomposite on PLA probed by Thermomechanical analysis (TMA). TMA is a technique to monitor the physical deformation of object under a constant load, while varying the temperature. For example, in this case, the elongation of the PLA nanocomposite samples (clay content = 0, 5 and 15 wt%) were measured by imposing a 9.8 mN load, while varying the temperature from 35 to 140 °C at a rate of 10 °C per a minute. The elongation of the samples starts when the temperature reaches glass transition temperature (T_g) of PLA, i.e. approximately 60 °C (Zhang et al., 2010). Then it gradually increases with the increase of temperature and it finally reached constant levels at the close of the observation period, indicating that the observed plastic deformation is closely related to glass-to-rubber transition of the amorphous component of PLA. It is also noted that the samples results in the different levels of elongation depending on the clay content. For example, the neat PLA sample shows 14.4 % of elongation. In contrast, the PLA-nanocomposite including 15 wt % of clay ends up with 9.1 % of elongation. The leveling off of the elongation indicates the formation of a network structure due to the presence of physical crosslinks created by the crystalline domain.

Although such observation effectively detects the macroscopic changes in the mechanical properties caused by the presence of clay particles, additional fundamental molecular level understanding of the reinforcement mechanism is also desired. Spectroscopic method should become an important tool to probe the phenomena at the molecular level.

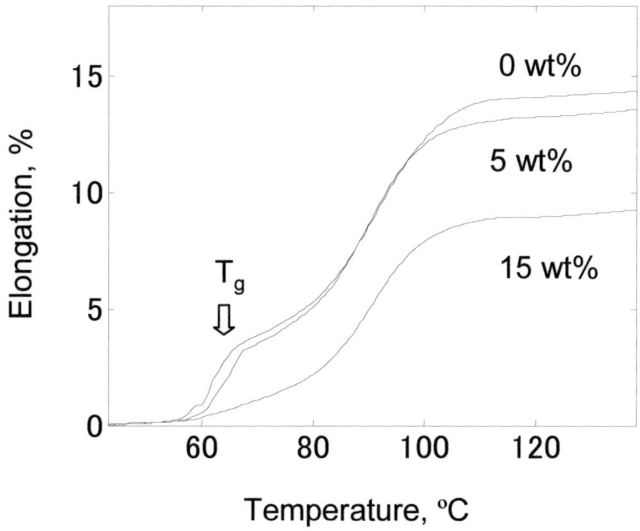

Fig. 5. Physical property of PLA samples proved by TMA.

3.2 PALAFAC analysis of NMR spectra of PLA nomocomposites

The temperature-dependent NMR spectra of the PLA samples collected under the varying temperature from 20 to 80 °C are shown in Fig. 6. Cross polarization-magic angle spinning (CP-MAS) NMR experiments were carried out on a Varian 400 NMR system spectrometer operated at 100.56 MHz for ^{13}C resonance with a cross polarization contact time of 2 ms (Fawcett, 1996). A zirconium oxide rotor of 4 mm diameter was used to acquire the NMR spectra at a spinning rate of 15 kHz. Each sample was packed into a 4 mm cylinder-type MAS rotor. A set of temperature-dependent NMR spectra were obtained under varying ambient temperature from 20 to 80 °C at every 20 °C step. The heating rate was approximately 10 °C per an hour.

Samples of semicrystalline polymers prepared from their melt possess complex supermolecular structure consisting of crystalline lamellae embedded in an amorphous matrix (Wunderlich, 1980). PLA essentially undergoes highly convoluted transition process, when temperature and its constitution are altered. These transitions include the melting of ordered molecular segments, as well as the grass-to-rubber transition and other relaxation of process of the amorphous component (Zhang et al., 2005; Meaurio *et al.*, 2006).

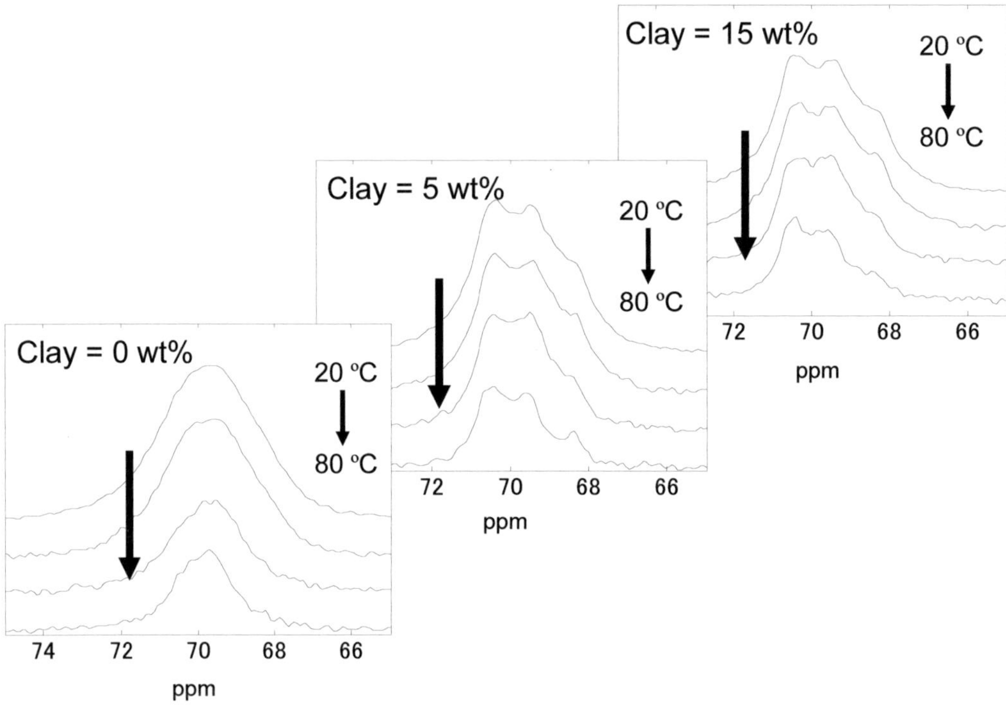

Fig. 6. Temperature-dependent CP-MAS NMR spectra of neat PLA and PLA nanocomposite samples.

The CP-MAS technique is ideal for the observation of ^{13}C spectra of solid samples. Since the local environment of a chemical group in solids are generally rigid, this leads to further considerations for crystallography or, more generally, molecular packing (Fawcett, 1996). The CPMAS NMR study of semicrystalline PLA samples is often complicated by the presence of overlapped contributions from coexisting crystalline and amorphous. For example, the unimodal peak observed around 69.5 ppm is assignable to CH structure which represents mobility of the main chain of the PLA (Tsuji *et al.*, 2010; Kister *et al.*, 1998). It is noted that the peak intensity gradually decreases with the increase of the temperature. This may be explained as the decrease in the cross polarization efficiency by the change in the molecular dynamics during the heating. Thus, the variation of the spectral intensity here reflects the structural alternation of PLA induced by the temperature.

More importantly, careful comparison of the samples reveals that the main feature of the NMR spectra of the three samples looks somewhat different. For example, the temperature-dependent NMR spectra of the PLA nanocomposite including 15 wt% clay provides specific three peaks at 70.5, 69.5 and 68.4, indicating the presence of the crystalline structure in the sample (Tsuji *et al.*, 2010; Kister *et al.*, 1998). When the sample has no clay in the system, these crystalline peaks are disappeared and compensated by the development of seemingly unimodal peak probably assigned to the amorphous of PLA (Tsuji *et al.*, 2010; Kister *et al.*, 1998). This indicates that the presence of the clay substantially influences supermolecular structure of the PLA. Consequently, it is very likely that the change in the spectral feature of the three-way data is closely related to temperature and clay content of the system. Thus, in turn, the fully detailed analysis of the data provides an interesting opportunity to probe the nature of the PLA nanocomposite by elucidating the variation of the NMR spectral intensity induced by the each perturbation with PARAFAC trilinear decomposition.

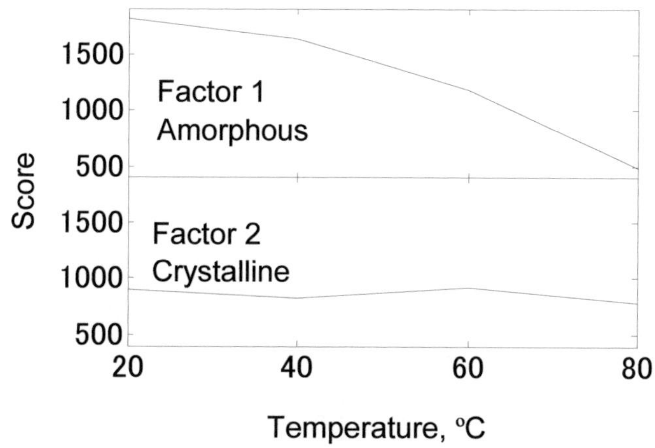

Fig. 7. Score vectors in score matrix **A** representing thermal behaviours of amorphous and crystalline components in PLA samples.

Fig. 7, 8 and 9 show results obtained from **A**, **B** and **C** matrices derived from PARAFAC analysis of the three-way NMR spectral data collected under varying temperature and clay content, respectively. Two major factors are indicated here, reflecting the fact that there are two species present in the system. One of the important benefits derived from PARAFAC decomposition of the multi-way data is the ability to rationally clarify the effect of the applied perturbations. For example, the matrix **A** represents abstract information on the temperature-induced behavior of the PLA under the influence of the clay content. In contrast, the matrix **C** holds essential information on the spectral intensity variation induced by the addition of the clay under the influence of the temperature. The matrix **B** contains loading vectors which provides chemical or physical interpretation to the patterns observed in the score matrices **A** and **C**.

It is noted that the loading vector of the first component of the matrix **B** (Fig. 8) resembles the spectral feature of the amorphous component of PLA. The loading vector of the second component of the matrix **B** shows characteristic three peaks assignable to crystalline component of the PLA. Thus it is most likely that the second factor represents thermal behaviours of the crystalline components in PLA samples.

Once the assignments for the loading vectors are established, it becomes possible to provide chemically meaningful interpretation to the score matrices **A** and **C** representing the dynamic behaviour of the components induced by the perturbations. For example, the score vector of the first factor in the matrix **A** represents the temperature-induced behaviour of the amorphous component of the PLA. On the other hand the score vector of the second factor means that of the crystalline component of the PLA. It is noted the score vector of the amorphous components exhibits obvious decrease with the temperature and such decrease becomes significant when the temperature exceeds its T_g. In contrast, the change in the score value of the crystalline component is small, indicating no major variation during the heating process.

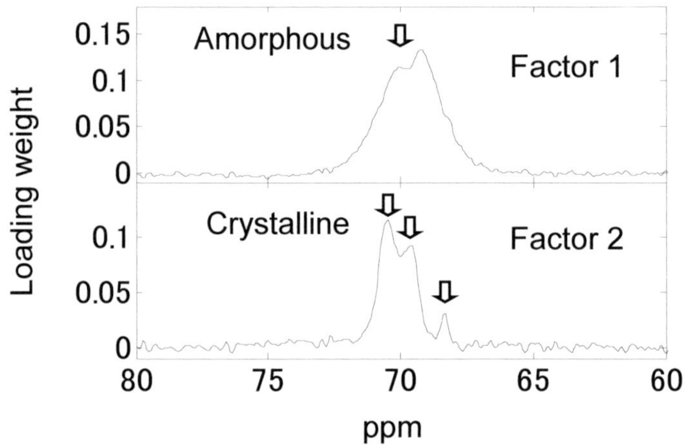

Fig. 8. Loading vectors in score matrix **B** representing thermal behaviours of amorphous and crystalline components in PLA samples.

The predominant variation of the amorphous component in the temperature region is explained as its glass-to-rubber transition. The change induced in the temperature region is associated with the Micro-Brownian motion of the PLA polymer segment. At a low temperature the amorphous regions of a polymer are in the glassy state. In this state the molecules are frozen on place. They may be able to vibrate slightly, but do not have any segmental motion. When the polymer is heated up to reach its T_g, the molecules can start to wiggle around to become rubbery state. Such segmental motion predominantly occurs in amorphous region of PLA while such motion is strongly restricted in systematically folded crystalline lamellae structure. Thus, it is very likely the observed change of the amorphous is related to glass-to-rubber transition of the amorphous component.

Now it is important to point out again that the predominant elongation in the TMA occurred around T_g. This elongation behaviour agrees well with the thermal behaviour of the amorphous component observed in the score matrix **A**. It thus suggests the physical elongation of the samples is essentially associated with the glass-to-rubber transition mainly occurred in the amorphous region.

It also becomes possible to provide the detailed interpretation to the pattern observed in the matrix **C** representing the clay-induced behaviours of amorphous and crystalline components in the PLA samples. The gradual decrease of the score of the first factor can be explained as the decrease of the amorphous component and the change in the sore of the first factor corresponds to the increase in the crystalline component by the addition of the

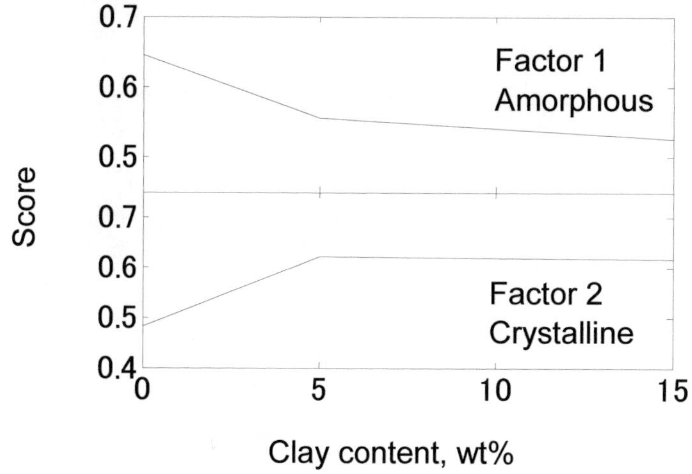

Fig. 9. Score vectors in score matrix **C** representing clay-induced behaviours of amorphous and crystalline components in PLA samples.

clay. It seems that the decrease in the amorphous is compensated by the development of the crystalline structure. In other words, the clay increases the frequency of the spontaneous nucleation of the PLA crystals.

PARAFAC trilinear model of the three-way NMR data of the PLA nanocomposites reveals that the crystalline and amorphous structures of the PLA nanocomposites undergo different transition under the heating. Namely, the change in the micro-Brownian motion of the polymer segments mainly occurs in the amorphous region. In addition, the different variations between the crystalline and amorphous component suggest the different effects of the presence of clay particles on them, i.e. nucleating effect of the clay. The decrease in the amorphous portion should result in the reduction of the structure undergoing the glass-to-rubber transition. Such variation of the crystallinity agrees well with the decreased elongation observed in the TMA. For example, in Fig. 5, the level of the elongation starting around T_g clearly decreases with the inclusion of the clay.

This hypothesis is also clearly supported with corresponding transmission electron microscope (TEM) images and differential scanning calorimetry (DSC) results of the PLA nanocomposite sample. Fig. 10 represents the TEM images of the PLA nanocomposite sample including 15 wt% clay. For example, in Fig. 10(a), one can see that the clay is broadly distributed over the PLA matrix. On the other hand, Fig. 10(b) reveals that some parts of the interlayer gallery is obviously extended, suggesting the insertion of the PLA polymer into the clay layers, namely intercalation.

DSC curves of the PLA samples, represented in Fig. 11, clearly show the presence of glass transition temperature around 60 °C. It is important to point out that this glass-to-rubber transition of amorphous component agrees well with the change in the elongation observed in the TMA. More importantly, the samples also provide obvious crystallization peak around 110 °C. The crystallization peak shows gradual increase by the inclusion of the clay content, suggesting quantitative increase in the amount of the crystalline structure. Thus, it

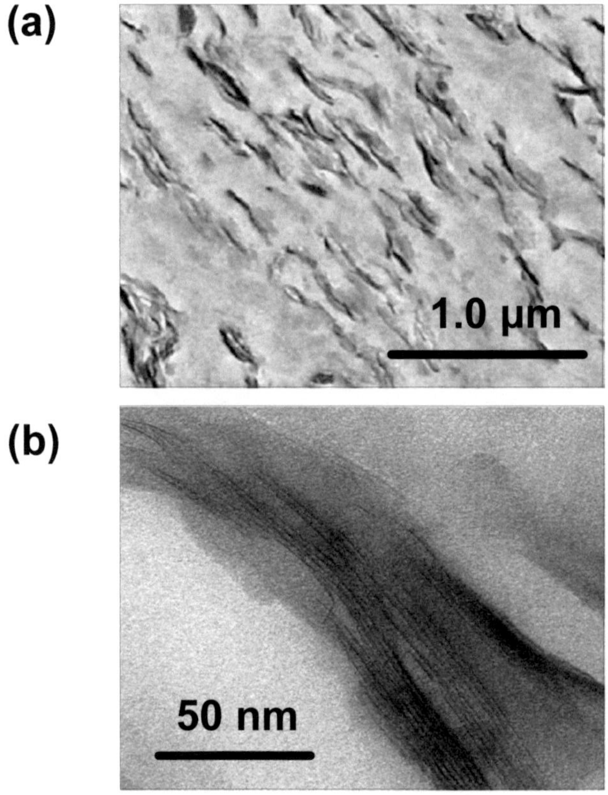

Fig. 10. TEM images of PLA nanocomposite sample.

is very likely that the clay works as the nucleating agent to increase the frequency of the spontaneous nucleation of the PLA crystals.

All the results put together, it provides overall picture of the system. When the clay is dispersed in the PLA matrix, the PLA polymer located at interlayer or around surface layer of the clay develops crystalline structure more frequently. The generation of the crystalline structure of PLA is compensated by the decrease of the amorphous content. This should decrease the structural portion substantially undergoing glass-to-rubber transition above T_g. Thus, in turn, it restricts the elongation of the samples during the heating process under a certain level of load. Consequently, it is demonstrated that PARAFAC analysis of the three-way data of the PLA nanocomposite samples effectively elucidates the mechanisms of the improvement of the mechanical property by the clay. By carrying out detailed band position shift analysis of the three way data of the temperature- and clay- dependent NMR spectra of the PLA samples, it becomes possible to extract chemically meaningful information concerning the variation of the crystalline structure closely associated with the nanocomposite system.

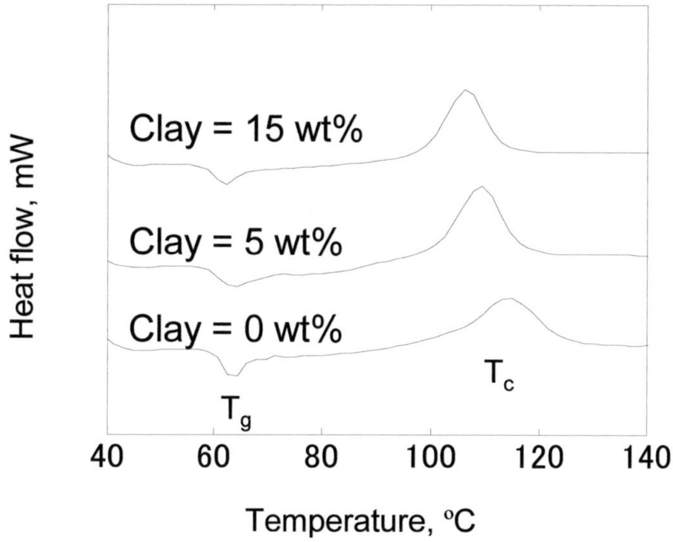

Fig. 11. DSC curves of neat PLA and PLA nanocomposite samples.

3. Conclusion

The basic background of PARAFAC and its practical example based on the temperature-dependent NMR spectra of the PLA nanocomposite samples are presented. The central concept of PARAFA decomposition of multi-way data lies in the fact that it can condense the essence of the information present in the multi-way data into a very compact matrix representation referred to as scores and loadings. Thus, while the score and loading matrices contain only a small number of factors, it effectively carries all the necessary information about spectral features and leads to sorting out the convoluted information content of highly complex chemical systems.

The effect of PLA nanocomposite is studied by the PARAFAC analysis of the temperature-dependent NMR spectra of several PLA nanocomposite samples including different clay contents. The PARAFAC analysis for the three-way data of the PLA nanocomposites revealed that the crystalline and amorphous structures of the PLA nanocomposites substantially undergo different transition under the heating. Namely, the change in the micro-Brownian motion of the polymer segments mainly occurs in the amorphous region when the PLA samples are heated up to their T_g. It also revealed that clay potentially works as nucleating effect of the clay. Namely, it increases the frequency of the spontaneous nucleation of the PLA crystals. Thus, in turn, the change in the population of the rigid crystalline and rubbery amorphous provides the improvement of the physical property. Consequently, it is possible to derive in-depth understanding of the PLA nanocomposites.

4. Acknowledgment

A part of this work was financially supported by NEDO "Technological Development of Ultra-hybrid Materials" Project.

5. References

Amigo, J. M., Skov, T., Coello, J., Maspoch, S. & Bro, R. (2008) Solving GC-MS problems with PARAFAC2. *TrAC Trends in Analytical Chemistry,* Vo. 27, No. 8, pp. 714-725

Awa, K., Okumura, T., Shinzawa, H., Otsuka, M. & Ozaki, Y. (2008). Self-modeling Curve Resolution (SMCR) Analysis of Near-infrared (NIR) Iimaging Data of Pharmaceutical Tablets. *Analytica Chimica Acta,* Vol. 619, No. 1, pp. 81-86

Bro, R. (2004). PARAFAC. Tutorial and applications. *Chemometrics and Intelligent Laboratory Systems,* Vol. 37, No. 2, pp. 149-171

Bro, R. & de Jong, S. (1997). A fast non-negativity constrained linear least squares algorithm for use in multi-way algorithms. *Journal of Chemometrics,* Vol. 11, No. 5, pp. 393-401

Bro, R. & Sidiropoulos, N. (1998). Least squares algorithms under unimodality and non-negativity constraints. *Journal of Chemometrics.* Vol. 12, No. 4, pp. 223–247

Bro, R., Viereck, N., Toft, M., Toft, H., Hansen, P. I. & Engelsen, S. B. (2010). Mathematical chromatography solves the cocktail party effect in mixtures using 2D spectra and PARAFAC. *TrAC Trends in Analytical Chemistry,* Vol. 29, No. 4, pp. 281-284

Cervantes-Uc, J. M., Espinosa, J. I. M., Cauich-Rodriguez, J. V., Avila-Ortega, A., Vazquez-Torres, H., Marcos-Fernandez. A. & San Roman, J. (2009). TGA/FTIR studies of segmented aliphatic polyurethanes and their nanocomposites prepared with commercial montmorillonites. *Polymer Degradation and Stability,* Vol. 94, No. 10, pp. 1666-1677

Christensen, J., Miquel Becker, B. & Frederiksen, C. S. (2005). Fluorescence spectroscopy and PARAFAC in the analysis of yogurt. *Chemometrics and Intelligent Laboratory Systems,* Vol. 75, No. 2, pp. 201-208

Ebrahimi, D., Kennedy, D. F., Messerle, B. A. & Hibbert, D. B. (2008). High throughput screening arrays of rhodium and iridium complexes as catalysts for intramolecular hydroamination using parallel factor analysis. *Analyst,* Vol. 133, No. 6, pp. 817-822

Fawcett, A. H. (1996). *Polymer Spectroscopy,* John Wiley & Sons, ISBN 0471960292, West Sussex, UK

Jiang, J.-H., Šašic, S., Yu, R.-Q. & Ozaki, Y. (2003). Resolution of two-way data from spectroscopic monitoring of reaction or process systems by parallel vector analysis (PVA) and window factor analysis (WFA): inspection of the effect of mass balance, methods and simulations. *Journal of Chemometrics,* Vol. 17, No. 3, pp. 186-197

Jiang, J.-H., Liang, Y. & Ozaki, Y. (2004). Principles and methodologies in self-modeling curve resolution. *Chemometrics and Intelligent Laboratory Systems,* Vol. 71, No. 1, pp. 1-12

Katti, K. S., Sikdar, D., Katti, D. R., Ghosh, P. & Verma. D. (2006). Molecular interactions in intercalated organically modified clay andclay–polycaprolactam nanocomposites: Experiments and modeling. *Polymer,* Vol. 47, No. 1, pp. 403-414

Kister, G., Cassanas, G. & Vert, M. (1998). Structure and morphology of solid lactide-glycolide copolymers from ^{13}C n.m.r., infra-red and Raman spectroscopy. *Polymer,* Vol. 39, No. 15, pp. 3335-3340

Meaurio, E., Zuza, E., López-Rodríguez. N. & Sarasua, J. R. (2006). Conformational Behavior of Poly(L-lactide) Studied by Infrared Spectroscopy. *Journal of Physical Chemistry B,* Vol. 110, No. 11 pp. 5790-5800

Rinnan, Å. & Andersen, C. M. (2005). Handling of first-order Rayleigh scatter in PARAFAC modelling of fluorescence excitation–emission data. *Chemometrics and Intelligent Laboratory Systems,* Vol. 76, No. 1, pp. 91-99

Shinzawa, H., Iwahashi, M., Noda, I. & Ozaki. Y. (2008a). Asynchronous Kernel Analysis for Binary Mixture Solutions of Ethanol and Carboxylic Acids. *Journal of Molecular Structure,* Vol. 883-884, No. 30, pp. 27-30

Shinzawa, H., Iwahashi, M., Noda, I. & Ozaki, Y. (2008b). A Convergence Criterion in Alternating Least Squares (ALS) by Global Phase Angle. *Journal of Molecular Structure,* Vol. 883-884, No. 30, pp. 73-78

Shinzawa, H., Jiang, J.-H., Iwahashi, M., Noda, I. & Ozaki, Y. (2007). Self-modeling Curve Resolution (SMCR) by Particle Swarm Optimization (PSO). *Analytica Chimica Acta,* Vol. 595, No. 1-2, pp. 275-281

Shinzawa, H., Awa, K., Kanematsu, W. & Ozaki, Y. (2010). Multivariate Data Analysis for Raman Spectroscopic Imaging. *Journal of Raman Spectroscopy,* Vol. 40, No. 12, pp. 1720-1725

Smilde, A., Bro, R. & Geladi, P. (November 2004). Multi-way Analysis: Applications in the Chemical Sciences. John Wiley & Sons, ISBN: 0471986911, West Sussex, UK

Suguna Lakshmi, M., Narmadha, B. & Reddy, B. S. R. (2008). Enhanced thermal stability and structural characteristics of different MMT-Clay/epoxy-nanocomposite materials. *Polymer Degradation and Stability,* Vol. 93, No. 1, pp 201-213

Tsuji, H., Kamo, S. & Horii, F. (2010). Solid-state ^{13}C NMR analyses of the structures of crystallized and quenched poly(lactide)s: Effects of crystallinity, water absorption, hydrolytic degradation, and tacticity. *Polymer,* Vol. 51, No. 10, pp. 2215-2220

Van Benthem, M. H., Lane, T. W., Davis, R. W., Lane, R. D. & Keenan, M. R., (2011). PARAFAC modeling of three-way hyperspectral images: Endogenous fluorophores as health biomarkers in aquatic species, *Chemometrics and Intelligent Laboratory Systems,* Vol. 106, No. 1, pp. 115-124

Wang, Z.-G., Jiang, J.-H., Ding, Y.-J., Wu, H.-L. & Yu, Ru-Qin., (2006). Trilinear evolving factor analysis for the resolution of three-way multi-component chromatograms. *Analytica Chimica Acta,* Vol. 558, No. 1-2, pp. 137-143

Wu, Y., Yuan, B., Zhao, J.-G. & Ozaki, Y. (2003). Hybrid Two-Dimensional Correlation and Parallel Factor Studies on the Switching Dynamics of a Surface-stabilized Ferroelectric Liquid Crystal. *Journal of Physical Chemistry B,* Vol. 107, No. 31, pp. 7706–7715

Wunderlich, B. (1980). *Macromolecular Physics: Vol. 2 Crystal Nucleation, Growth, Annealing,* Academic Press, New York, USA

Zhang, J., Li, C., Duan, Y., Domb, A. J. & Ozaki, Y. (2010). Glass transition and disorder-to-order phase transition behavior of poly(l-lacticacid) revealed by infrared spectroscopy. *Vibrational Spectroscopy,* Vol. 53, No. 2, pp. 307–310

Zhang, J., Sato, H., Tsuji, H., Noda, I. & Ozaki, Y. (2005). Differences in the CH_3···O=C interactions among poly(L-lactide), poly(L-lactide)/poly(D-lactide) stereocomplex, and poly(3-hydroxybutyrate) studied by infrared spectroscopy. *Journal of Molecular Structure.* Vol. 735–736, No. 14, pp. 249–257

14

Application of Chemometrics to the Interpretation of Analytical Separations Data

James J. Harynuk, A. Paulina de la Mata and Nikolai A. Sinkov
Department of Chemistry, University of Alberta
Canada

1. Introduction

Interesting real-world samples are almost always present as mixtures containing the analyte(s) of interest and a matrix of components that are irrelevant to answering the analytical question at hand. Additionally, the compounds comprising the matrix are usually present in far greater abundance (both number and concentration) than the analytes of interest, making quantification or even detection of these analytes difficult if not impossible.

When tasked with these types of samples, analysts turn to some form of separations technique such as gas or liquid chromatography (GC or LC) or capillary electrophoresis (CE) so that individual components in each sample may be quantified. More recently, more complex analytical questions are being probed, for example profiling blood or urine to identify a disease state or ascertaining the geographic origin of a food/beverage sample. These tasks often go beyond the simple quantification of one or two analytes in a sample. For these and other similar questions, separations scientists are turning more often to chemometric tools as a means of visualizing and interpreting the rich data that they obtain from their separations systems.

Here we present a brief overview of separations approaches, with a focus on the data that are derived from different methods and on phenomena in the separations approach that lead to challenges in data interpretation. This is followed by a discussion of approaches that exist for the chemometric interpretation of separations data, specific challenges that arise in the chemometric treatment of these data, and solutions that have been implemented to deal with these challenges.

1.1 Separations techniques

Chromatography is widely used for the separation, purification, and analysis of mixtures. In general, analytes contained in either a gaseous or liquid mobile phase are flowed past a stationary phase which is usually confined within a column. Depending on the chemistries of the analytes and the conditions of the separation (mobile/stationary phase compositions, temperature, etc.) different compounds will partition between the two phases to varying degrees. The separation arises due to this differential partitioning, with analytes which associate weakly with the stationary phase passing through the column more quickly than those with a greater affinity for the stationary phase (Miller, 2005; Cazes, 2010).

There are many types of chromatography, with the most common being liquid chromatography (LC) where analytes partition between a mobile liquid phase and an immobile stationary phase, and gas chromatography (GC) where the mobile phase is a gas and the stationary phase is a solid or more often a viscous, liquid-like polymer. There are numerous modes for LC separations, including for example reverse-phase (RPLC), normal-phase (NPLC), ion (IC), size exclusion (SEC), and hydrophilic interaction (HILIC) to name a few. From a point of view of chemometric data interpretation and the discussion in this chapter, all of these LC separations generate data which are equivalent. In any chromatographic separation, the sample is delivered to the inlet of the column while the outlet is connected to a detector, which records a continuous signal. The detector response rises and then falls to baseline based on the analyte flux passing through it, ideally generating one separate peak with an approximately Gaussian shape for each individual analyte. Assuming that the conditions for repeat analyses are not changed, the peak for a given analyte will appear at the same time in every analysis, with the peak area/height being proportional to the quantity of analyte present in a sample (Poole, 2003; Miller, 2005).

Another separations technique which is popular for some samples is capillary electrophoresis (CE). Here, an electric field applied across a fused silica capillary containing a buffer induces motion of the buffer and analytes in the sample. The CE separation is dependent on differential mobilities of analytes in the solution in the presence of the electric field. This difference in mobilities is based on the fact that different analytes have different charges and sizes in solution. While the separation mechanism of CE is fundamentally different from the chromatographic mechanism, the data are a series of peaks recorded as a function of time. Consequently, the same tools can be applied to data from a CE separation, and similar concerns exist for the interpretation of these data (Poole, 2003; Miller, 2005). For ease of readability, and because chemometrics are more often applied to chromatographic data than electrophoretic data, we will often refer to a chromatogram in this chapter. This could equally be an electropherogram; when considering the application of chemometric techniques to separations data whether the origin is electrophoretic or chromatographic is largely irrelevant.

When tasked with incredibly complex samples, analysts are now turning more and more frequently to so-called comprehensive multidimensional separations (e.g.: GC×GC, LC×LC, CE×CE) (Liu & Phillips, 1991; Erni & Frei, 1978; Michels et al., 2002). In these techniques, the mixture of compounds is sequentially separated by two different separation mechanisms. In the case of GC×GC, for example, a sample might be separated first on an apolar column, followed by a polar column. The exact workings of comprehensive multidimensional separations are beyond the scope of this work, and are discussed elsewhere (Górecki et al., 2004; Cortes et al., 2009; François et al., 2009; Kivilompolo et al., 2011; Li et al., 2011). However, these techniques are gaining in popularity, and are capable of separating exceedingly complex mixtures comprising thousands of individual compounds. Due to the vastly improved separation power of these techniques, the data are much more information-rich, and without some form of chemometric treatment it is essentially impossible to do more than scratch the surface of the information contained therein.

1.2 Separations data

The detector signal from a separations experiment, when plotted vs. time, yields a series of (ideally) Gaussian peaks, each representing one compound in the sample. Acquisition speed is one consideration for a chromatographic detector: it must be sufficient to faithfully record the profile of each compound as it passes through the detector. In order to obtain an accurate peak profile, the minimum number of acquisition points required across a peak is 10. Thus, the required speed of the detector is intrinsically linked to the nature of the separation. In separations where the base width of the peaks are on the order of 5 s, a data rate of 2 Hz would be acceptable, but when peak widths are 100-200 ms, as in GC×GC, then detector rates on the order of 50-100 Hz are required for quantitative analysis.

From a point of view of chemometric analysis of separations data, another important consideration is whether the detector is univariate or multivariate. Univariate detectors, such as the flame ionisation detector, or single-wavelength UV-visible spectrometer, record only one variable as a function of time, generating data which take the form of a vector of instrument response. Other detectors, typically mass spectrometers and multi-channel spectroscopic instruments, can be operated such that they record a multivariate response. Data from these instruments comprise an array of signal responses with each row representing a time when a response was recorded, and each column representing a variable that was recorded (e.g.: detector wavelength, ion mass-to-charge ratio). To the chemometrician, it is immediately obvious that there are numerous advantages to collecting multivariate chromatographic data; however, it is worth noting that most of this advantage has been by and large ignored by chromatographers. Typically, only the profile of a single variable vs. time would be used to selectively quantify an analyte, or the detector response across all channels at a given time used to help identify a peak.

One other aspect of raw separations data is the sheer number of variables measured for each sample. When a univariate detector is used for a 15 min separation, operating with an acquisition speed of 10 Hz, the data will be a vector of 9000 individual measurements per sample. If a multivariate detector is employed instead, for example a mass spectrometer operating over a 30-300 m/z mass range, this number increases to 2 439 000 individual variables arranged in a 9000 × 271 array per sample! In the case of GC×GC-MS analyses, which are typically 60 min in length but have a high-speed MS collecting data at rates of ~100 Hz, there are on the order of 100 million data points collected for each sample.

2. Challenges with chromatographic data

Variations in analytical separations data are, in principle, no different from those derived from any other instrument; being based on both chemical and non-chemical aspects of the analysis. All relevant information will be contained within the chemical variations and any chemometric approach to interpreting chromatographic data must be capable of identifying relevant chemical variation while minimizing the effects of irrelevant chemical and non-chemical variations. Sources of irrelevant chemical variation include matrix peaks, here defined as any chemical source of signal introduced with the sample, but having no bearing on the conclusions drawn from the data. Additionally, there is background signal which can for example derive from changes in mobile phase concentration which influence detector signals in LC or chemical "bleed" signatures from stationary phases as they degrade in GC. Non-chemical variations include, for example, baseline drift (for non-chemical reasons), retention time shifts (due to minor fluctuations in operating conditions), and electronic noise. These may easily interfere with the relevant chemical information, degrading model performance and the validity of results (de la Mata-Espinosa et al., 2011a). Figure 1 presents an overlay of several LC chromatograms of similar samples exemplifying the challenges of baseline drift and retention time shifts. One of the major challenges in handling chromatographic data using chemometric tools is appropriate pre-processing to remove as many non-chemical and irrelevant chemical variations as possible from the data set.

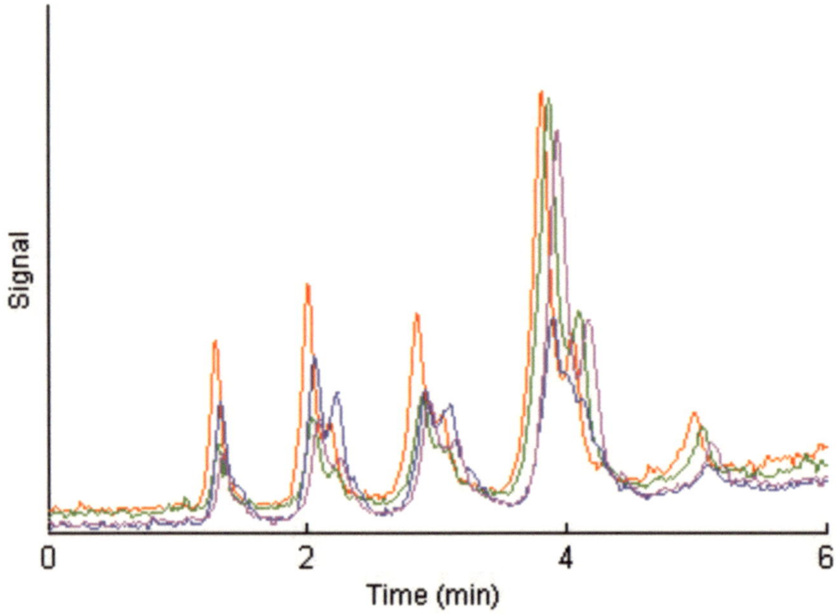

Fig. 1. LC chromatograms of edible oils showing a high degree of variation in baseline.

Initial efforts into the application of statistical and chemometric tools to chromatographic data were accomplished using data that were processed to provide a list of detected, integrated peak areas or heights (or the calibrated concentrations for known compounds). However, the trend in recent years has turned towards the direct chemometric interpretation of raw chromatographic signals (Watson et al., 2006; Johnson & Synovec, 2002). The reason for this trend is that many errors can occur during integration of raw signals (Asher et al., 2009; de la Mata-Espinosa et al., 2011b). By applying chemometric tools directly to the raw data, many of these errors can be avoided. Of course, when working with the raw data, other issues become more important, most notably retention time shifts and the population of available variables.

2.1 Baseline and noise

Baseline variations, such as noise and drift, are due to small changes in experimental conditions, for example changes in detector response due to the mobile phase gradient in LC separations or increased levels of stationary phase bleed at higher temperatures in temperature-programmed GC. Other sources of noise and drift could include changes in detector response as its components age, contamination of solvents or gases, and of course electronic noise (which is minimal in modern chromatographic systems).

Chemometric approaches to handling chromatographic data should incorporate baseline correction of some form. When raw chromatographic data are processed, the method of baseline correction and its importance are generally obvious to the analyst. In the case where integrated peak tables are used, this is often done automatically by the chromatographic software with little consideration by the analyst, even though the manner in which the baseline is calculated will significantly influence the determination of peak areas/heights.

2.2 Retention time shifts

In all separations, retention times of peaks can easily shift by a few seconds from one analysis to the next. This is not much of an issue with simple samples having only a few peaks which are then integrated prior to chemometric analysis. However, retention times of peaks are used for identifying the compounds. With complex separations, unstable retention times may result in unreliable peak identification, making comparisons from one run to the next impossible. When comparing raw data this is even more important as one must ensure that the peak for a given component is always registered in the exact same position in the data matrix so that the algorithms will recognize the signals correctly.

The causes of retention time shifts depend on the separations technique being used. In GC, peaks may shift due to degradation of the stationary phase, decreasing retention times over time; build-up of heavy matrix components which foul the column, effectively changing the chemistry of the stationary phase; minor gas leaks which alter the flow rate; or even matrix effects on the evaporation rate in the injector, affecting the rate of mass transfer to the column. In LC, peak shifts may be due to small fluctuations in mobile phase chemistry from one run to the next; temperature fluctuations which in turn affect solvent viscosity and solute diffusion coefficients, altering the kinetics as well as the thermodynamics of the separation; or degradation / fouling of the stationary phase of the column. CE is the technique most prone to drastic shifts in migration time, due to the instability of the electroosmotic flow in the capillary (Figure 2). Electroosmotic flow depends on the applied voltage, the buffer concentration and composition, and is incredibly sensitive to the surface chemistry of the capillary. The act of analyzing a sample by CE will often have a minor, possibly irreversible effect on the capillary surface, resulting in a change in the migration time of an analyte.

Shifts in retention times are minimized by proper instrument maintenance, precise control of instrumental conditions or by using approaches such as retention time locking in GC to account for variations in instrument performance (Etxebarria et al., 2009; Mommers et al., 2011) and relative retention times in CE. Even with these approaches, some retention time shifting will occur and require more advanced alignment techniques for correction prior to chemometric analysis.

2.3 Incomplete separation

Another challenge with the interpretation of chromatographic data is incomplete separation of peaks. If two or more compounds have similar retention characteristics under a given set of separation conditions, they will not be completely resolved, as evidenced by the peak clusters in Figure 1. In these cases, apportioning the signal between the different compounds

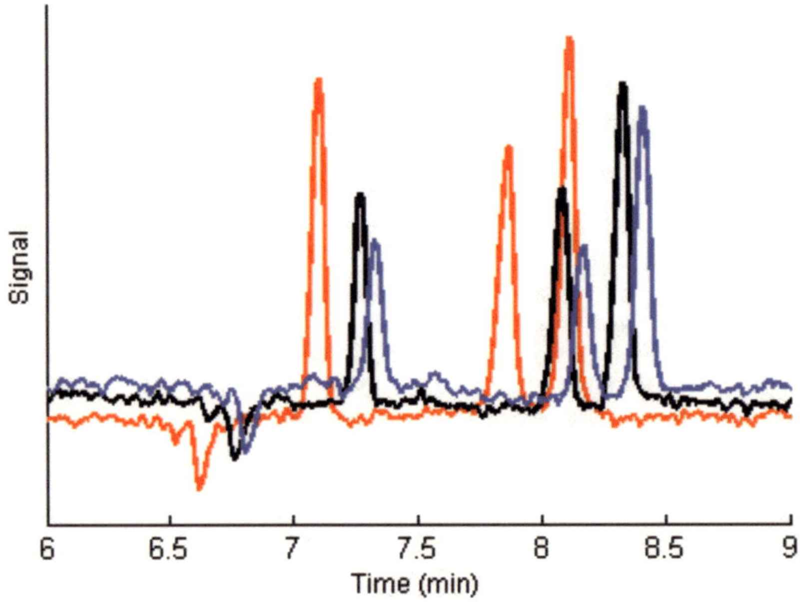

Fig. 2. CE of substituted benzenes showing extreme misalignment.

becomes a challenge, especially for univariate signals. The general approach used for these cases is one of deconvolution: decomposing the analytical signal to determine the contribution of each coeluting compound, or to determine the contribution of the compound of interest, disregarding the remaining data.

2.4 Data overload

As shown in Section 1.2, raw chromatographic signals present an overabundance of data to the analyst. This poses several challenges. From a practical point of view, attempts to construct a chemometric model using the entirety of the data set could easily exceed the capabilities of the computer system being used. More fundamentally, if the raw data are considered, the number of variables measured for each sample will vastly outnumber the number of samples available in the data set. These overdetermined systems can defeat many chemometric techniques due, for example, to collinear variables. Finally, for most chromatograms, especially multidimensional ones, only a small fraction of the data points actually contain meaningful signal. Most of the signal is due to background noise or irrelevant matrix components. Consequently, the raw data must somehow be reduced in size prior to chemometric analysis. This is typically achieved via a feature selection approach, as discussed in Section 3.3.3.

3. Pre-processing steps for chromatographic data

3.1 Baseline correction

The aim of baseline correction is to separate the analyte signal of interest from signal which arises due to changes in mobile phase composition or stationary phase bleed and signal due to electronic noise. Several baseline correction methods have been proposed in literature, with the two most common approaches being to fit a curve to the data and subtract this value from the signal, and modeling the baseline to exclude it using factor models (Amigo et al., 2010).

Curve fitting is the classical approach used in virtually all commercial software packages provided by vendors of separations equipment. The algorithms used in this approach fit a polynomial function across segments of the chromatogram using regions where no analyte peaks elute to determine the coefficients of the polynomial and then interpolating the background signal for regions where peaks are eluting. The functions are usually first-order polynomials; however, higher-order polynomials or a series of connected first-order polynomials are also used in some situations. Having determined the equation of the background signal, the fitted line is then subtracted from the signal (Brereton, 2003; Gan et al., 2006; Kaczmarek et al., 2005; Zhang et al., 2010; Persson & Strang, 2003; Eilers, 2003). Correction of the baseline using curve fitting is demonstrated in Figure 3.

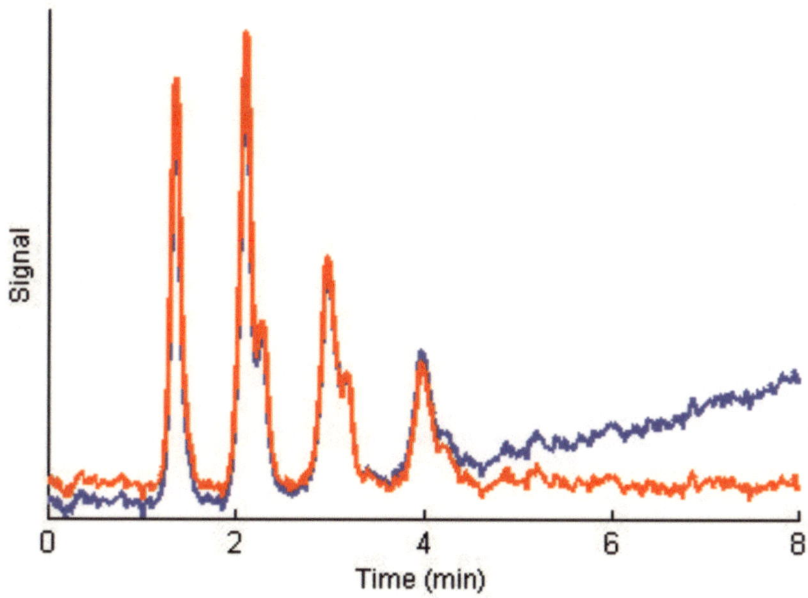

Fig. 3. An LC chromatogram before (blue) and after (red) baseline correction.

The approach of using models such as parallel factor analysis (PARAFAC) for background correction is analogous to the use of these approaches for deconvoluting coeluting peaks. As these models are more often used for this purpose than for simple background correction, they will be discussed in more detail in Section 3.3. These approaches often rely on having a multivariate signal and are applied to the chromatogram or more typically small selected regions where a single analyte elutes. The result of applying these deconvolution techniques for background correction is essentially the deconvolution of a single analyte peak, with the background noise making up the error matrix (Amigo et al., 2010). These approaches are generally more powerful and likely result in better quality analytical data, but they are not widely used in separation science. The reason for this is likely historical as these tools have only recently become available to the separation sciences, while the classical curve fitting approach is well established, works with univariate detectors, and performs well in most practical situations.

3.2 Alignment of separations data

The retention times of analytes in separations fluctuate from one analytical run to the next and, in order for chemometric techniques to be applied to separations data, these fluctuations must be corrected during pre-processing. This ensures that the signal from each analyte in each analysis is correctly registered within the data matrix to be processed. There are essentially two approaches to this problem: integrated peak tables, or mathematical warping and alignment of the raw signal.

3.2.1 Peak tables

Integrated peak tables are the simplest way to ensure that analytical separations data are properly aligned for chemometric processing. In order to use this approach, one must be able to reliably assign a unique identifier to each peak in each sample of the data set, and ensure that the same compound is identified with the same identifier in each sample. It should be noted that while the compound name is an obvious identifier, a series of labels such as *Unknown x*, where *x* is a numerical identifier would also be acceptable in the event that compound names were unknown, so long as compounds are matched correctly. Rather than identifying peaks by retention time, one could use relative retention times or retention indices in order to adjust for slight variations in the retention times of peaks. Algorithms for aligning peak tables exist and perform well, so long as some peaks can be easily and reliably matched across all chromatograms (Lavine et al., 2001).

The challenges with this approach stem from its reliance on integrated peak tables. Thus, any integration errors due to poorly-resolved peaks or peaks that are missed due to falling outside of integration parameters in the software will impact any subsequent analysis.

3.2.2 Raw signal alignment

Alignment of raw chromatographic signals prior to chemometric processing is more complex than the alignment of peak tables. In addition to the three more popular algorithms that will be presented below, there are several others that have been developed (Yao et al., 2007; Toppo et al., 2008; Eilers, 2004; Van Nederkassel et al., 2006). In deciding which approach to use, one of the first questions to be answered is if the analysis is to be qualitative or quantitative. This is because some alignment methods can distort peaks, affecting their quantification. Some of the more common algorithms include correlation optimized warping (COW) (Nielsen et al., 1998; Tomasi et al., 2004), correlation optimized shifting (coshift) (Van den Berg, 2005), and a piecewise peak-matching algorithm (Johnson et al., 2003).

In instances where there are non-systematic peak shifts, COW is a popular algorithm. COW relies on stretching or compressing segments of a sample signal such that the correlation coefficient between it and a reference signal is maximized for each interval. Care must be taken with the selection of the input parameters to avoid significant changes in peak shapes as this approach to the warping of the chromatogram has been shown to affect peak areas, leading to poor quantitative conclusions (Nielsen et al., 1998; Tomasi et al., 2004).

A fast and simple alignment algorithm is coshift. This algorithm is useful when data only require a single left-right shift in retention time. The entire data matrix is shifted in one direction or the other by a set amount, maximizing the correlation between a target and the data matrix that required alignment. The single shifting value for the entire data matrix is a weakness, especially for chromatographic data where peaks can shift in different directions and to different extents in a single file. To handle this, an algorithm termed icoshift (interval-correlation-shifting) has been derived from coshift. Icoshift aligns each data matrix to a target by maximizing the cross-correlation between the sample and the target within a series of user-defined intervals (Savorani et al., 2010). The use of multiple intervals permits the alignment of separations data where shifts of different magnitudes and directions occur. These alignment algorithms have been used successfully for both one-dimensional data (de la Mata-Espinosa, 2011a; Liang, 2010; Laursen, 2010) and two-dimensional data, with some modifications (Zhang, 2008). It is important to note that the shifting of chromatograms using coshift or icoshift does not lead to distortions of peak shape, and consequently does not introduce errors into quantitative results.

The piecewise peak matching approach (Johnson et al., 2003) provides another avenue for chromatographic alignment. In this approach, peaks are identified in a target signal to which all other signals will be aligned. The algorithm then identifies peaks within the sample signals located within predetermined windows of the peaks in the target. Peaks within windows are deemed to come from the same compound, and matched. The chromatograms are aligned by stretching or compressing the regions between peak apexes. A variant of this algorithm can be used when MS data are available. In this case, the mass spectrum at the apex of each peak in the target signal is compared to the mass spectrum of each peak within a set window on the sample signal and peaks are matched if their spectra have a high enough match quality (Watson et al., 2006). A general scheme for peak alignment using this approach is described in Figure 4. Depending on the number and relative positions of the peaks in chromatograms matched using this approach, peak shapes may be altered, possibly affecting quantitative results.

One of the biggest challenges for all alignment algorithms is that they depend on the data to be aligned being reasonably similar in terms of both matrix and analyte peaks. In some instances this will not be the case. In our laboratory, we have observed this when analyzing arson debris where the matrix and analytes form an incredibly complex and variable chromatogram from one sample to the next. A similar situation can be easily imagined when processing samples of biological origin. One solution to this issue is to add markers to every sample prior to the separation step in the analysis. These markers should be easily identifiable within the samples, even under conditions where they coelute with matrix components; should occur in multiple, evenly distributed locations along the chromatogram, and should not occur natively in the samples. One choice is a series of deuterated compounds which, with MS detection, are trivial to identify even in a complex mixture (Sinkov et al., 2011b). One additional benefit is that these compounds can act as internal standards if quantitative results are desired.

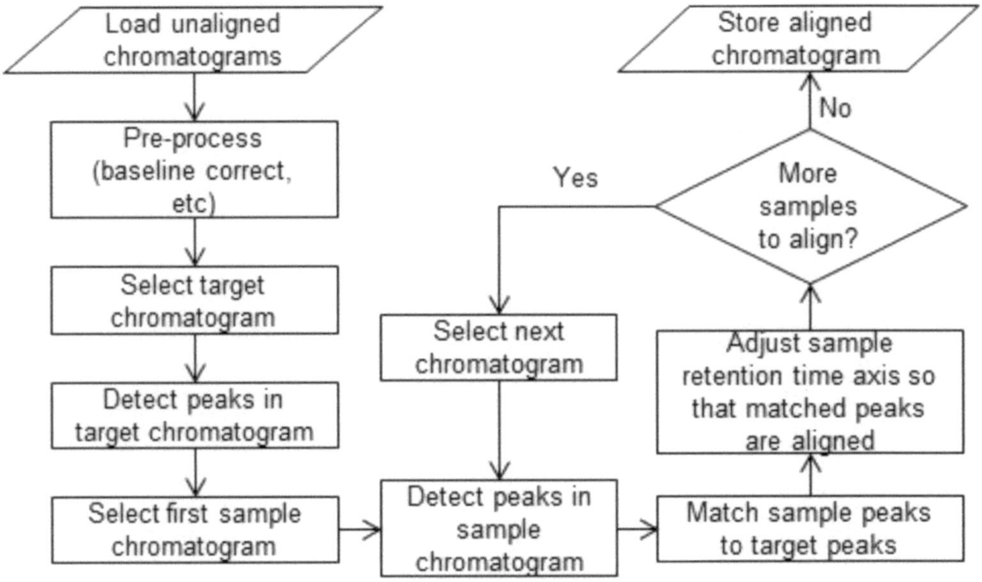

Fig. 4. Flowchart for target-based chromatographic alignment, adapted from (Johnson et al., 2003).

3.3 Deconvolution of overlapping peaks

The central issue in deconvolution is depicted in Figure 5. The instrument response is represented as a black solid line which is the sum of the four dashed, coloured peaks. Ideally, the four signals should be individually quantified. This is a common problem for analytical separations, even those of relatively simple mixtures. Some of these issues may be solved by changing the experimental conditions or using characteristic features (wavelengths or ions) of the coeluting analytes and a multivariate detector to selectively detect and quantify them. However, in many cases this is insufficient and more advanced techniques must be used. The strategies used for deconvolution depend heavily on whether the detector signal is univariate or multivariate.

3.3.1 Deconvolution of univariate signals

In the case of univariate signals, one is typically limited to using univariate curve-fitting analyses where a number of Gaussian or modified Gaussian curves are determined such that the sum of these curves fits the experimentally observed cluster of peaks (Felinger, 1994). In these approaches, only a small window of chromatographic data (one peak cluster) should be processed at a time, and constraints such as fixed peak widths, shapes, unimodality, and non-negativity are often required to ensure the validity of the solution.

To solve a univariate deconvolution problem, approaches such as evolving factor analysis (EFA) (Maeder, 1987) or multivariate curve resolution (MCR) (Tauler & Barceló, 1993), among others (Vivó-Truyols et al., 2002; Sarkar et al., 1998; Kong et al. 2005) can be used. When these approaches are used with univariate data, the variables to be solved for are the number, positions, and abundances of each of the peaks that make up the signal.

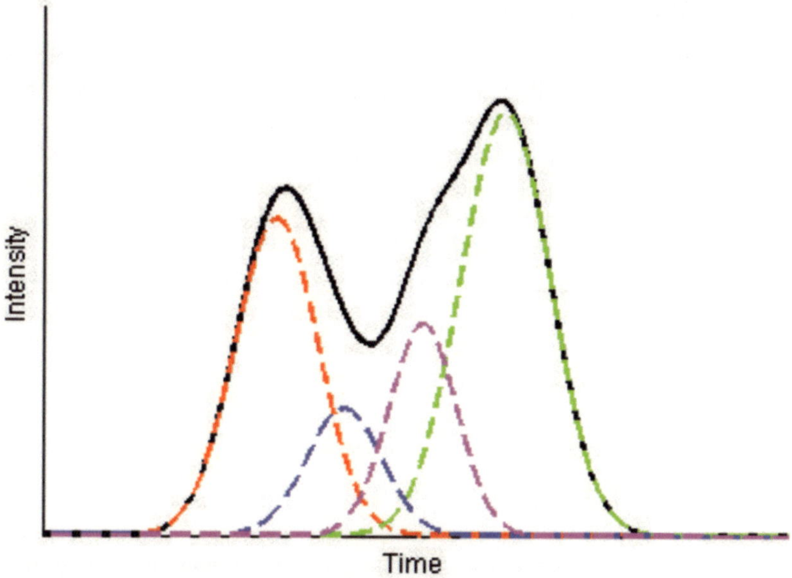

Fig. 5. Deconvolution of overlapping peaks. The black, solid trace represents the analytical signal observed at the detector, which is the sum of the four peaks represented by dashed lines.

Multivariate curve resolution is widely applicable to separations data and is one of the most common approaches (Franch-Lage et al., 2011; Marini et al., 2011, de la Mata-Espinosa et al., 2011a). The aim of this technique is to determine the number of components present in a sample and the contribution of each component to the sample. In performing MCR, the concentration and response profiles for each analyte are obtained, providing a qualitative and semi-quantitative overview of the components in an unresolved mixture without *a priori* knowledge of the mixture composition.

3.3.2 Deconvolution of multivariate signals

When multivariate detectors are used for separations, the additional dimension of information can be exploited to aid in deconvolution. MCR and EFA can also be used with multivariate data. In the case of MCR, the experimental matrix is decomposed into a matrix of concentration vs. time profiles (deconvoluted peaks) and pure spectral profiles of each compound. Knowledge of the number of components contributing to the signal in the region being deconvoluted is useful to guide the process and improve the results (de Juan & Tauler, 2006), though strictly speaking it is not required.

Parallel factor analysis (PARAFAC) (Harshman, 1970; Bro, 1997; Amigo et al., 2010) is a technique that is ideally suited for interpreting multivariate separations data. PARAFAC is a decomposition model for multivariate data which provides three matrices, **A**, **B** and **C** which contain the scores and loadings for each component. The residuals, **E**, and the number of factors, *r*, are also extracted. The PARAFAC decomposition finds the best trilinear model that minimizes the sum squares of the residuals in the model through a procedure of alternating least squares.

The biggest advantage of using PARAFAC over other models is the uniqueness of the solution; PARAFAC is less flexible and uses fewer degrees of freedom, being a more restricted model. However, its unique solution reflects actual pure analyte profiles in both the time dimension and the spectral dimension. Thus, the results of PARAFAC analysis on a cluster of overlapping multivariate peaks provide both qualitative and quantitative data where the deconvoluted signals appear as analyte peaks. One restriction to the use of PARAFAC is that the data must be trilinear (Bro, 1997; Amigo et al., 2010). In the case of chromatographic techniques with a multivariate detector, the dimensions are retention time, detector signal, and samples. In the case of comprehensive multidimensional separations, such as GC×GC, PARAFAC considers retention in the two dimensions and the samples as the three dimensions.

3.4 Feature selection

High data acquisition rates combined with the length of time required for many separations results in a large number of data points collected for a given separation (see Section 1.2). In many situations, most of the data are collected when no analytes are eluting from the system, and represent background signal when only mobile phase is reaching the detector. In the case of spectroscopic and especially mass spectral detectors, at a given point in time, many of the recorded data in this dimension will not contain useful information, even when an analyte of interest is eluting. Furthermore, many components in the mixture can be completely irrelevant to analysis (Johnson & Synovec, 2002; Sinkov & Harynuk, 2011a). Consequently, only a small portion of separations data is potentially useful. It is also well known that any model will be heavily influenced by the specific variables that are included in its construction (Kjeldahl & Bro, 2010).

The inclusion of irrelevant data is detrimental to the model because the mathematics attempt to account for variations observed in these irrelevant variables. Consequently the model is forced to model noise, resulting in a decrease in its predictive ability. Worse yet, the model could fit the data well and provide a seemingly useful prediction, until cross-validation shows otherwise. Finally, the inclusion of extraneous variables increases the demands on the computer system being employed, making model construction slower, or in some cases outright impossible. Thus, prior reduction of separations data to a manageable size is crucial. Figure 6 depicts situations where either too few or too many variables were used to model a system.

One common manner to achieve data reduction is to use a table of integrated peaks instead of raw chromatographic data. This has the advantage of reducing the number of variables to those compounds included in the peak list, removing baseline noise and, if the analyst knows which exact peaks to use, removing signal from irrelevant compounds. Problems with this approach include the restriction to identified compounds, which may or may not include all of the information required for modeling, and integration errors that skew results. Finally, even with an error-free comprehensive peak table, the analyst must still perform feature selection since many peaks will undoubtedly be irrelevant to the analysis.

In the case of multivariate detection, it can be advantageous to monitor only one or a few channels (wavelengths, ions, etc.) as this will selectively detect only a portion of the analytes, allowing the analyst to avoid many interfering species while greatly reducing the size of the data. However, in these cases the analyst must know exactly what signals to use and runs the risk of missing important features of the data encoded in the channels that were ignored. Further, using this approach destroys much of the multivariate advantage that can be realized through using these more complex (and expensive) detection strategies.

Objective feature selection techniques generally have two steps: variable ranking, and variable selection. Objective variable ranking techniques such as analysis of variance (ANOVA) (Johnson & Synovec, 2002), the discriminating variable test (DIVA) (Rajalahti et al., 2009a, 2009b), and informative vectors (Teofilo et al., 2009) have the distinct advantage that variables are ranked based on a mathematically calculable "perceived utility" and not on subjective analyst perception. In essence, the data are given the chance to inform the user of what is relevant and what is likely noise, providing an approach that can be generalized to any set of analytical data.

ANOVA is an effective method when the goal is to discriminate between classes of samples. ANOVA calculates the F ratio for each variable: the ratio of between-class variance to within-class variance. If the F ratio for a given variable is high, it is deemed to be more valuable for describing the difference between classes. Once the F ratio has been calculated for every data point in the chromatogram, the variables can be ranked in order of decreasing F ratio. A chemometric model is then constructed using a fraction of variables having the highest F ratio. One significant advantage of ANOVA is that the algorithm can be written with memory conservation in mind and thus is easily applied to data sets with very large numbers of samples and variables (hundreds or thousands of samples, each containing millions of variables). Consequently, it can be easily applied to a set of GC-MS chromatograms across the entire chromatogram, something that is difficult for other feature ranking approaches.

DIVA is a feature ranking technique that aids feature selection prior to chemometric analysis (Rajalahti et al., 2009a, 2009b). This approach involves the creation of a PLS-DA model using all candidate variables. Projecting this PLS-DA model onto a new single LV yields what is termed a target projected (TP) model (Rajalahti et al., 2009a). From this, the ratio of explained variance to residual variance for each variable in the TP model provides its selectivity ratio, upon which variables are ranked (Rajalahti et al., 2009a, 2009b; Kvalheim, 1990; Kvalheim & Karstang, 1989). DIVA produces a ranking that is slightly different than that produced by ANOVA, though a direct comparison on chromatographic data has not yet been performed to our knowledge.

Once variables have been ranked, those to be included in the model must be selected. This is generally achieved by constructing a model using a forward-selection or backwards elimination approach, in an attempt to maximize some metric of model quality. Model quality can be assessed based on several metrics such as mean correct classification rates (Rajalahti et al., 2009b) or the degree of separation between classes of samples in principal component (or latent variable) space, for example using either a Euclidian distance-based metric (Pierce et al., 2005) or a metric that accounts for size and shape of clusters (Sinkov & Harynuk, 2011a).

The one exception to the rank-and-select approach are genetic algorithms (Yoshida et al., 2001), though due to the sheer number of variables present in a typical separation, these are not often used on the raw separations data as arriving at the optimal number and combination of variables is computationally inefficient and uncertain.

Sometimes, several feature selection methods are used for a given analysis. For example, an analyst might reduce chromatogram to a peak table, selecting a series of candidate variables of interest and then perform further variable ranking and optimization on the integrated peak table, especially in the case of multidimensional separations where hundreds, if not thousands of compounds can be resolved (Felkel et al., 2010).

Finally, cross-validation is extremely important, especially when processing raw separations data and using a feature ranking approach such as ANOVA. As discussed previously, raw separations data contain on the order of 10^5 to 10^6 data points for each sample. In these cases of overdetermined systems it is entirely possible that some combinations of variables containing only noise will, by random chance, indicate a difference between samples. When handling raw separations data, a good approach to avoid this problem is to break the data set into three separate sets: a training set to construct the model, an optimization set to optimize data processing parameters (such as alignment and feature selection), and finally a test set to determine if the optimized model has any meaning (Brereton, 2007). Of course this does require that one collect data for a large number of samples so that a representative population of samples exists for each of the three subsets of data.

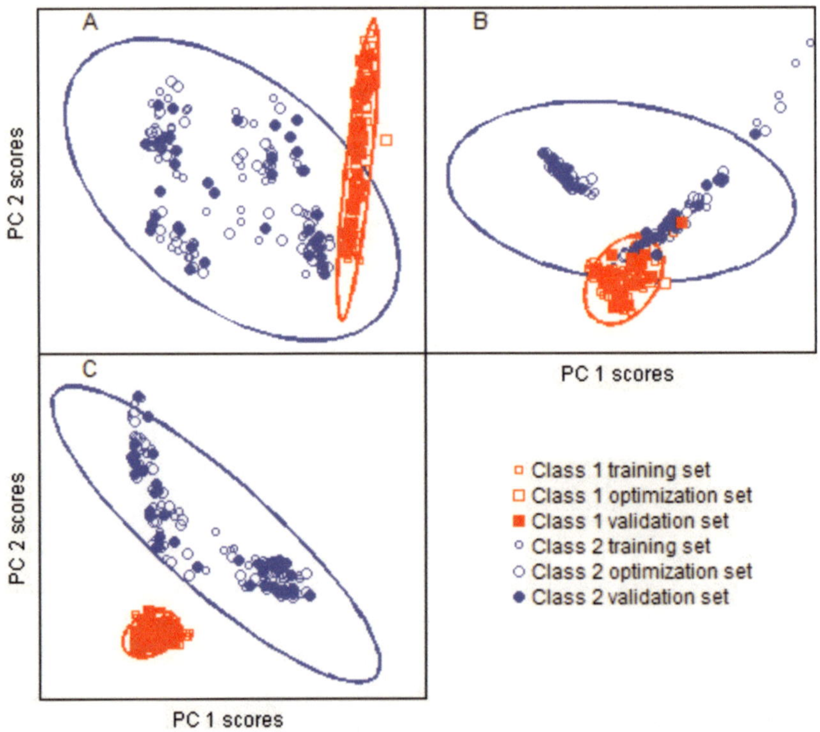

Fig. 6. Models constructed from the same data set using different numbers of top-ranked variables. (A) Too few variables; (B) Too many variables; (C) Optimal number of variables.

4. Applications and examples

After applying the appropriate pre-processing, different chemometric techniques can be applied according to the aim of the study. Pattern recognition is one of the chemometric methods most used in analytical chemistry and this is true for separations data. Pattern recognition can be generally divided into two classes: exploratory data analysis and unsupervised and supervised pattern recognition (Otto, 2007; Brereton, 2007).

Exploratory data analysis aims to extract important information, detect outliers and identify relationships between samples and its use is recommended prior to the application of other chemometric techniques. Examples of the use of exploratory data analysis tools applied to separations data include principal component analyisis (PCA) (de la Mata-Espinosa et al., 2011a; Ruiz-Samblas et al., 2011) and factor analysis (Stanimirova et al., 2011).

Unsupervised pattern recognition techniques uncover patterns within a data set without *a priori* class assignment of samples. Here, the objective is to find patterns in the data which allow grouping of similar samples using, for example, cluster analysis which has been applied to separations data by Reid et al. (2007). When supervised pattern recognition is used, the classes of samples in a training set are known and used to calibrate a model, which is then used to predict class assignments of unknown samples. Some examples of which are linear discriminant analysis (LDA), and partial least squares-discriminant analysis (PLS-DA) (de la Mata-Espinosa et al., 2011b; Zorzetti et al., 2011; Sinkov et al., 2011b). In a study performed by Sinkov et al., two alignment techniques for chromatographic data were compared. The data comprised raw GC-MS chromatograms of simulated arson debris where some samples contained different types of gasoline weathered to different extents spiked into debris samples which themselves exhibited a high degree of variability in their chemical composition. The goal was to build a PLS-DA model that could correctly classify debris samples based on whether or not they contained gasoline (Figure 7). As can be seen, the alignment algorithm used has a direct impact on the quality of the predictions. In Figure 7A, there are multiple false positives, false negatives, and ambiguous samples. In Figure 7B, all samples are classified correctly and there are no ambiguous samples.

Another example of applying chemometrics to separations data is depicted in Figures 8 and 9. Here, interval PLS (iPLS) was applied to blends of oils in order to quantify the relative concentration of olive oil in the samples (de la Mata-Espinosa et al., 2011b). iPLS divides the data into a number of intervals and then calculates a PLS model for each interval. In this example, the two peak segments which presented the lower root mean square error of cross validation (RMSECV) were used for building the final PLS model.

As mentioned in Section 3.3.2, PARAFAC is a chemometric tool for multidimensional data treatment. The scores and loadings obtained with PARAFAC can be used in two-way models for data exploration and quantitative analysis (Vosough et al., 2010). When small deviations in trilinearity exist within the data, usually due to relatively small shifts in retention time in the case of separations data, a modified version of PARAFAC called PARAFAC2 is recommended for use (Bro et al., 1999).

Like PARAFAC, PARAFAC2 decomposes raw data into loading and score matrices but without the imposition of trilinearity as in PARAFAC. Even without this constraint, the PARAFAC2 model preserves the property of uniqueness that is so advantageous with PARAFAC. Thus, analyte profiles and concentrations can be estimated by PARAFAC2 even if chromatographic alignment is not perfect (Amigo et al., 2008; Skov et al., 2009).

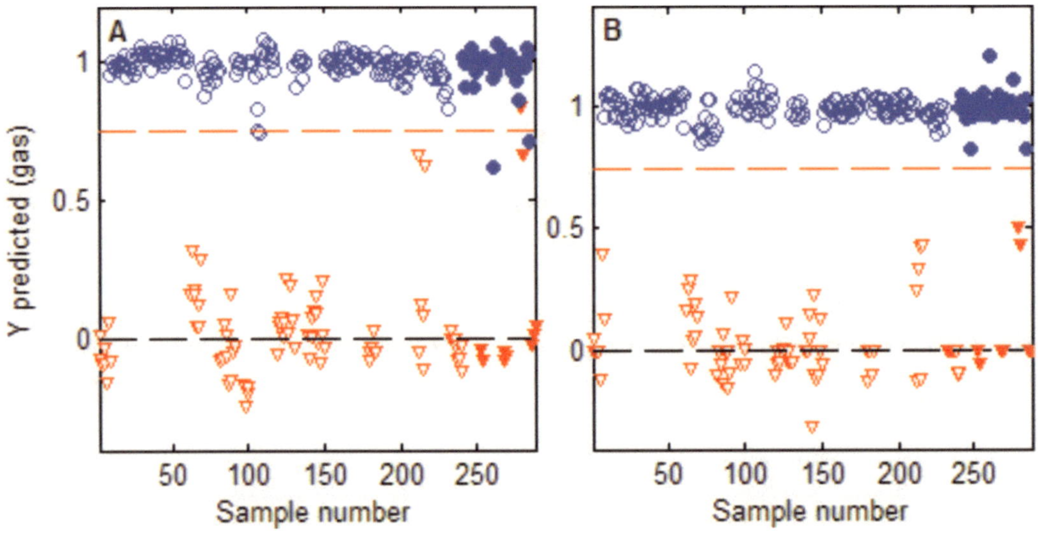

Fig. 7. PLS-DA Models for identifying gasoline in simulated arson debris derived from the same raw data, but aligned with different techniques. (A) Feature-based alignment; (B) Deuterated alkane ladder – based alignment. All other treatment and model construction algorithms were the same in both cases. Hollow markers indicate data in the training set while filled markers indicate data in the validation set. Circles represent debris containing gasoline while triangles represent gasoline-free debris. Reprinted from Sinkov et al., 2011b, with permission.

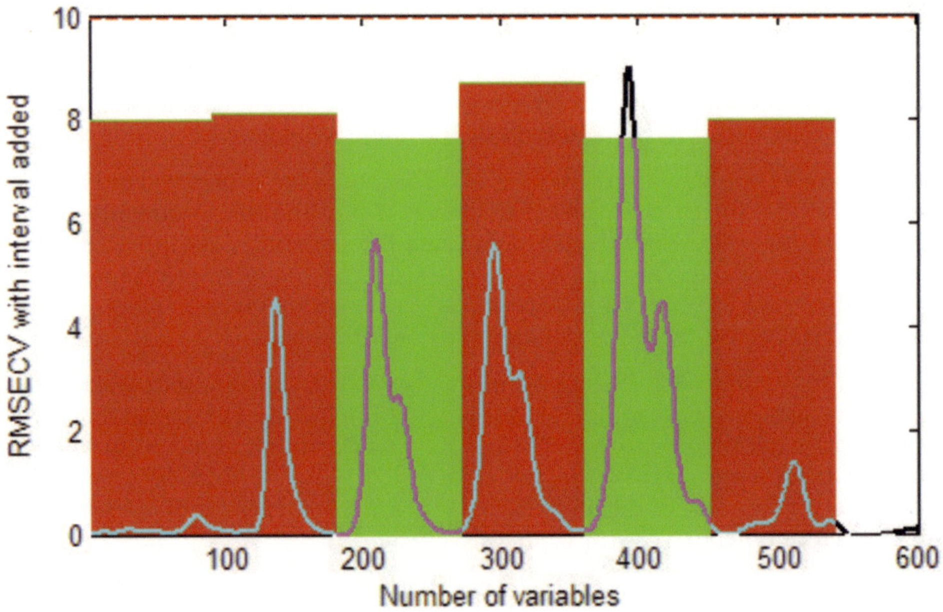

Fig. 8. Feature selection using iPLS. Segments in green showed lower RMSECV and were thus used to construct the final model. Reprinted from de la Mata-Espinosa et al., 2011b, with permission.

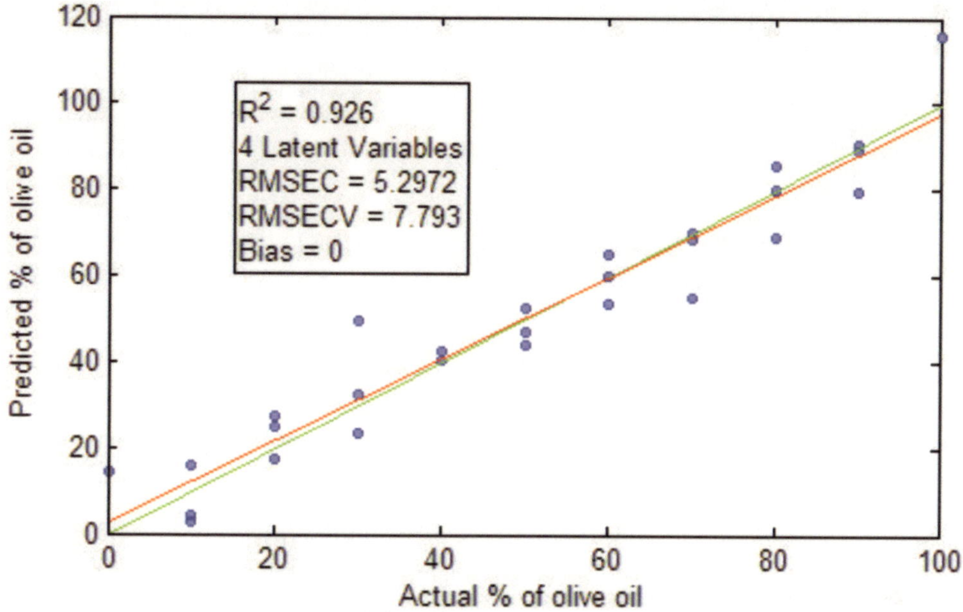

Fig. 9. Predicted vs. actual % olive oil using PLS model constructed based on results in Figure 8. Reprinted from de la Mata-Espinosa et al., 2011b, with permission.

5. Conclusions

The analyst must choose from a plethora of methods for processing separations data, a potentially daunting task. It is our hope that this review will help chromatographers entertaining thoughts of applying chemometrics to their data understand what they must consider when choosing how to prepare their data. Likewise, it is hoped that we have informed chemometricians of some of the specific challenges associated with the processing of chromatographic data and the origins of those limitations. In the development of a chemometric model for the interpretation of separations data, there are numerous opportunities for missteps that will exclude key information from the model and/or generate meaningless results. However, when due care is taken there are also many opportunities to apply chemometric techniques to transform the rich data generated by these powerful analytical tools into valuable information effectively and efficiently.

6. References

Amigo, J.M.; Skov, T.; Bro, R.; Coello, J. & Maspoch, S. (2008). Solving GC-MS problems with PARAFAC2. *Trends in Analytical Chemistry*, Vol.27, No.8, (September 2008), pp. 714-725, ISSN 0165-9936

Amigo, J.M.; Skov, T. & Bro, R. (2010). ChroMATHography: solving chromatographic issues with mathematical models and intuitive graphics. *Chemical Reviews*, Vol.110, No.8, (May 2010), pp. 4582-4605, ISSN 1520-6890

Asher, B.J.; D'Angostino, L.A.; Way, J.D.; Wong, C.S. & Harynuk, J.J. (2009). Comparison of peak integration methods for the determination of enantiomeric fraction in environmental samples. *Chemosphere*, Vol.75, No.8, (May 2009), pp. 1042-1048, ISSN 0045-6535

Brereton, R.G. (2003). *Chemometrics Data Analysis for the Laboratory and Chemical Plant,* Wiley, ISBN 0-474-78977-8, UK

Brereton, R.G. (2007). *Applied Chemometrics for Scientists*, John Wiley & Sons Inc., ISBN 978-0-470-01686-2, Toronto, Canada

Bro, R. (1997). PARAFAC. Tutorial and applications. *Chemometrics Intelligent Laboratory Systems,* Vol.38, No.2, (October 1997), pp. 149-171, ISSN 0169-7439

Bro, R.; Andersson, C.A.& Kiers, H.A.L. (2009). PARAFAC-Part II. Modeling chromatographic data with retention times shifts. *Journal of Chemometrics,* Vol.13, No.3-4, (May-August 1999), pp. 295-309, ISSN 0886-9383

Casez, J. (2010). *Encyclopaedia of Chromatography,* (3rd ed.) CRC Press, ISBN 1-4200-8483, Florida, USA

Cortes, H.J.; Winniford, B.; Luong, J. & Pursch, M. (2009). Comprehensive two dimensional gas chromatography review. *Journal of Separation Science,* Vol.32, No.5-6, (March 2009), pp. 883-904, ISSN 1615-9306

de Juan, A. & Tauler, R. (2006). Multivariate Curve Resolution (MCR) from 2000: Progress in concepts and applications. *Critical Reviews in Analytical Chemistry,* Vol.36, No.3-4, (2006) pp. 163-176, ISSN 1040-8347

de la Mata-Espinosa, P.; Bosque-Sendra, J.M.; Bro, R. & Cuadros-Rodríguez, L. (2011a). Discriminating olive and non-olive oils using HPLC-CAD and chemometrics. *Analytical and Bioanalytcial Chemistry,* Vol.399, No.6, (February, 2011), pp. 2083-2092, ISSN 1618-2650

de la Mata-Espinosa, P.; Bosque-Sendra, J.M.; Bro, R. & Cuadros-Rodríguez, L. (2011b). Olive oil quantification of edible vegetable oil blends using triacylglycerols chromatographic fingerprints and chemometric tools. *Talanta,* Vol.85, No.1, (July 2011), pp. 183-196, ISSN 0039-9140

Eilers, P.H.C. (2003). A perfect Smoother. *Analytical Chemistry,* Vol.75, No.14, (July 2003) pp. 3631-3636, ISSN 0003-2700

Eilers, P.H.C. (2004). Parametric Time Warping. *Analytical Chemistry,* Vol.76, No.2, (January 2004), pp. 404-411, ISSN 0003-2700

Erni, F. & Frei, R.W. (1978). 2-Dimensional column liquid-chromatographic technique for resolution of complex mixtures. *Journal of Chromatography,* Vol.149, (February 1978), pp. 561-569 ISSN 0021-9673

Etxebarria, N.; Zuloaga, O.; Olivares, M.; Bartolomé. L.J. & Navarro, P. (2009). Retention-time locked methods in gas chromatography. *Journal of Chromatography A*, Vol.1216, No.10, (March 2009), pp. 1624-1629 ISSN 0021-9673

Felinger, A. (1994). Deconvolution of overlapping skewed peaks. *Analytical Chemistry,* Vol.66, No.19, (October 1994), pp. 3066-3072, ISSN 0003-2700

Felkel, Y.; Dorr, N.; Glatz, F. & Varmuza, K. (2010). Determination of the total acid number (TAN) of used gas engine oils by IR and chemometrics applying a combined strategy for variable selection. *Chemometrics and Intelligent Laboratory Systems,* Vol. 101, No. 1, (March, 2010), pp. 14-22 ISSN 0169-7439

Franch-Lage, F.; Amigo, J.M.; Skibsted, E.; Maspoch, S. & Coello, J. (2011). Fast assessment of the surface distribution of API and excipients in tablets using NIR-hyperspectral imaging. *International Journal of Pharmaceutics,* Vol.441, No.1-2, (June 2011), pp. 27-35, ISSN 0378-5173

François, I.; Sandra, K. & Sandra, P. (2009). Comprehensive liquid chromatography: Fundamental aspects and practical considerations—A review. *Analytica Chimica Acta,* Vol.641, No.1-2, (May 2009), pp. 14-31, ISSN 0003-2670

Gan, F.; Ruan, G. & Mo, J. (2006). Baseline correction by improved iterative polynomial fitting with automatic threshold. *Chemometrics and Intelligent Laboratory Systems,* Vol.82, No.1 (May 2006), pp. 59-65, ISSN 0169-7439

Górecki, T.; Harynuk, J. & Panić, O. (2004). The evolution of comprehensive two-dimensional gas chromatography, *Journal of Separation Science,* Vol.27 (2004) pp. 359-379, ISSN 1615-9306

Harshman, R.A. (1970). Foundations of the PARAFAC procedure: models and conditions for an 'exploratory' multimodal factor analysis. *UCLA Working Papers Phonet.* Vol 16, (1970), pp. 1-84

Johnson, K.J. & Synovec, R.E. (2002). Pattern recognition of jet fuels: comprehensive GC×GC with ANOVA-based feature selection and principal component analysis. *Chemometrics and Intelligent Laboratory Systems,* Vol.60, No.1-2, (January 2002), pp. 225-237, ISSN 0169-7439

Johnson, K.J.; Wright, B.W.; Jarman, K.H. & Synovec, R.E. (2003). High-speed peak matching algorithm for retention time alignment of gas chromatographic data for chemometric analysis. *Journal of Chromatography A,* Vol.996, No.1-2, (May 2003), pp. 141-155, ISSN 0021-9673

Kaczmarek, K.; Walczak, B.; de Jong, S. & Vandeginste, B.G.M. (2005). Baseline reduction in two dimensional gel electrophoresis images. *Acta Chromatographica,* Vol.15 (2005), pp. 82-96, ISSN 1233-2356

Kivilompolo, M.; Pol, J. & Hyotylainen, T. (2011). Comprehensive two-dimensional liquid chormatography (LC×LC): A review. *LC GC Europe,* Vol.24, No 5 (May 2011), pp. 232-+, ISSN 1471-6577

Kjeldahl, K. & Bro, R. (2010). Some common misunderstandings in chemometrics. *Journal of Chemometrics,* Vol.24, No.7-8, (July-August, 2011), pp. 558-564, ISSN 0886-9383

Kong, K.; Ye, F.; Guo, L.; Tian, J. & Xu, G. (2005). Deconvolution of overlapped peaks based on the exponentially modified Gaussian model in comprehensive two-dimensional gas chromatography, *Journal of Chromatography A,* Vol.1086, No.1-2 (September 2005) pp. 160-164, ISSN 0021-9673

Kvalheim, O.M. & Karstang, T.V. (1989). Interpretation of latent-variable regression models. *Chemometrics and Intelligent Laboratory Systems,* Vol.7, No.1-2, (December 1989), pp. 39-51, ISSN 0169-7439

Kvalheim, O.M. (1990). Latent-variable regression models with higher-order terms: An extension of response modelling by orthogonal design and multiple linear regression. *Chemometrics and Intelligent Laboratory Systems,* Vol.8, No.1, (May 1990), pp. 59-67, ISSN 0169-7439

Lavine, B.K.; Brzozowski, D.; Moores, A.J.; Davidson, C.E. & Mayfield, H.T. (2001). Genetic algorithm for fuel spill identification. *Analytica Chimica Acta,* Vol.437, No.2, (June 2001), pp. 233-246, ISSN 0003-2670

Laursen, K.; Frederiksen, S.S.; Leuenhagen, C. & Bro, R. (2010). Chemometric quality contrl of chromatographic purity. *Journal of Chromatography A,* Vol.1217, No.42 (October 2010), pp. 6503-6510, ISSN 0021-9673

Li, Y.H.; Wojcik, R & Dovichi, N.J. (2011). A replaceable microreactor for on-line protein digestion in a two dimensional capillary electrophoresis system with tandem mass spectrometry detection. *Journal of Chromatography A*, Vol.1218, No.15 (April 2011), pp. 2007-2011, ISSN 0021-9673

Liang, Y.; Xie, P. & Chau, F. (2010). Chromatographic fingerprinting and related chemometric techniques for quality control of traditional Chinese medicines. *Journal of Separation Science,* Vol.33, No.3 (February 2010), pp. 410-421, ISSN 1615-9314

Liu, Z. & Phillips, J.B. (1991). Comprehensive 2-dimensional gas-chromatography using a modulator interface. *Journal of Chromatographic Science,* Vol.29, No.6 (June 1991), pp. 227-231, ISSN 0021-9665

Maeder, M. (1987). Evolving factor analysis for the resolution of overlapping chromatographic peaks. *Analytical Chemistry,* Vol.59, No.3, (February 1987), pp 527-530, ISSN 0003-2700

Marini, F.; D'Aloise, A.; Bucci, R.; Buiarelli, F.; Magri, A.L. & Magri, D. (2011), Fast analysis of 4 phenolic acids in olive oil by HPLC-DAD and chemometrics, *Chemometrics and Intelligent laboratory systems,* Vol.106, No.1, (March 2011), pp. 142-149, ISSN 0169-7439

Michels, D.A.; Hu, S.; Schoenherr, R.M.; Eggertson, M.J. & Dovichi, N.J. (2002), Fully automated two-dimensional capillary electrophoresis for high sensitivity protein analysis, *Molecular & Cellular Proteomics,* Vol.1, No.1, (January 2002), pp. 69-74, ISSN 1535-9476

Miller, J.M. (2005). *Chromatography: concepts and contrasts,* (2nd ed.) Wiley, ISBN 0471472077, Hoboken, USA

Mommers, J.; Knooren, J.; Mengerink, Y.; Wilbers, A.; Vreuls, R. & van der Wal, S. (2011). Retention time locking procedure for comprehensive two-dimensional gas chromatography. *Journal of Chromatography A*, Vol.1218, No.21 (May, 2011), pp. 3159-3165 ISSN 0021-9673

Nielsen, N-P.; Cartensen, J.M. & Smedsgaard, J. (1998). Aligning of single and multiple wavelength chromatographic profiles for chemometric data analysis using correlation optimised warping. *Journal of Chromatography A,* Vol.805, No.1-2 (May 1998), pp. 17-35 ISSN 0021-9673

Otto, M. (2007). *Chemometrics,* Wiley-VCH, ISBN 978-3-527-31418-8, Weinheim, Germany

Persson, P.O. & Strang, G. (2003). Smoothing by Savitzky-Golay and Legendre filters, In: *Mathematical Systems Theory in Biology, Communications, Computation and Finance,* Rosenthal J. Gilliam D.S., pp. 301-315, IMA Vol. Math. Appl., 134, Springer, ISBN 978-0387-40319-9, New York, USA

Pierce K.M.; Hope J.L.; Johnson K.J.; Wright B.W. & Synovec R.E. (2005). Classification of gasoline data obtained by gas chromatography using a piecewise alignment algorithm combined with feature selection and principal component analysis. *Journal of Chromatography A*, Vol.1096, No.1-2, (November 2005), pp. 101-110, ISSN 0021-9673

Poole, C.F. (2003). *The Essence of Chromatography*, (1st ed.), Elsevier, ISBN 0444501983, Amsterdam, The Netherlands

Rajalahti, T.; Arneberg, R.; Berven, F.S.; Myhr, K.M.; Ulvik, R.J. & Kvalheim, O.M. (2009a). Biomarker discovery in mass spectral profiles by means of selectivity ratio plot. *Chemometrics and Intelligent Laboratory Systems*, Vol. 95, No. 1, (January 2009), pp. 35-48, ISSN 0169-7439

Rajalahti, T.; Arneberg, R.; Kroksveen, A.C.; Berle, M.; Myhr, K.M. & Kvalheim, O.M. (2009b). Discriminating Variable Test and Selectivity Ratio Plot: Quantitative Tools for Interpretation and Variable (Biomarker) Selection in Complex Spectral or Chromatographic Profiles. *Analytical Chemistry*, Vol. 81, No. 7, (April 2009), pp. 2581-2590, ISSN 0169-7439

Reid, R.G.; Durham, D.G.; Boyle S.P.; Low, A.S. & Wangboonskul, J. (2007). Differentiation of opium and poppy straw using capillary electrophoresis and pattern recognition techniques. *Analytica Chimica Acta*, Vol.605, No. 1, (December 2007), pp. 20-27, ISSN 0003-2670

Ruiz-Samblas, C.; Cuadros-Rodriguez, L.; Gonzalez-Casado, A.; Rodriguez Garcia, F.D.P; de la Mata-Espinosa, P.; Bosque-Sendra, J.M. (2011). Multivariate analysis of HT/GC-(IT)MS chromatographic profiles of triacylglycerols for classification of olive oil varieties, *Analytical and Bionalytical Chemistry*, Vol.399, No.6 (February 2011), pp. 2093-2103, ISSN 1618-2642

Sarkar, S.; Dutta, P.K. & Roy, N.C. (1998). A blind-deconvolution approach for chromatographic and spectroscopic peak restoration, *IEEE transactions on instrumentation and measurement*, Vol.47, No.4 (August 1998), pp. 941-947, ISSN 0018-9456

Savorani, F.; Tomasi, G. & Engelsen, S.B. (2010). Icoshift: A versatile tool for the rapid alignment of 1D NMR spectra. *Journal of Magnetic Resonance*, Vol.202, No.2, (February 2010), pp. 190-202 ISSN 1090-7807

Sinkov, N.A. & Harynuk, J.J. (2011a). Cluster resolution: A metric for automated, objective and optimized feature selection in chemometric modeling. *Talanta*, Vol.83, No.4, (January 2011), pp. 1079-1087, ISSN 0039-9140

Sinkov, N.A.; Johnston, B.M.; Sandercock, P.M.L. & Harynuk, J.J. (2011b). Automated optimization and construction of chemometric models based on highly variable raw chromatographic data. *Analytica Chimica Acta*, Vol.697, No.1-2, (July 2011), pp. 8-15, ISSN 1873-4324

Skov, T.; Hoggard, J.C.; Bro, R. & Synovec, R.E. (2009). Handling within run retention time shifts in two-dimensional chromatography data using shift correction and modeling. *Journal of Chromatography A*, Vol.1216, No.18, (May 2009), pp. 4020-4029, ISSN 0021-9673

Stanimirova, I.; Boucon, C. & Walczak, B. (2011). Relating gas chromatographic profiles to sensory measurements describing the end products of the Maillard reaction. *Talanta*, Vol.83, No 4, (January 2011), pp. 1239-1246, ISSN 0039-9140

Tauler, R. & Barceló, D. (1993). Multivariate curve resolution applied to liquid chromatography-diode array detection. *Trends in Analytical Chemistry*, Vol.12, No.8, (1993), pp. 319-327, ISSN 0165-9936

Teofilo, R.F.; Martins, J.P.A. & Ferreira, M.M.C. (2009). Sorting variables by using informative vectors as a strategy for feature selection in multivariate regression.

Journal of Chemometrics, Vol.23, No.1-2, (January-February 2009), pp. 32-48, ISSN 0886-9383

Tomasi, G.; Van den Berg, F. & Andersson, C. (2004). Correlation optimized warping and dynamic time warping as preprocessing methods for chromatographic data. *Journal of Chemometrics*, Vol.18, No.5, (May 2004), pp. 231-241, ISSN 0886-9383

Toppo, S.; Roveri, A.; Vitale, M.P.; Zaccarin, M.; Serain, E.; Apostolidis, E.; Gion, M., Mariorino, M. & Ursini, F. (2008). MPA: A multiple peak alignment algorithm to perform multiple comparisons of liquid-phase proteomic profiles. *Proteomics*, Vol.8, No.2, (January 2008), pp. 250-253 ISSN 1615-9861

Van den Berg, F.; Tomasi, G. & Viereck, N. (2005). Warping: investigation of NMR pre-processing and correction, In: *Magnetic Resonance in Food Science: The Multivariate Challenge*, Engelsen, S.B., Belton, P.S., Jakobsen, H.J., pp. 131-138, Royal Society of Chemistry, ISBN 0854046488, Cambridge, UK

Van Nederkassel, A.M.; Dazykowski, M.; Eilers, P.H.C. & Vander Heyden, Y. (2006). A comparison of three algorithms for chromatograms alignment. *Journal of Chromatography A*, Vol.118, No.2 (June 2006), pp. 199-210 ISSN 0021-9673

Vivó-Truyols, G.; Torres-Lapasió, J.R.; Caballero R.D. & García-Alvarez-Coque, M.C. (2002). Peak deconvolution in one-dimensional chromatography using a two-way data approach. *Journal of Chromatography A*, Vol.958, No.1-2, (June, 2002), pp. 35-49, ISSN 0021-9673

Vosough, M.; Bayat, M. & Salemi, A. (2010). Matrix-free analysis of aflatoxins in pistachio nuts using parallel factor modeling of liquid chromatography diode-array detection data. *Analytica Chimica Acta*, Vol.663, No.1, (March 2010), pp. 11-18. ISSN 0003-2670

Watson, N.E.; VanWingerden, M.M.; Pierce, K.M.; Wright, B.W. & Synovec, R.E. (2006). Classification of high-speed gas chromatography-mass spectrometry data by principal component analysis coupled with piecewise alignment and feature selection. *Journal of Chromatography A*, Vol.1129, No.1, (September, 2006), pp. 111-118, ISSN 0021-9673

Yao, W., Yin, X. & Hu Y. (2007). A new algorithm of piecewise automated beam search for peak alignment of chromatographic fingerprints. *Journal of Chromatography A*, Vol. 1160, No.1-2, (August 2007), pp. 254-262. ISSN 0021-9673

Yoshida H.; Leardi R.; Funatsu K. & Varmuza K. (2001) Feature selection by genetic algorithms for mass spectral classifiers. *Analytica Chimica Acta*, 446, 1-2, (November 2001), pp. 485-494, ISSN 0003-2670

Zhang D.; Huang, X.; Regnier, F.E. & Zhang, M. (2008). Two-dimensional correlation optimized warping algorithm for aligning GC×GC-MS data. *Analytical Chemistry*, Vol.80, No.8 (April 2008), pp. 2664-2671, ISSN 0003-2700

Zhang, Z.M.; Chen, S. & Liang, Y.Z. (2010). Baseline correction using adaptive iteratively reweighted penalized least squares. *Analyst*, Vol.5 (February 2010), pp. 1138-1146, ISSN 0003-2654

Zorzetti, B.M.; Shaver, J.M. & Harynuk, J.J. (2011). Estimation of the age of a weathered mixture of volatile organic compounds. *Analytica Chimica Acta*, Vol.694, No.1-2, (May 2011), pp. 31-37, ISSN 0003-2670